Historic Concrete
background to appraisal

Edited by

James Sutherland, Dawn Humm and Mike Chrimes

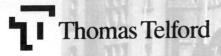
Thomas Telford

Published by Thomas Telford Publishing, Thomas Telford Ltd, 1 Heron Quay, London E14 4JD. URL: http://www.thomastelford.com

Distributors for Thomas Telford books are
USA: ASCE Press, 1801 Alexander Bell Drive, Reston, VA 20191-4400, USA
Japan: Maruzen Co. Ltd, Book Department, 3–10 Nihonbashi 2-chome, Chuo-ku, Tokyo 103
Australia: DA Books and Journals, 648 Whitehorse Road, Mitcham 3132, Victoria

First published 2001

Also available from Thomas Telford Books

Manual of numerical methods in concrete. MYH Bangash. ISBN 07277 2942 6
Owen Williams (The Engineer's Contribution to Contemporary Architecture series). D Yeomans and D Cottam. ISBN 07277 3018 5
Innovations in concrete. David Bennett. ISBN 07277 2005 8
Structural detailing in steel. MYH Bangash. ISBN 07277 2850 4

A catalogue record for this book is available from the British Library

ISBN: 0 7277 2875 X

Typeset by Gray Publishing, Tunbridge Wells
Printed and bound in Great Britain by MPG Books, Bodmin

List of contributors

R.D. ANCHOR, *Anchor Consultants, Birmingham*

D.A. BRUGGEMANN, *Principal Engineer, Halliburton Brown & Root*

M.N. BUSSELL, *Specialist consultant in historic structures, formerly Senior Research and Development Engineer, Ove Arup & Partners*

M. CHRIMES, *Head Librarian, Institution of Civil Engineers*

DR M.H. GOULD, *formerly honorary senior research fellow, Queen's University, Belfast*

K.J. HOLLOCK, *Engineer, Halliburton Brown & Root*

D. HUMM, *Construction Information Consultant*

L.B. HURST, *Consultant, Hurst Pierce & Malcolm*

P.B. MORICE, *Emeritus Professor of Civil Engineering, University of Southampton*

SIR ALAN MUIR WOOD, *President, International Tunnelling Association, past president Institution of Civil Engineers Institution, Consultant, Halcrow Group Limited*

F. NEWBY, *former senior partner, F.J. Samuely & Partners*

B.N. SHARP, *Consultant, Halcrow Group Limited*

G.P. SIMS, *Principal Consultant, Halliburton Brown & Root*

W.J.R SMYTH, *Consultant, Ove Arup & Partners*

G. SOMERVILLE, *Consultant, British Cement Association*

R.J.M. SUTHERLAND, *former partner and consultant, Harris & Sutherland*

H. TOTTENHAM, *Consultant, formerly Professor of Structural Engineering, University of Southampton*

F. WALLEY, *Consultant, Ove Arup & Partners*

J. WEILER, *Consultant Architectural Historian*

G. WEST, *Consultant, formerly Transport Research Laboratory*

A. WITTEN, *Information Scientist, formerly Chief Librarian, Institution of Structural Engineers*

Frank Newby

It was with great sadness that the Editors learnt of the death of one of the contributors, Frank Newby, on 10 May 2001, before this book to which he contributed so much was published.

Although Frank wrote only one chapter, he contributed far more than that, both through sharing his knowledge of concrete design, and through the time he dedicated to making the 1996 exhibition 'Revolution or Evolution', which partly inspired this book, a reality.

Frank was one of the leading structural engineers of his generation, a Fellow of the Royal Academy of Engineering, Fellow and Gold Medallist of the Institution of Structural Engineers, Hon FRIBA and for 32 years senior partner in F.J. Samuely and Partners.

These bald statistics give no idea of the character of this friendly, creative, very generous but in many ways self-effacing man. Of his many interests official status certainly was not one. He did not wish to be a president or chairman of anything but give him something to design or an intricate problem to solve and he would blossom.

1   Introduction

James Sutherland

It is common knowledge that the dome of the Pantheon in Rome is made of concrete but few realize that the Liver Building (Figure 1.1), which has dominated the Liverpool waterfront since about 1909 is a pioneer example of reinforced concrete framing. Even fewer people know of the Eldon Street flats, also in Liverpool and even more notable as pioneers, being built in large panel precast concrete in 1905.[1] Here the structure (Figure 1.2) is remarkably similar to that 'invented' as system building in the 1950s although perhaps more robust. The first flat slab concrete floors in Britain are normally attributed to Owen Williams in the 1930s more than 20 years after the independent invention of flat slab construction in Switzerland and America. Recently an example (Figure 1.3) has been found in Britain dated 1919.[2] Could there be an even earlier one? These are just cases of

Figure 1.1   The Royal Liver Building, Liverpool completed in 1909. Early high-rise concrete framing.

information lost or ignored. Concrete, it seems, is not seen as a suitable subject for historical study, yet in the last 100 years it has probably had a greater impact on our surroundings, and indirectly on our way of life, than any other material. One wonders why this influence is not more widely recognized. Apart from any cultural interest, there are very practical reasons for studying the history of concrete.

In Britain we have a vast stock of concrete structures of varying ages. Frequently these need to be appraised, repaired or altered, but relevant drawings or other records seldom exist and concrete is probably the most totally opaque of all structural materials. With steel, timber and masonry one can generally see, and even measure, what is there but with concrete it is often hard to judge from outside even whether it is mass concrete or reinforced or prestressed. It might not be structural at all but just protecting a steel frame. To find out more one can use ultrasonic testers and covermeters, drill holes or chip away surfaces but, even so, the results may be puzzling. The best starting point for appraisal is some knowledge

a

b

Figure 1.2   Concrete flats in Eldon Street, Liverpool. Pioneer large panel precast concrete construction of 1905. (a) Transporting units from the precasting yard. (b) Erecting wall units, with floor units to follow. (c) The completed building. Reproduced with the permission of the Liverpool Record Office, Liverpool Libraries and Information Services.

Figure 1.2c

*Figure 1.3   Bryant &
May's factory of 1919
with possibly the earliest
flat-slab concrete
construction in Britain.*

of what is likely to be found at different dates and with different types of structure. This is where history — the theme of this book — comes in.

The book arose out of a set of papers on different aspects of the history of reinforced and prestressed concrete published by the Institution of Civil Engineers in 1996.[3] Concurrently there was also an exhibition called 'Revolution or Evolution' and a half-day meeting. All feedback pointed strongly to the desirability of publishing an edited and expanded version of the papers in book form, the thinking being that engineers who may be asked to make appraisals are more likely to remember, find and refer to a book than to look out back numbers of a journal. Likewise architects, surveyors, historians of construction and general readers could well find a book but be less likely to search for papers in a journal which, at first, might not seem relevant to their interests. Time will show whether this thinking was correct.

The format of the book

For the most part the book deals with concrete in Britain, but with references to discoveries and works in other countries where these have influenced practice in Britain. The first 14 chapters are edited versions of the original papers with much of the statistical and tabulated material transferred to appendices for easy reference. A further five chapters have been added covering military applications of concrete and its use in tunnels, in roads and pavements, in water-retaining structures and in dams, thus making good some major omissions on civil engineering.

No rigid starting date has been taken in the different chapters, the authors deciding on significant periods in each case. For foundations, military works and water-retaining structures the treatment goes back to the middle of the 19th century, or marginally before, to include mass concrete in fortifications, reservoirs and the bases of gas holders. Also for the development of cements and for patent flooring systems the effective starting time is around 1850 while for reinforced concrete in Britain it is seen as about 1900 and for prestressing the middle of the Second World War.

Effectively there are no finishing dates and generally no attempt has been made to cover present practice. Reinforced concrete continues to find favour for most types of structure but here the concentration is on the 'working stress' era up to around 1970, with a strong emphasis on out-dated reinforcing details and only a glancing reference to limit state thinking and present codes. There is little on water-retaining structures like water towers and swimming pools after 1920 because by this time reinforced concrete was well established and techniques were similar to those for other concrete structures. With prestressed concrete the principles have remained constant since its introduction in the 1940s but the size of tendons and anchorages has increased and the detailing has developed. As with reinforcing bars, patent piling systems and early prestressing anchorages and ducts are recorded in some detail in the appendices.

It would never be possible in one book to cover every historical aspect of the structural use of concrete. Even with the number of chapters expanded beyond the range of the original papers there must still be gaps. For instance, silos, bunkers, pipelines, sewers, masts and fencing are not specifically referred to. Also the architectural treatment of concrete, the development of precasting (especially in relation to quality of finish), system building, semi-structural concrete cladding and artificial stone are all very relevant both to the historian and to engineers carrying out appraisals but are hardly touched upon in the main chapters. Likewise the problems with concrete, which have received so much publicity in the last 40 years are only covered incidentally. Some notes on these 'neglected' topics are given below.

Visual aspects of concrete

In Britain concrete was seen at first as a purely structural material, perhaps at its most impressive on a large scale in bridges and also in utilitarian structures where appearance was not thought to matter, as in the case of most jetties and industrial plants. Except for artificial or reconstructed stone, which is really an ersatz material mimicking natural masonry, the really creative approach to the appearance of concrete did not come until the 1930s when rough-boarded shuttering and various forms of exposed aggregate finish were introduced. Maxwell Ayrton's architectural detailing to Twickenham Bridge (1928–33) is a good example.[4] It included bush-hammering, brushing off the surface cement when green and, in the case of the breakwaters, a hammered ribbed finish later dubbed as 'Elephant House' after the London Zoo building of 1962–65.

It was after the Second World War that carefully designed exposed concrete surfaces became really popular, especially in buildings. The positions of construction joints and tie bolts were controlled, with the latter often emphasized, and, in the

Figure 1.4   Visually dominant concrete of the 1960s (Essex University Library). (a) Precast and in-situ *concrete all designed to be exposed (rough-boarded and fractured rib finishes). (b) Detail of ribbed finish to fascia beams during 'fracturing'.*

case of rough-boarded formwork, the width, surface finish and irregularity of the boards were all carefully set out. This care may be seen as a softening of the 'honest' or 'brutal' architectural functionalism of the period. The Hayward Gallery and the National Theatre on the South Bank of the Thames are well-known examples, the finishes at the National Theatre being visually particularly successful. With such care what may have been seen at first as a method of saving money rapidly became just the reverse, the extra cost often greatly exceeding that of all but the most expensive applied finishes.

Apart from rough-boarded finishes, exposed aggregate surfaces also became popular in the 1950s and 1960s and were exploited with varying success, the best again being planned with great care and with joints emphasized.

Also after 1950 precasting of concrete was recognized not just as a means of saving time but of improving the appearance of structural framing and of concrete cladding. Quality is easier to control with precast than with *in-situ* concrete and, at worst, individual units can be rejected without too much disruption. Here the visual problem lies largely in controlling the widths of joints and their alignment.

The Library at Essex University (Figure 1.4) is virtually a dictionary of the varieties of the techniques, which were used in the 1960s to achieve fine integral finishes and the virtual elimination of applied coatings on buildings of fundamentally dominant form. Structural 'honesty' had by then become near to a religion.

This is not the place to argue the aesthetic principles behind exposed concrete surfaces. However it must be admitted that much of the public dislike of concrete, which became strong in the 1970s and 1980s, has been due to poor and uneven weathering with the occasional crack or rusty streak due to exposed reinforcement or iron in the aggregate. The big misconception has been that there are some types of concrete finish, which, even if expensive, are permanent. In many cases simple cleaning on a regular basis may be all that is needed for good maintenance but, where some actual defects like spalling have occurred, it is very difficult to make repairs to concrete which are visually acceptable in the long term. Clever craftsmen can hide patching initially but there is a tendency for evidence of it to reappear and become more and more obvious year after year. Also after some years sharp edges may be blurred by acid rain and aggregate exposed irregularly. In many cases owners of buildings with exposed concrete finishes have opted for painting — as with stucco — but this does undoubtedly alter the original image and of course needs to be repeated regularly.

The misconception that there are permanent finishes to concrete which need no maintenance is only matched by the misunderstanding over the durability of reinforced concrete as a structural material.

The durability of concrete

Writing in a pioneer textbook published in 1904, C.F. Marsh stated that the durability of reinforced concrete was 'well established' and that the cost of maintaining it was 'nil'.[5] This myth has continued for a surprisingly long time. It may not be quite dead yet. If applied to mass concrete or to reinforced concrete wholly protected from damp, Marsh's statement would be near to the truth but much reinforced concrete is exposed to the weather or to internal leaks in buildings or bridge decks.

Engineers always seem to have felt a strong urge to reinforce concrete sections even where this is not necessary to resist calculated forces and often both designers and those carrying out the construction have taken far too cavalier an attitude to cover to reinforcement. In Chapter 5, M.N. Bussell enlarges on the lack of adequate guidance on cover and on durability (at least in the period up to the 1930s). With precast units such as columns and mullions much of this reinforcement was doubtless put in to prevent damage during handling and in the case of *in-situ* concrete to control shrinkage. Figure 1.5 shows two sections through precast reinforced mullions and sills as illustrated in a publication first issued in 1918.[6] Here the cover seems impossibly small by any standards and it is doubtful whether the reinforcement was needed at all. If actually used, one must wonder

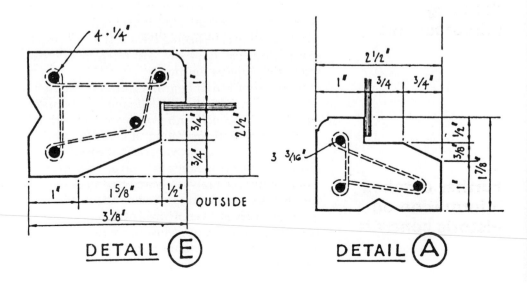

Figure 1.5   Example of very small concrete cover to reinforcement as advocated in a book first published in 1918.[6]

if this detail led to rust and splitting? It is hard to be sure. Fence posts are a good example of the variable performance of reinforced concrete with low covers.

Lack of cover to reinforcement has not been the only cause of deterioration in concrete. In Chapter 6, Somerville discusses several aspects of durability with useful references, emphasizing in particular our blindness to the reduced protection of steel as carbonation takes place. This was not recognized in practice until the 1970s although scientifically it was known half a century earlier.

In spite of many shortcomings with past practice much excellent reinforced and prestressed concrete exists today but we should not assume that it will not need maintenance in the future. Appraising engineers would do well to emphasize this continuing need when reporting to clients, even where visually there appear to be no defects at present. Everyone expects timber, iron and steel to need maintenance. Concrete cannot be considered immune.

System building in concrete

There was a strong movement towards concrete building systems in the 1950s and early 1960s. This was partly ideological and partly because of a real need to build quickly after the War using the minimum amount of steel. The systems all depended on prefabrication. Intergrid and Laingspan, initially intended for schools and both based on prestressing, were used quite extensively for office blocks as well. Individual system-built houses in precast concrete such as the Airey house date mainly from 1945 but some were built in the 1930s.[7] In the 1950s and 1960s several large-panel precast systems were developed in Britain, or imported from the Continent, for use in housing blocks up to 25 storeys high.[8] The fashion for high-rise housing was short lived on social grounds and, structurally, large-panel precast systems went out of favour following the partial collapse of the Ronan Point tower in 1968.[9] Nevertheless many such buildings remain, all the taller ones having been strengthened. Apart from full system building, precast wall panels with insulation sandwiched between concrete skins have been commonly used on all types of building from the 1950s onwards.

It is not always easy to understand from visual inspection how these system buildings were designed and especially how the concrete units were joined. Here engineers may find it useful to consult the archive of trade literature referred to at the end of this introduction.

Postscript on detailing reinforcement

It is interesting to speculate on the impact of reinforced concrete on those working on site around 1900 and soon after. The most uninitiated site worker would have had a feeling for how a beam or column of timber or steel should behave, even if without understanding the distribution of the stresses. However with reinforced concrete it must at first have been far from obvious what the bars were doing and what was important about the way in which they were placed. New conventions for detailing had to be developed. Figure 1.6 shows the evolution in the representation of reinforcement from a highly pictorial form for a complex detail early in the century to a typical stylized CAD output, which must be incomprehensible to anyone outside the industry, but presents no problem to the steelfixer today.

Further detailed information on historic concrete

Some useful references to many aspects of structural concrete, especially those not covered in individual chapters of this book, can be found in the Appendix. Journals such as *Concrete and Constructional Engineering* and *Concrete* are a mine of technical information as are the books in the *Concrete Series* by Concrete Publications Ltd, which are listed in this Appendix. These are just examples. There are also the journals or proceedings of the engineering institutions and general

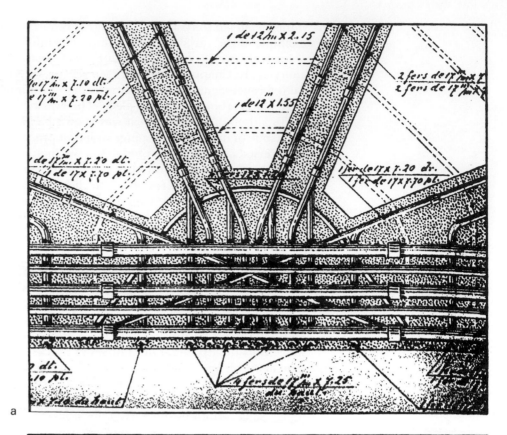

a

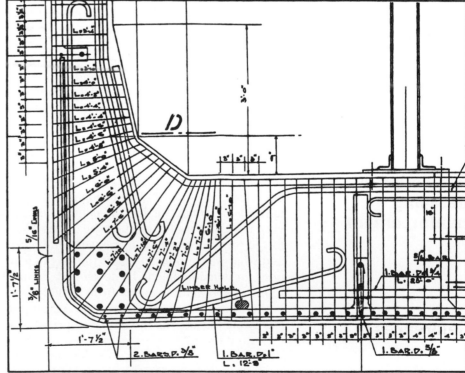

b

Figure 1.6   The development of methods of defining reinforcement on drawings. (a) Pictorial detailing of c. 1900 (Hennebique). (b) Formalized pictorial detailing of c. 1910. (c) Stylized CAD detailing of the 1990s.

engineering and architectural journals which are listed with their periods of publication.

As well as published accounts of structures and structural forms, there is a useful archive of 'real' reinforced concrete drawings dating back to 1903.

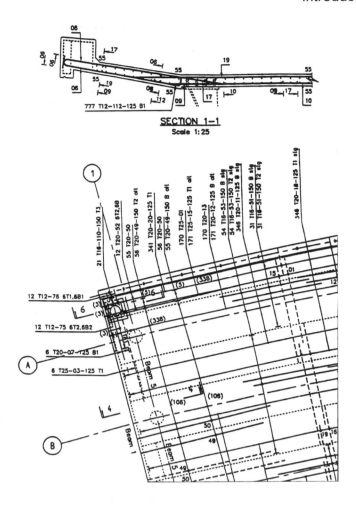

Figure 1.6c

This was built up by the Institution of Civil Engineers together with the Concrete Society and is held at the Institution, partly as a historical record but also to help appraising engineers to know what they might expect or to understand better what they have found.

Thanks are due to all those who have already provided this historical material and, in particular, to L.G. Mouchel & Partners for the gift of many drawings and especially for a number of complete dossiers on their early projects with calculations and other information as well as working drawings. Further additions to this archive would be welcome. Finally trade brochures are being gathered on early reinforcing methods, on prestressing anchorages, on precast concrete building systems and on patent piling systems and held by the Institution of Civil Engineers while the Science Museum has a growing collection of related hardware.

References

1. Report of City Engineer, City of Liverpool. Concrete Dwellings, Eldon Street: Liverpool 28 April 1905; Patent Specification No. 6115 A.D.1901 (John Alexander Brodie); Moore, R., An early system of large-panel building. RIBA J., 1969, 383–86.
2. Yeomans, D., Construction since 1900: Materials. Batsford, 1997: 121–23.
3. Historic Concrete. Proc. Instn Civ. Engrs Structs & Bldgs, 1996, 116, 255–480.
4. Gray, W.S., Childe, H.L., Concrete Surface Finishes, Renderings and Terrazzo. London, Concrete Publications Ltd., 1935.
5. Marsh, C.F., Reinforced Concrete, Archibald Constable, 1904.
6. Lakeman, A., Concrete Houses and Small Garages, 4th edn. London, Concrete Publications Ltd., 1918, Rewritten 1949.

7. Post-War Building Studies No. 1. House Construction. HMSO, 1944; Post-War Building Studies No. 23. House Construction Second Report. HMSO, 1946.

8. Diamant, R.M.E. (in collaboration with the Architect and Building News). Industrialised Building 50 International Methods. London, Illiffe Books Ltd., 1964 & 1965.

9. Ministry of Housing and Local Government. Report on the Inquiry into the Collapse of Flats at Ronan Point, Canning Town. HMSO, 1968.

2 The innovative uses of concrete by engineers and architects

Frank Newby

Synopsis

Concrete as originally exploited has little tensile strength, but with the development of reinforced concrete at the end of the 19th century, a composite material capable of taking tensile stresses was available to engineers. Taking an international view, this chapter looks at the way in which engineers and architects chose to exploit this and the later developments such as prestressed concrete, illustrating innovative structural and architectural uses in a historical context.

Introduction

In general terms concrete is an agglomerate of coarse material such as crushed stone, gravel, or broken bricks or tiles, and sand mixed with a binder and water which in time forms a hard artificial stone. Mixed with sand, binders, such as lime, have been used from the earliest times to form mortars for brickwork. Similarly the first concrete was lime concrete, and as improved binders were developed concretes were classified by the type of binder (such as, pozzolana, natural cement, artificial or Portland cement and epoxy). Each type of concrete has its own properties and has been exploited in many different ways by architects and engineers.

Concrete as defined above has little tensile strength compared with its compressive strength. Towards the end of the 19th century, reinforcement of concrete to form a composite material capable of taking tensile stresses became feasible, and the age of reinforced concrete began. The original wrought iron reinforcement was replaced by steel in its different forms, not to mention copper and bamboo, and today the use of glass and carbon fibre reinforcement is in its infancy.

An alternative method of making concrete capable of taking tensile loads came in the 1930s with the advent of prestressing, where the concrete is precompressed so that tensile stresses do not occur under working conditions. This represented a fundamental change in structural thinking, and prestressed concrete can be considered as a new material with its own technology and uses.

The chief characteristic of concrete is that when freshly made it forms a plastic mass which requires containment and support until it has sufficiently hardened. Its form and appearance are therefore defined by the shape and surface texture of the mould, which are specified by the engineer or architect, and by the sequence of building, as determined by the contractor.

The three main structural requirements are first that the concrete construction will satisfactorily carry all loads into the ground or onto other supports, secondly that in underwater construction it will harden, and thirdly that it will retain its stability in a fire. Today concrete is also used as external weatherproof cladding, for limiting acoustic transmission and for providing thermal mass.

In order to illustrate the general development of concrete construction against which works in the UK can be assessed, examples showing innovative structural and architectural use are described in mainly chronological order.

Plain concrete

Although mortars and even weak concretes were developed in Egypt and earlier civilizations, it was in the Roman period that the first major use of concrete was made. Vitruvius in his *Ten books on architecture* describes 'a kind of powder from natural causes produces astonishing results from the neighbourhood of Vesuvius. This substance mixed with lime and small stones not only lends strength to buildings but when piers of it are constructed in the sea they set hard under water'. He was referring to pozzolana, as we know it today, a natural cement of volcanic origin. Whereas lime obtains its hardness from atmospheric exposure, pozzolana has silica and alumina inclusions which react with the lime even under water.

The earliest extant domes which used pozzolana concrete appeared in the roofs to baths in Pompeii in the second to first century BC. Being plastic, concrete can be poured into three-dimensional shapes and is much cheaper than cutting stonework. The surface of the domes reflects the rough timber shuttering.

In the reconstruction of Rome in the first century AD, much use was made of pozzolana concrete. When Hadrian took over construction, one of his first projects was that of constructing a civic temple, the Pantheon (Figure 2.1). This had a dome 43.3 m in diameter, some three times larger than any other built, and was completed in 128 AD. He appreciated the potential of pozzolana concrete and had

Figure 2.1   The Pantheon in Rome, 128 AD.

the courage and financial power to proceed with his revolutionary design. To minimize the weight of the dome he introduced five rows of coffers shaped to be seen from the ground. He also reduced the density of the pozzolana concrete as he built up to the eye of the dome. The external surface of the dome was stepped to facilitate casting against vertical shuttering. In describing the Pantheon, Rowland Mainstone[1] states, 'The Roman concrete dome ... permitted, for the first time, an architecture of large unencumbered interior spaces that could be experienced only from within'.

Although the properties of pozzolana had been known and exploited for centuries, it was not until John Smeaton's experiments in the mid-1750s that the reasons for these properties were understood. Smeaton's first civil engineering project, undertaken at the age of 32, was to rebuild the Eddystone Lighthouse, for which he needed hydraulic mortar which would set under water. He solved his immediate problem by using pozzolana, which he knew about from Vitruvius and from Bélidor's *Architecture hydraulique*, but his interest was roused. He therefore carried out a series of tests to discover what it was that made some limes, tarras and pozzolana hydraulic and established that it was the clay content alone that imparted this property. Details of these tests, which eventually led to the further widespread research into the production of hydraulic limes and cements, are given by Smeaton in his 1791 book on the lighthouse, and the history of these developments are detailed by Pasley[2] and by Vicat.[3]

'Roman cement', patented by Parker in 1796, which was a natural hydraulic cement made from calcining cement stones found in the chalk, set too quickly for use in foundations but was used extensively for watertight mortar. The development of 'Portland cement', the artificial cement patented by Aspdin in 1824, did not seriously advance until about 1859 when John Grant, engineer to the Metropolitan Board of Works, carried out a long series of tests before using it on a major civil engineering project, the London main drainage works.[4]

Lime concrete was specified in 1817 by the architect, Robert Smirke, for the foundations of the Millbank Penitentiary, London. Quicklime and gravel were mixed and, for no logical reason, dropped into a trench from a height of 6 ft and water added. He later used the same technique when underpinning and rebuilding the Customs House in 1827. In 1840, Charles Barry also specified a lime concrete for the foundations of his new Palace of Westminster, which was also dropped into place, this time from a height of 10 ft.

In France, François Coignet, who was active in the chemical industry, attempted in 1852 to build an exposed lime concrete walled factory following the ancient pisé system of construction. After overcoming difficulties with mixes and water content on this, his first encounter with concrete, he commissioned an architect to design a sophisticated four-storey house in St Denis, Paris, which had exposed concrete replicating a typical stone building. The intermediate floors had fireproof floors of timber beams encased in lime concrete, while the roof slab had iron joists encased in concrete.[5]

Because of its success he took out a British patent in 1855 (No. 2659) entitled 'Emploi de Béton' which gave details of his method of construction and contained an addendum emphasizing the complete sufficiency of concrete.

> Concrete walls need no facing materials such as stone etc. The hollow part of the mould in which the concrete is poured should have the form to be given to the mass whether the walls be plain or with projections such as cornices, string-courses ... or any kind of ornament.

In a submission to the organizing committee of the Universal Exhibition of 1855, where he wished to build a concrete house, he makes a telling remark. 'The reign of stone in building construction seems to have come to an end. Cement, concrete and iron are destined to replace it.'

Figure 2.2   Church at Le Vesinet, 1862.

Later he also wrote,[6]

> 'building cheap vaults of unusual spans for great public spaces... will not be a servile imitation of Roman works but will far surpass them in daring, elegance and economy'.

Following the success of his early works, Coignet became a building contractor, showing to the world by his many designs, particularly the church at Le Vesinet (Figure 2.2) in 1862, that mass concrete was an acceptable material for construction. His work was illustrated in the UK where Joseph Tall patented a shuttering system in 1865 (No. 822) which came to be widely used in the building of houses. In the second half of the 19th century concrete was the new material that excited architects, particularly for fireproof construction, and there are a surprising number of buildings still extant with exposed concrete facades.

The concrete finish frequently left a great deal to be desired, sometimes to the point where the concrete had to be covered with a concrete render or designed using coloured precast concrete tiles as permanent shuttering. Despite these problems, its fireproof property meant that many theatres were built in mass concrete with fireproof floors, notably the Royal English Opera House, London (now the Palace Theatre), which was begun in 1888.[7] The embedding of iron joists in concrete to provide fireproof flooring systems had flourished in the UK from 1844

Figure 2.3   Glenfinnan Viaduct, Scotland, 1898.

and these are described by Lawrance Hurst.[34] In the field of civil engineering there are no better examples than the Borrodale Bridge, with a single span of 127 ft 6 in and the 21-arch Glenfinnan Viaduct, which is 416 yd in length and up to 100 ft in height (Figure 2.3). Both these structures were designed in mass concrete by engineers Simpson and Wilson of Glasgow and built by 'Concrete Bob' McAlpine in 1898 for the West Highland Railway.

Development of reinforced concrete in Europe

Although Wilkinson[8] took out a patent for reinforced concrete in 1854 in the UK, the first practical appearance of the material was in France in 1855 with Lambot's 'ferciment' concrete boat. However, a far more important development took place in 1867 when Monier produced his reinforced concrete flower pots which led to his patent for structural reinforced concrete in 1877. At the same time Hyatt,[9] the inventor of glass prisms for pavement lights, had organized a number of tests on concrete beams reinforced with iron ties which were carried out by David Kirkaldy and found to be satisfactory. He also carried out fire tests to show that the steel and concrete acted in unison under heat.

In 1879 the astute contractor, Wayss, bought the German rights to Monier's patent, and in 1887 he published *Das System Monier (Eisengerippe met Cementumhülling) in seiner Anwendung aug das gesammte Bauwesen*. As often happened, the new system was first used in civil engineering, in this case for arched bridges, some of which have the elegance of Maillart. One of these, built in 1890, spanned 37.2 m (Figure 2.4).

François Hennebique,[10] a provincial building contractor working in Belgium, experimented with concrete reinforced with iron from about 1879, and in 1892 patented his system, giving his method of calculation and typical reinforcement details for the bending of beams and slabs. He accepted many commissions before moving to Paris in 1894, when he gave up contracting and supervised the training of concrete contractors instead. At first his engineers carried out the calculations in Paris, but later this was done by his agents in their own offices and sent to Paris for checking.

The building system was essentially a frame with infill floor slabs and external cladding in brick, stone, concrete or glass. Continuity of reinforcement through

Figure 2.4   Monier Bridge, 1890.

Figure 2.5   Hennebique's Tourcoing factory, 1895.

the haunched column and beam joints provided rigidity naturally and made can-tilevers easy to construct. Architects accepted these characteristics and developed their designs accordingly. As early as 1895 the spinning mill at Tourcoing, one of Hennebique's earliest projects, already clearly illustrates the new concrete archi-tecture (Figure 2.5). By the time he had produced his first building in the UK (the 1897 Weaver flour mill in Swansea with agent, L.G. Mouchel), several 100 projects had already been constructed in Europe.

The Paris Exhibition of 1900 revealed for the first time to a wide international audience the great potential of reinforced concrete both in architecture and in civil engineering, since it was used for a number of pavilions and other installa-tions such as bridges and sewage works. For the Petit Palais, Hennebique together with the architect Charles Girault, built a free-standing spiral slab staircase, a three-dimensional form which would be repeated many times. In 1904 Hennebique built a house for himself in Paris, the flamboyant design of which again demonstrated the role of concrete in domestic architecture. For its construction thin precast concrete units 14 in high were used as permanent shuttering. Hennebique published a house magazine, *Le béton armé*, which was circulated to all his regional offices giving details of the latest projects and information on new developments. In 1899 one of his engineers, Paul Christophe published the first textbook on reinforced concrete, *Le béton armé et ses applications.*

Many other systems were patented, mainly on the geometry and nature of the reinforcement, but only a few survived and were developed into the 20th century. These included those of Coignet and Considère, both of whom followed Hennebique in opening offices in the UK. These are described by Michael Bussell in Chapter 4.

Development of reinforced concrete in the USA

Natural cements had been produced in the USA since the early 1820s and Portland cement since 1871. As in France and the UK, unreinforced concrete walls and fireproof filler-joist floors were used, but in the USA concrete blocks with different finishes and colours became popular. Mass concrete received a boost when it was used for the foundations of the Statue of Liberty.

Ernest Ransome,[11] who first used reinforced concrete in America, began by running his father's San Francisco factory which was producing artificial stone. In 1844 he patented the use of cold twisted iron bars of square section as reinforcement, carrying out extensive tests to confirm his patent. In 1888 he used reinforced concrete for the main floor structure, which was supported on internal cast iron columns and external stone walls, but the following year the whole structure was in concrete. On another contract in 1889 he introduced and patented the ribbed floor construction still used today (Figure 2.6). He precast the beams and together with *in-situ* concrete slabs formed a composite construction. He went on to build many notable multi-storey factories and warehouses. The 16-storey Ingalls building in Cincinnati of 1902–1903 (Figure 2.7), then the world's tallest building, made use of Ransome's system. His system and that of the Trussed Concrete (Truscon) Co., which used the Kahn bar, were the most widely used in this early period, although the Monier and Hennebique systems had also been introduced.

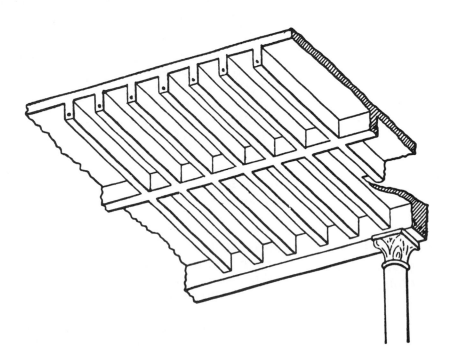

Figure 2.6 Ransome rib floor, 1889.

Figure 2.7   Ingalls building, 1903.

Earlier, in 1902, Orland W. Norcross, an engineer from Boston, patented a flat slab system of radial reinforcement from columns so as to eliminate connecting beams between columns for economy, having successfully built a structure on this principle. The first mature method of flat slab construction was the inclusion of an enlarged mushroom head first used by CAP. Turner in 1905–1906 and patented in 1908. He used four layers of reinforcement orthogonally and diagonally between columns. The omission of beams gave extra headroom or allowed the floor-to-floor height to be reduced, and it allowed services or partitions at a constant height as well as providing more light internally. Turner became as busy an engineer as Ransome and Truscon.

The work of Robert Maillart, 1872–1940

Robert Maillart[12] was trained at ETH in Zurich under Professor Ritter. He graduated in 1894 and started work with a contractor who was building Hennebique structures. He then moved to the Department of Works of Zurich, where he designed and built a mass concrete arched bridge which the city architect clad in stone. In 1901, with another contractor, he designed his first reinforced concrete bridge, a three-hinged arched structure spanning 38 m at Zuoz, in which he showed his appreciation of the character of reinforced concrete by joining together the

Figure 2.8   Tavanasa Bridge, 1905.

slabs and walls to form a monolithic box construction. The road slab cantilevered beyond the spandrel to form a marked shadow line, a feature which was to become typical of Maillart's own designs and of many bridges today. His professor, Ritter, checked the design and supervised a load test.

In 1902, at the age of 30, he set up his own design and contracting firm and successfully tendered for the bridge at Bilwil. Like Zuoz, it had two spans and was of similar construction. Maillart's moulding of the concrete at joints, a feature which also became a hallmark of his work, is of particular interest. As with Zuoz, small cracks occurred in the solid spandrels, so in 1905 for the 51 m span Tavanasa Bridge (Figure 2.8) he reduced the spandrel concrete to a minimum, expressing the three-hinged structure with elegance and minimum cost.

Maillart also studied the nature and economics of floor construction. In 1909 he patented his 'beamless deck' system after testing prototypes, and for a few years built many buildings, mainly warehouses, using the system in Switzerland (Figure 2.9), Russia and other parts of Europe. In contrast to Turner's system of 1908 in the USA which used four layers of mesh, Maillart only used two layers in each face of the slab with widely splayed column heads and octagonal or square columns, a more rational engineering solution and quite different from the frames of Hennebique.

During the 1914–18 war Maillart was marooned in Russia, but on his return he set up office again but as a consultant. His next development in bridge design was in 1923 when he built a deck-stiffened arch at Flienglibach. Here a thin arched slab was loaded through walls from the road deck, which had deep parapet beams of sufficient stiffness to minimize the deformation of the arch under unequal loading. Two years later he built the Valtschielbach Bridge (Figure 2.10) to a similar design.

His greatest bridge, also a deck-stiffened arch, was at Schwandbach in 1933, where the elliptical ground plan of the road deck is supported on a varying width thin arch by vertical walls to give a truly three-dimensional form for the whole (Figure 2.11). His structural virtuosity and architectural flair are also revealed in his unusual but logical roof to the train shed at Chiasso[13] with its organic shape and in his spectacular Cement Hall of 1939 built for the National Fair in Zurich.

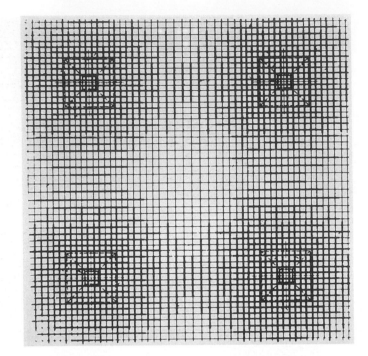

Figure 2.9   Maillart's flat slab, 1909.

Figure 2.10   Valtschielbach Bridge, 1923.

The work of Eugène Freyssinet, 1879–1962

Born 7 years later than Maillart, Freyssinet[14] studied reinforced concrete under Rabat in Paris. In 1905, at the age of 26, he took up his first professional job (with a provincial government department) at the period when Maillart was building the Tavanasa Bridge. In 1910 Freyssinet designed the Verdre Bridge with three spans: 225, 240 and 225 ft. Each span was a three-pinned arch with decentring jacks positioned horizontally at the crown. Due to creep of the concrete the crowns deflected 5 in, so he reinstated the jacks to relevel. For later bridges he turned to two-pin arches to reduce deflections and carried out further studies into creep and into concrete mix design to produce higher strengths.

Figure 2.11   Schwandbach Bridge, 1933.

Figure 2.12   Montluçon, 1915.

In 1914 Freyssinet left government service and became technical manager of a construction firm, Mercier Limousin et Cie, but had to spend four years in the army constructing many industrial projects in reinforced concrete which used a minimum weight of steel which was in short supply. Innovation and invention were readily accepted and Freyssinet introduced some of the first concrete shells ever built.

In 1915 he covered the glass works at Montluçon with a series of arches with concrete vaults in between (Figure 2.12). The following year, for the Schneider

Figure 2.13   Orly hangars, 1921.

Figure 2.14   Conoid shells, 1928.

factory at Le Creusot, he produced a continuous roof vault with external stiffeners and internal ties with the weight distributed onto external arches. For aircraft hangars he turned to concrete vaults, and for the two airship sheds at Orly in 1921 he produced thin corrugated vaults spanning 70 m, rising 50 m, and 200 m in length. These hangars were later seen by architects as great works of art (Figure 2.13). In 1928 he began to use conoid shells (Figure 2.14) for a number of projects such as the Austerlitz station in Paris, while his Plougastel Bridge, completed in 1930, broke all records with its three spans of 180 m (Figure 2.15). He described this project at the Liège conference of 1930, by which time, aged 50, he had retired from the firm to devote his time to the development of prestressed concrete, a system he had patented in 1928. However, it must be said that for the previous 40 years a number of engineers[15] had tried to invent a workable system.

During the Plougastel Bridge project, Freyssinet began to understand the nature and magnitude of creep, concluding that the losses in prestress would not be excessive if high strength concrete and high tensile steel were used. He succeeded in making hollow pretensioned, prestressed concrete lighting poles on which he spent much of his fortune, but to no avail as the market collapsed. In 1934 his luck changed when, at the last minute, he saved the Marine Terminal at Le Havre from collapse (it was settling at the rate of 1 in per month) by adding extra concrete

Figure 2.15 Plougastel Bridge, 1930.

between existing foundations and stressing them together to form a grillage which he supported on prestressed concrete piles. This was his first attempt to exploit his ideas on a large scale. Prestressing was accepted, and in 1936 he built his first bridge, of 19 m span and 4.9 m wide, over the dam at Portes-de-Feu. Very soon he was building bridges from prefabricated units and stressing them together. His most telling remark was that 'prestressing is a state of the mind' and singly it was the most far-reaching engineering invention of this century. Francis Walley has given an extensive review of the subject in Chpater 10.[36]

The architect and the new material — reinforced concrete

The first architect in France to develop concrete construction was Auguste Perret,[16] who in 1903 chose a concrete frame for a 10-storey block of flats in 25 Rue Franklin in Paris (Figure 2.16). Externally he emphasized the structure by cladding the frame with a plain ceramic tile and the infill with a decorative tile with a subtle difference in colour. Glass blocks, recently invented, also became part of the design. Although he worked in his father's building firm, the flats were constructed by others, but for a car showroom two years later he also acted as the contractor and exposed the concrete frame to produce a functional classical facade (Figure 2.17).

After his father's death in 1905, he and his brother renamed the firm Perret Frères and within a short time became specialists in reinforced concrete. They were appointed to construct the theatre in the Champs Elysées in 1911 to a design by Henri van de Welde. Perret criticized the proposed structure and ended up redesigning it with plans signed by the official architect. Because of the confined site on which to build two auditoria (the larger seating 2000 people), the structure was brilliantly rationalized. He went on to explore structural expression in his industrial buildings and the interior of the Esden clothing factory in Paris in 1919 gave a new dimension to the possibilities of reinforced concrete. Three years later in 1922 he took on the design and building of a minimum cost church at Le Raincy (Figure 2.18) where his use of pierced precast concrete panels for external walls together with exposed concrete barrel vault roofing expressed a true concrete aesthetic.

Figure 2.16   Rue Franklin, Paris, 1903.

Figure 2.17   Car showroom Paris, 1905.

In the UK the earliest buildings in which architects exploited the nature of concrete were the 1907 Lion Chambers in Hope Street, Glasgow (Figure 2.19) and the 1909 Liver Building in Liverpool, which was 15 storeys high, both built by the Hennebique Company. In the USA Frank Lloyd Wright thought that concrete would bring a new architecture and it is interesting to see how, in 1906, he designed

Figure 2.18   Church at Le Raincy, 1922.

Figure 2.19   Lion Chambers, Glasgow, 1907.

his Unity Church in Oak Park, Chicago, where he used monolithic walls with elaborate mouldings (Figure 2.20). He turned to using concrete blocks as permanent shuttering, the blocks, some of which used white cement, being made in plaster moulds. Many houses in California were to be built with this technique.

In the early 20th century in Germany the engineer–contractors, Wayss & Freytag, and Dyckerhoff & Wildmann, were busy constructing many industrial buildings with a variety of roof forms. In 1913 Dyckerhoff & Wildmann, with the architect Max Berg, built an impressive exposed concrete Centennial Hall in Breslau (Figure 2.21) with a dome 213 ft in diameter supported on a series of arches and clearly exhibiting the plasticity of concrete. At the same time a new movement

Figure 2.20   Unity Temple, Oak Park, Chicago, 1906.

Figure 2.21   Centennial Hall, Breslau, 1913.

in architecture emerged, influenced by the huge new exposed concrete industrial structures such as warehouses and silos which were the result of scientific method and economy. Gropius, Le Corbusier and Mies van der Rohe were the major exponents. In 1911 Walter Gropius led the way with his Fagus factory in Anfeld a.d.Leine, which has a reinforced concrete frame with stub cantilevers on which the steel and glass skin was supported. He used a similar construction in 1925 for the Bauhaus in Dessau. Le Corbusier spent a short time in Perret's office in 1909–10 and in 1915 published a sketch of his ferro-concrete skeleton for dwellings which formed the basis of later designs by his contemporaries. A good survey of architectural design in concrete was given by T.P. Bennett in 1927.[17]

At the first international conference on reinforced concrete in Liège in 1930,[18] Dischinger of Dyckerhoff & Wildman gave details of the development of shells from his firm's dome at Jena in 1924 to their cylindrical shell of 1926. Predicted future forms included northlight shells and square domes similar to those at Reval (now Tallinn), built in 1916,[19] and at Brynmar, constructed much later in 1951. In the discussion Freyssinet talked about his experience with concrete shells.

One engineer who was absent from the Liège conference was Sir Owen Williams,[20] who had started work in 1911 with the British office of the US Indented Bar-Engineering Co. He quickly moved to the Trussed Concrete Steel Co. and became chief estimating engineer at the age of 23. During the 1914–18 war he designed aircraft and concrete ships. In 1920 Owen Williams set up as a consultant and was appointed structural engineer for the Wembley Exhibition of 1924, and he later built a number of bridges with Maxwell Ayrton as architect. Williams's flair for concrete design is readily seen in his Findhorn, Spey, Crubenmore and Loch Alvie Bridges on the A9 in Scotland (Figure 2.22). In 1929 at the age of 39 he registered as an architect and took on the responsibility for the design of a whole

Figure 2.22   A9 bridges by Owen Williams, 1925–26.

building, not just the structure. His Boots 'wets' building in Nottingham of 1930–32 with its large-scale economic flat slab structure and its standard industrial curtain wall on the perimeter was admired by Modern Movement architects. Although he followed this with various minimum structure buildings for the Daily Express, Peckham Health Centre and others, the Boots 'wets' is still his best achievement. Like Maillart, he took especial care in moulding the concrete at structural joints such as the abutments of his bridges and the column heads of his flat slabs.

The International Congress of Modern Architecture (CIAM), set up in 1928, brought together modern architects such as Gropius, Le Corbusier and like-minded architects in the UK such as Wells Coates and Joseph Emberton. Conferences discussed the emerging society and how architects could best serve its needs. In 1931 at the age of 30 Berthold Lubetkin[21] came to England from Russia via Paris and formed the architectural practice, Tecton, which was given the commission to design the Gorilla House at London Zoo. Lubetkin had studied reinforced concrete in Berlin and Paris, where he met Le Corbusier and Mies van der Rohe, and had attended Perret's Atelier. While he was designing the Gorilla House, the Paris office of Christiani & Neilsen, engineering contractors who specialized in reinforced concrete works, suggested that he should meet Ove Arup, their chief designer in London. This was the beginning of a long and fruitful collaboration between Lubetkin, considered the leading modern architect, and Arup, the engineer–contractor. Arup moved to the contractors J.L. Kier in 1933 as director and chief designer so that, after 11 years of civil engineering experience, he could also tender for the construction of buildings and thus associate with Modern Movement architects.

Lubetkin's earliest buildings were his best. For the block of flats, Highpoint, Arup persuaded Lubetkin to use external load-bearing walls and floor slabs supported on spine beams so that there was less structural obstruction in the flats than with the typical two-way frame. However, some structural gymnastics were necessary to maintain a standard architectural grid of columns at ground-floor level. Sliding formwork, originally developed for silo construction, was innovatively used here by Arup, the contractor, for the exposed concrete external walls of a domestic building, although later there were problems with weathering owing to their 5 in thickness.

The Penguin Pool at London Zoo was Lubetkin's second project and came to symbolize modern reinforced concrete architecture. With structural daring Lubetkin elegantly intertwined two spiral ramps, first hinted at by the spiral staircases in Paris. Arup worked closely on the project and approved the structural form of the final proposal.

The calculations for the ramps were complex and were carried out by Felix Samuely,[22] who also suggested that the section should be trapezoidal, a change which was agreed to. Samuely had just arrived from Berlin where he had been a co-partner in the consulting firm of Samuely & Berger, set up by him in 1929 at the age of 27. Samuely worked with Kier for nine months before again setting up a partnership, this time with Helsby and Hannam. They were to be the structural consultants of Bexhill Pavilion, with the architects Mendelssohn (with whom Samuely had worked in Berlin) and Chermayeff. Being an independent consultant with experience in concrete and, more importantly, in the new technology of welded steelwork, he attracted the Modern Movement architects from 1934 to the outbreak of war. Of particular note were the Wells Coates flats in Brighton and Kensington (Figure 2.23) and the shallow 100 ft diameter concrete dome at Folkestone for Pleydell Bouverie. Like Arup, Samuely was dedicated to the integration of structure in architecture and appreciated that the role of the engineer was to give the architect the confidence to create a structure for his building. He was senior lecturer in structures at the Architectural Association, a member of the MARS Group and joint editor of their *Master Plan of London*, published in the *Architectural Review* 1942.

Figure 2.23   Palace Gate flats, Kensington, 1938.

Arup was to some extent impeded by staying with Kier because as a contractor he had to tender for work rather than having the freedom to offer his services to architects as an independent consultant. Nevertheless, he joined the Architectural Association and the MARS Group, where he followed his intellectual pursuits and formed close friendships with a wide circle of architects. The mathematician and structural analyst Ronald Jenkins joined Kier in 1935, 2 years after the design of the Penguin Pool, and followed Arup when he left to set up his own contracting firm, Arup & Arup, in 1938.

Eduardo Torroja, 1889–1961

Torroja was one of the most creative engineers of this centry. His contribution to the understanding of building structures and their development is comparable to that of Maillart with his bridges. Frank Lloyd Wright considered him the creator of organic structures and architecture.

Torroja's three-dimensional concept of structures can be clearly seen in the roof of a hospital clinic built in 1926, one of his first projects as a consulting engineer. The structure is similar to a top hat on an octagonal plan, with a compression ring on top and a wide brim in tension. His search for economic and pure structures is evident in his design, also in 1926, for a portal-framed bridge in a cutting, where the interconnecting slabs between the frames were located in positions where they acted as compression flanges. Nervi was to use the same idea in his design for the sophisticated lecture theatre in the Unesco building in Paris in 1952.

Torroja, the son of a geometer and architect, was brought up in the Spanish atmosphere of brick or tiled vaults which became even more extravagant after the introduction of Portland cement. Working with an architect, his first notable shell in 1933 was a 3½ in thick 156 ft span dome at Algeciras supported on eight columns interconnected by pretensioned ties (Figure 2.24). A more interesting shaped shell with a span of 180 ft was used for a pelota court (Fronton Recoletos) in Madrid in 1935. Here he combined latticing in the glazed areas with a solid 3¼ in thick slab, elegantly solving the problem set by the architect Zuaso. The structural design was confirmed by a model test carried out at the University of Madrid, where Torroja was professor.

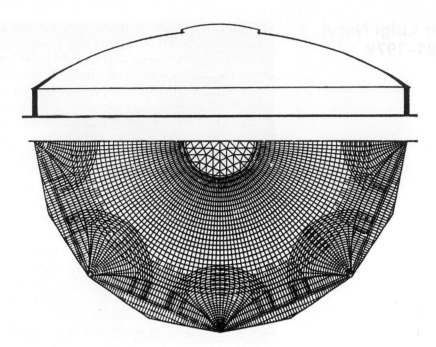

Figure 2.24   Dome at Algeciras, 1933.

Figure 2.25   Aqueduct at Alloz, 1940.

Many examples of his work could be given, but one in particular exemplifies the graceful simplicity of his designs. This is the prestressed concrete aqueduct at Alloz built in 1940 (Figure 2.25), where the concrete is maintained in constant compression for watertightness. In his book *Philosophy of Structures*,[23] which should be essential reading for all engineering and architectural students, he makes a telling statement: 'The process of visualizing or conceiving a structure is an art — it is motivated by an inner experience, by intuition.'

Pier Luigi Nervi, 1891–1979

Having had a good academic training and excellent experience with concrete construction, Nervi set up his own contracting firm at the age of 30. It enabled him to conceive, design, develop and build his own structures. In 1926 he built a cinema in Naples which had a three-dimensional top-hat structure, 30 m in diameter, similar to Torroja's hospital roof construction of the same year.

Built in 1930, the staircase tower at his Florence Stadium, with its spiral flights stiffened by a reverse spiral beam, is not only economic but it is also concrete sculpture (Figure 2.26). His first large ribbed shell structure was the hangar built for the Italian air force in 1935 (Figure 2.27). This majestic building of *in-situ* concrete covers a column-free area of 350 ft by 132 ft. A model test was used in the design. A similar hangar, built in 1940, shows a bolder and simpler support system. The cost and time of construction were reduced by using precast concrete latticed units which had projecting reinforcement subsequently welded for continuity.

For long-span roofs, Nervi appreciated that a lightweight structure was beneficial. He first proposed the use of thin slabs of ferro-cement, cement mortar reinforced with superimposed layers of mesh and small diameter bars, for an exhibition building at the Rome World Fair of 1939. Ferro-cement, which is in fact a development of Lambot's ferciment of 1847, has a high degree of elasticity and resistance to cracking. No formwork is required and architectural or organic shapes are easy to form. During the war, Nervi was commissioned to build boats using ferro-cement and tests showed that such construction was possible. However, its first actual use only came in 1945 when Nervi built a 165 t yacht with a hull thickness of 3.5 cm. Its first use in building construction followed in 1946 when he built a small warehouse for his firm. After much testing by Professor Oberti, he used it on some large-scale roof structures, including the 94 m span arched Exhibition Hall in Turin of 1948–49 (Figure 2.28).

This roof was made up from ferro-cement inverted trough units 2.5 m long and 1.45 m high and with a wall thickness of under 5 cm sitting end to end and

Figure 2.26   Florence Stadium staircase, 1930.

Figure 2.27   Air force hangar, 1935.

Figure 2.28   Exhibition Hall, Turin, 1949.

connected with *in-situ* concrete chords. The system worked well and Nervi went on to use ferro-cement for all his later roof structures and for permanent formwork on his next innovative structural form, isostatic ribbed floor slabs (Figure 2.29), where each rib carries load to the columns according to its relative stiffness. Their positioning is determined by the designer and therein lies the art

Figure 2.29   Isostatic slab, Gatti wool factory, Rome, 1953.

of the engineer. Nervi's art is to be seen in all his structures from the Turin Exhibition Hall to the 1957 Sports Palace in Rome and has been well documented by him.[24] What is of interest is the work he carried out as a consulting engineer with architects on buildings. Although parts of the structures are exposed and are recognizably Nervi's, he seems to have had little influence on the overall architectural form, which may confirm that a brilliant structure does not necessarily produce good architecture. However, Nervi has left us with a rich selection of structures and writings to study.[25]

Postwar construction in Europe

In the continent innovative use was made during the war of prestressed concrete units for the construction of bombproof U-boat pens and for motorway bridges, while in the UK the Mulberry harbours were outstanding.[26] The main problem for the building industry of postwar Europe was the rapid reconstruction of houses, schools and factories, while for the civil engineers it was the construction of bridges. Skilled craftsmen and structural materials — steel and timber — were in short supply, so that industrialized system building of large, precast concrete floor units and wall panels was used extensively for multistorey housing. Precast and prestressed concrete floor and long-span roof systems, reinforced concrete shells and prestressed concrete bridges all made their appearance, together with new, larger capacity mechanical handling equipment such as tower and mobile cranes.

Architects had to learn how to use the new structural forms and to express the aesthetics of large panel construction. They exposed a wide range of aggregates, tooled the concrete or used special plastic linings to the moulds, and it was found that white cement and sand became an acceptable colour when wet. Prestressed concrete floor units spanned up to 50 ft and gave architects a greater flexibility in the planning of buildings, while the introduction of 30–40 ft span *in-situ* concrete continuous coffered flat slabs cast onto plastic pans allowed the concrete to be left exposed or simply painted.

In 1946, after 24 years of contracting, Ove Arup set up as a consulting engineer, forming his partnership in 1949. Because of his concern for the integration

Figure 2.30 Hatfield Technical College, 1950.

of structure in architecture, on which he had widely written and lectured, and particularly because of his intellect and personality, he attracted architect clients and outstanding engineers.

Ove Arup & Partners' answer to multi-storey housing was the *in-situ* concrete box-frame construction with industrialized shuttering systems, though they turned to prefabrication for other structures. At the 1951 Festival of Britain, a showcase for structural ingenuity, they built a continuous prestressed concrete footbridge and a prestressed concrete two-way grid roof. The partnership have been consultants on many prestigious concrete-framed buildings such as Coventry Cathedral and the Sydney Opera House, on which Peter Rice gained his practical experience before becoming the leading engineer-architect of the last 20 years.

In 1950 Felix Samuely built his first and perhaps his most innovative prestressed concrete structure. For the heavily loaded floors of a printing factory in Bristol, he designed precast concrete three-hinged frames with prestressed concrete ties. For ancillary buildings he used prestressed concrete planks as reinforcement for continuous composite floors. At the same time his structure for Hatfield Technical College had site-cast large three-hinged roof frames and trussed beams for the composite floor construction (Figure 2.30). He instigated further research on composite construction[27] and ended up with a series of standard designs which were checked by full-scale tests similar to those carried out by Ransome in 1889 to confirm his composite ribbed construction.

For roofs, prestressed concrete beams and trusses reached unprecedented spans in industrial buildings. In the UK A.J. Harris, who had worked with Freyssinet and who ran the Prestressed Concrete Co., was responsible for the innovative design of the BEA hangar at London Airport in 1951 (Figure 2.31). Secondary roof units, made up from precast units post-tensioned together, span 110 ft and are supported on 150 ft span primary post-tensioned *in-situ* concrete box beams.[28]

Figure 2.31   BEA hangar, London Airport, 1951.

Figure 2.32   BOAC hangar, London Airport, 1955.

At the same time Owen Williams was designing a hangar for BOAC (Figure 2.32) with a 336 ft span external *in-situ* concrete arch with counterbalances supporting a 140 ft roof structure. On the adjacent site he won a limited competition to build one of the first stayed roofs in the UK. This was a hangar with double cantilevers of 110 ft with raking steel ties.

That the development of concrete shell structures across Europe was spectacular may be seen in the many projects illustrated in the proceedings of the first and second symposia on concrete shell roof construction in London in 1952 and Oslo in 1957. Peter Morice and Hugh Tottenham have written on early shell development, and Robert Anchor describes Twisteel's extensive construction in the UK of north-light shells for warehouses and industrial buildings in the 1950s. British Reinforced Concrete Ltd also built standard shells, but it was to the independent consulting engineers, such as Ronald Jenkins of Ove Arup & Partners, that architects went to discuss their roofing proposals.

Figure 2.33   Assembly hall,
Wigan, 1949.

Figure 2.34   Clifton
Cathedral, Bristol, 1965.

The type and size of shells evolved with notable structures by Heinz Isler in Switzerland and by Esquillan in Paris. Esquillan's immense CNIT exhibition hall in Paris, completed in 1958, is a double shell dome covering a triangular area and supported at the corners 218 m apart.

In the UK Samuely built his only barrel-vault roof for a factory in Bristol in 1948. For economy and to provide a wider range of architectural forms, he turned to

Figure 2.35   MODA building, Jeddah, 1978.

folded plate roofs (or 'Faltwerke' as it was known in Germany). Here the shell roof is made up from a series of planes rather than a curve. These could be made either of *in-situ* concrete or of a composite construction of precast units with the shell reinforcement in the *in-situ* concrete screed (Figure 2.33). To eliminate top shuttering for steep slopes he prestressed solid units together.

For a school in London in 1949 he not only used a concrete folded plate for the balcony structure of the assembly hall, but supported it on a star beam[29] which distributed the load to three columns. This innovative system was used to advantage by F.J. Samuely & Partners for Clifton Cathedral (Figure 2.34) in 1965, six years after Samuely's death at the early age of 56. Their roof construction in Jeddah, intended to carry helicopters (Figure 2.35), clearly shows the influence of Nervi's isostatic floor design.

The work of Felix Candela (b. 1910)

Any history of shell construction would be incomplete without a description of the work of Felix Candela,[30] an architect by training but with an interest in mathematics and structures. In 1936 he was about to leave Spain on a travelling scholarship to visit Dischinger & Finsterwalder and study concrete shells when the Civil War broke out. He joined the Republicans and in 1939 was deported to Mexico. After a few years of varied experience with contractors, he built for himself his first shell in 1949, an experimental ctesiphon shell where drooped hessian between arches was sprayed with concrete.

Two years later, at the age of 41, Candela set up a contracting organization with his brother to build shell roofs in which he did the architectural and engineering design. In 1951 the firm began with a startling building, the Cosmic Rays Pavilion (Figure 2.36), where the brief called for a concrete slab with a maximum thickness of 5/8 in. Candela re-read Aimond's paper on hypars,[31] written in 1936, and did relatively simple calculations for its design. The building was instantly acclaimed by architects and he went on to build cylindrical shells, folded plates and conoids before concentrating on hypars. Because he was responsible for the construction, he could take risks and experiment. He did not agree with elastic design of shells and developed his own method of calculation, based on ultimate load design.

Figure 2.36   Cosmic Rays Pavilion, Mexico City, 1951.

Figure 2.37   Iglesia de la Virgen Milagrosa, 1955.

His output was prodigious and each structure was a development of the last and showed a complete mastery of the hypar form. Of particular note is the Iglesia de la Virgen Milagrosa built in 1955 (Figure 2.37).

Long-span bridges and tall buildings

The need for new bridges was more acute in Europe than in Britain. Freyssinet first built prestressed concrete bridges in France in 1942 and after the war continued with many over the River Marne using prefabricated units, by which time prestressing was beginning to be widely accepted by engineers. Many

Figure 2.38   Lake Maracaibo Bridge, 1957–62.

systems of post-tensioning, following Freyssinet's cone anchorage, were developed and described at the various world conferences on prestressed concrete, as were the varied uses of the system. Fritz Leonhardt was particularly prominent in the development of prestressed concrete bridges and has written on the aesthetics of bridges. In Italy, Riccardo Morandi built his first prestressed concrete bridge in 1950. Twelve years later he crossed Lake Maracaibo, Venezuela,[32] with a 5½ mi long bridge with maximum spans of 771 ft which uses a double cantilever system with prestressed concrete inclined ties (Figure 2.38).

Cable-stayed bridges increased the economic spans of concrete and each year greater distances are being bridged as reliable concrete strengths increase. Amongst these, the Ganter Bridge in Switzerland (Figure 2.39) built in 1980 by Christian Menn, a pupil of Maillart, is much acclaimed. The outer spans are curved on plan and the diagonal ties were stressed after the concrete had been poured. The main span is 174 m and the piers rise to 150 m from the valley floor.

The development of tall reinforced concrete buildings, following Ransome's 16-storey Ingalls building of 1903–1904, is best seen in the USA. Concrete service cores, enclosing lifts and staircases, provided the stiffness to withstand wind loads transmitted to them by the floor slabs, which left the external framework to carry vertical loads only. In 1953 in Chicago, the Marina City apartment block, circular on plan, reached 60 storeys using this system. At the same time Myron Goldsmith[33] was preparing his thesis on the appearance of tall concrete buildings (Figure 2.40). He envisaged a system whereby windloads would be carried on a massive external framework, with wide-centred columns and storey-height beams which would also carry intermediate floors. Some buildings of this form appeared later in Europe. Goldsmith was working with Skidmore, Owings & Merrill (SOM) in Chicago and teaching at IIT where Mies van der Rohe of the Bauhaus was professor. Goldsmith and Fazlur Kahn, a structural engineer also at SOM, carried on

Figure 2.39   Ganter Bridge, Switzerland, 1980.

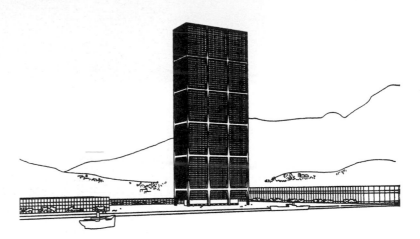

Figure 2.40   Myron Goldsmith's thesis on the appearance of tall concrete buildings, 1953.

with research at IIT on tall buildings from which the idea of shell and core construction evolved. Their Brunswick Building in Chicago (Figure 2.41) is a good example. As the core is relatively stiffer than the external frame near the ground, wind loads are gradually transferred to the core so that the external frame only carries vertical loads. While he was working with Bruce Graham at SOM on the design of the Two Shell Plaza in Houston (Figure 2.42), Kahn studied the distribution of load in a regular external framework of a shell and core system sitting on wide centred columns. He used a powerful computer to determine the sizes of columns so that they were all equally stressed. This indicated that the visual appearance of a shell core system could be exploited by the architect and that it would be possible to analyse many different configurations of external columns.

Also in Chicago, which has excellent sand and aggregates, high-strength concretes are being developed and used. Thus the maximum possible height of buildings is continually increasing and external latticed concrete frameworks are beginning to compete with tall steel construction. However, the tallest concrete construction is not used for buildings but for the latest North Sea oil rigs. This new problem which has been facing civil engineers is comparable with that faced by Brunel and Stephenson in the 19th century when designing long-span bridges

Figure 2.41 Brunswick
Building, Chicago, 1966.

Figure 2.42 Two Shell Plaza,
Houston, 1972.

Figure 2.43   Brent Field oil rig, 1991.

to carry heavy locomotives. Varied designs for rigs in steel and in concrete, each breaking new ground and advancing technology, have been built, but their designers go unnoticed by society. For the Brent Field, Norwegian contractors have just installed a 1500 ft high rig in reinforced concrete (Figure 2.43). Further historic details of the construction is given in Chpater 13.[37]

Conclusions

This short survey of the development of concrete highlights the roles of the engineer and architect. Construction under water and fireproofing of buildings were their respective technical problems from the Roman period until the discovery of natural cement and the important invention of Portland cement. Portland cement concrete was able to carry substantial compressive loads, which made it suitable for foundations, load-bearing walls and fireproof floors. Reinforcing Portland cement concrete at the end of the 19th century produced a revolutionary new structural material which brought about the scale of construction we see today.

Its innovative use by engineers came very early with Maillart and his Tavanasa Bridge of 1905. His mathematical and structural prowess led him to envisage intuitively a concrete box construction for a three-hinged arch bridge. He had his own contracting firm and so was able to carry the risk of both design and construction. In 1909 he patented a fully tested flat slab building construction. His breadth of vision and inventiveness may also be seen in his deck-stiffened arch bridge system and his Chiasso roof truss, while his Cement Hall in Zurich shows his artistic virtuosity.

Eugène Freyssinet was also an engineer and contractor whose innovative uses of concrete in early shells and bridges led him to his momentous concept of pre-stressing. Nervi and Candela followed in the footsteps of Maillart and Freyssinet in that they were contractors which allowed them the freedom to develop their ideas. Nervi invented ferro-cement for lightweight long-span roof construction, and his artistic flair and writings increased the perspective of design in concrete. Candela, an architect by training, illustrated the many possible forms of hyperbolic paraboloid shells.

Eduardo Torroja and Fazlur Kahn were both structural consultants who made significant innovative uses of concrete but also contributed to a better understanding of the nature of structures. Kahn made use of a powerful computer, the evolving tool for structural analysis, to develop his various schemes for tall buildings, with encouragement from architects at SOM in Chicago.

From Hadrian onwards, architects have always been interested in the latest technology. During the nineteenth century they used cements for renders and plain concrete to imitate stonework and also to solve the problem of fire resistance. Perret in Paris pioneered the exposed external framed building, while Frank Lloyd Wright in 1906 built his first concrete-walled Unity Temple in Chicago. In Germany in 1913 the 64 m Breslau dome surpassed that of the Pantheon in size. The Modern Movement architects made use of concrete as a framework onto which they hung cladding or as a perforated wall, void of decoration. As reinforced concrete was a new material without a tradition of knowledge or experience, architects had to turn to specialist engineers for advice. When shells and then prestressed concrete appeared, the advances in their use for building were made by close partnerships of mutually appreciative architects and engineers.

Architect–engineer collaboration is essential for concrete architecture to progress, but for civil engineering the architect-engineer Santiago Calatrava, who spans both disciplines, may have the answer. This short study shows that historically those civil engineers with artistic flair have been responsible for many of the major innovations in the use of concrete. Surely then, engineering students should be encouraged to develop their artistic ability and a study of history would given them a sense of presence in the ever-changing construction industry.

References

1. Mainstone, R.J., Developments in Structural Form. MIT Press: London, 1975.
2. Pasley, C.W., Observations on Limes, Calcareous Cements, Mortars (etc.). John Weale: London, 1938.
3. Smith, J.T., A Practical and Scientific Treatise on Calcareous Mortars and Cements (translated from L.J. Vicat's Résumé des connaissances positives). John Weale: London, 1837.
4. Grant, J., Experiments on the Strength of Cement, Chiefly in Reference to the Portland Cement Used in the Southern Main Drainage Works. Spon: London, 1875.
5. Collins, P., Concrete. The Vision of a New Architectury. Faber & Faber: London, 1958, 27–28.
6. Coignet, F., Bétons Agglomérés Appliqués à l'art de Contruire. E. Lacroix: Paris, 1861: 196
7. Collins, P., Concrete. The Vision of a New Architecture. Faber & Faber: London, 1958: 54.
8. Brown, J.M., W.B. Wilkinson (1819–1902) and his place in the history of reinforced concrete. Trans. Newcomen Soc., 1966–97, XXXIX, 129–142.
9. Hyatt, T., An Account of Some Experiments with Portland-Cement-Concrete Combined with Iron, as a Building Material. Chiswick Press: London, 1877.
10. Cusack, P., François Hennebique: The specialist organisation and the success of ferro concrete 1892–1909. Trans. Newcomen Soc., 1984–85. LVI, 71–86.
11. Banham, R., A Concrete Atlantis. MIT Press: London, 1986.
12. Billington, D., Robert Maillart's Bridges. The Art of Engineering. Princeton University Press: Princeton, 1979.
13. Billington, D., Robert Maillart and the Art of Reinforced Concrete. MIT Press: Cambridge, Massachusetts, 1990.
14. Ordonez, J.A.F., Eugène Freyssinet, 2C edn., Barcelona, 1979.
15. Abeles, P. W., The Principles and Practice of Prestressed Concrete. Crosby Lockwood: London, 1949.
16. Collins, P., Concrete. The Vision of a New Architecture. Faber & Faber: London, 1958.
17. Bennett, T. P., Architectural Design in Concrete. Ernest Benn: London, 1927.

18. 1930 Premier Congrès International du Béton et du Béton Armé (Mémoires, résumés, comptes rendus et discussions). La Technique Des Travaux: Liège, 1932.
19. Christiani, Neilsen. Twentyfive Years of Civil Engineering 1904–1929. Copenhagen, 1929.
20. Cottam, D., Owen Williams. Architectural Association (Works III), London, 1986.
21. Allen, J., Berthold Lubetkin. RIBA Publications: London, 1992.
22. Higgs, M., Felix Samuely 1902–59. Archit. Assn J., 1960, LXXVI, No. 843, 2–31.
23. Torroja, E., Philosophy of Structures. University of California Press, 1967.
24. Nervi, P.L., Structures — the Works of Pier Luigi Nervi. F.W. Dodge: New York, 1957.
25. Nervi, P.L., Aesthetics and Technology in Building. Harvard University Press, 1966.
26. The Civil Engineer in War. Institution of Civil Engineers: London, 1948.
27. Samuely, F.J., Composite construction. J. Instn Civ. Engrs, 1952, Feb., 222–59.
28. Harris, A.J., Hangars at London Airport. J. Instn Struct. Engrs, 1952, XXX(10), 226–35.
29. Samuely, F.J., Space frames and stressed skin construction. RIBA J., 1952, 166–73.
30. Faber, C., Candela. The Shell Builder. The Architectural Press: London, 1963.
31. Aimond, F., Etude Statique des Voiles Minces en Paraboloide Hyperbolique Travaillant Sans Flexion. Publications IABSE: Zurich, Vol. IV, 1936.
32. Boaga, G., Boni, B., The Concrete Architecture of Riccardo Morandi. Tiranti: London, 1965.
33. Goldsmith, M., Buildings and Concepts. Rizzoli: New York, 1987.

3 Concrete and the structural use of cements in England before 1890

Lawrance Hurst

Synopsis

The years between the invention of Roman cement in 1796 and 1890, by when Portland cement had effectively replaced all its competitors as the binder for mortar and concrete, also saw the development of concrete and its increasing use for foundations and for fireproof flooring. This chapter reviews that development and the use of cements in buildings by reference particularly to the Royal Institute of British Architects' Transactions which were the forum for discussion because building design was the preserve of architects, generally unassisted by engineers.

The next chapter takes over the story with the advent of reinforced concrete building frames by the specialists who had crossed the Channel and the Atlantic to take out English patents.

Introduction

During the 19th century most buildings were designed and drawn in an architect's or surveyor's office, and their drawings included all necessary details of the structure. The 1909 London County Council (General Powers Act)[1] was the first statutory document to lay down floor loadings, permissible stresses and a basis of design and hence it was not until then that consulting engineers started to be concerned with building structures. During the period covered by this chapter, the Institution of Civil Engineers is not the place to seek guidance to common practice, but rather the Royal Institute of British Architects. In those days the Institution of Civil Engineers was concerned with engineering works, leaving all aspects of buildings to the RIBA.

Certain types of buildings were designed by engineers, such as railway stations, but they were in those cases generally acting as architects, and parts of building were designed by specialist engineers — R.M. Ordish and J.W. Grover designed the Royal Albert Hall dome for Major General Scott, himself an engineer, but acting in that case as an architect. The remainder of the ironwork, the masonry, the foundations and the fireproof flooring were however designed and detailed in Scott's own office. Architects undoubtedly had assistance with their structural designs but that assistance is seldom identified, unless it was obtained from the supplier of the iron beams or the fireproof flooring or the reinforced concrete components. Foundations, masonry, unreinforced concrete and similar non-specialist components were detailed by the architect, any design necessary being either based on experience of not unsatisfactory behaviour or by reference to published formulae, which were generally derived empirically from tests.

It is notable that the first paper in Volume 1 of the Transactions of the Institute of British Architects is George Godwin's prize essay upon *the Nature and properties of concrete, and its application to construction up to the present period*,[2] which is followed by notes on the concrete foundations at Westminster New Bridewell[3] and on concrete underpinning at Chatham Dockyard[4] and later on M.I. Brunel's experiments on reinforced brickwork.[5] These papers and others from the Transactions of the RIBA are referred to later in this chapter.

Binders and aggregates

Cements

Today we know what is meant by the terms concrete, mortar and cement: unless qualified by additional words, cement means Portland cement, mortar means a mixture of Portland cement and sand, and concrete is made with coarse and fine aggregates and Portland cement. Before and during most of the 19th century such certainty did not exist, and until 1790 the distinction between cements which set or did not set under water, termed hydraulic or non-hydraulic, was not known, or if known was not really appreciated. Any limes burnt from limestone with a small clay content that would have had some hydraulic properties were unlikely to be appreciated because limes were invariably slaked and that would have killed those hydraulic properties at birth. Mortars and renders which needed hydraulic properties, perhaps for laying bricks in frosty weather or particularly exposed walls, incorporated terras or crushed brick to give a quicker set.

John Smeaton's experiments to discover a suitable mortar for use in the construction of the Eddystone Lighthouse were reported in his Narrative, first published in 1791,[6] where he set out the properties of 'water cements' and distinguished between cements which set by chemical action, under water, and those which merely hardened, by carbonation and evaporation, and which dissolved when placed in water. He deduced that ground volcanic rock — either puzzolana (now usually spelt pozzolana) from Italy or terras (also known as trass and terrace) from Germany via Holland — were needed as a constituent of a water cement. His recipes never came into general use, because in 1796, James Parker patented his Roman cement.[7,8] Parker's Roman cement, also known as Frost's, was a quick setting artificial hydraulic cement, 'artificial' meaning manufactured, not 'natural' as were the mined volcanic hydraulic cements. It was used for water-proof renders and for mortar for laying bricks in wet conditions during the first half of the 19th century, until such time as true Portland cement became available in 1845–50. Other patent cements were also used during the same period principally for exterior renders designed to imitate stone and for water-proofing.[9,10]

'Cement mortar', as used by Sir Robert Smirke in the brickwork above the lime concrete underpinning to Custom House in 1826 and as used by Mark Isambard Brunel in the Thames Tunnel was made with Roman cement.[11] The remarkable watertightness of the Thames Tunnel is a tribute not only to the workmanship but also to the use of Roman cement, without which it would not have been possible. Robert Stephenson made a rule never to use lime mortar in the arches of tunnels, but to build them with (Roman) cement exclusively[12] (Figure 3.1). Robert Smirke also used three or four courses of brickwork laid in Roman cement as chain bond in rubble walling at Maidstone County Court House.[13]

Mortar or render made with Roman cement is generally a brownish colour, and is much harder than the white or dirty grey lime mortar of the period. It was used for precast embellishments to be incorporated with stucco and for mortars where its quick setting or strength or hydraulic properties were needed. It was however seldom mixed with fine and coarse aggregates to make concrete, because it set too quickly. References to Roman cement are to be found up to 1880, when it was still significantly cheaper than Portland cement, but not much evidence of its use is to be found in those later years.

Portland cement was the term used by Joseph Aspdin in his patent of 1824, but it was not until about 1845 that I.C. Johnson, the manager of the cement works at Swanscombe, Kent, burnt the new materials at a high enough temperature to produce what we now know as Portland cement,[14] which only slowly replaced Roman cement in mortars and renders, because it was much more expensive, but with a few exceptions it was not mixed with aggregates to make concrete for use in buildings until about 1865.

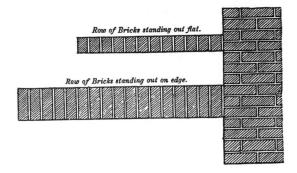

Row of Bricks standing out flat.

Row of Bricks standing out on edge.

Figure 3.1   A method of testing and comparing cements used by Pasley (p. 79 in Ref. 11) and others . Neat (Roman) cement was used and each brick held up by pressing the point of a trowel firmly against it for 2 or 3 min. The next brick was similarly set 6 or 7 min later. Francis White & Co., the cement manufacturers, repeatedly struck out 29 bricks in this way before the joints gave way.

In 1850 Dobson in his *Rudimentary Treatise on Foundations and Concrete Works*,[15] which is basically on engineering works, not buildings, is quite clear — 'concrete is made of gravel, sand and ground lime, mixed together with water' and 'beton' is concrete made with hydraulic lime. Neither Parker nor Frost nor Roman cement appear in the index. Others define concrete as a combination of aggregates, cement and water mixed at the same time, but beton meant mixing the sand, cement and water together first before adding it to the coarse aggregates. Concrete figures largely in the discussion following a paper read to the RIBA in 1875 on *New materials and recent inventions connected with building*[16] but in many instances it is not apparent if Portland cement or lime was used as a binder, such was the confusion of terminology in general use at that time. It is therefore necessary to be particularly cautious in making deductions from descriptions of 'concrete' or 'cement' in 19th century references. In some particular uses gypsum was used as the binder for mortars in concretes — for flooring in the Nottingham area and for concrete in Dennett's floors.

Aggregates

Whilst sand was initially mixed with lime or cement, Roman or Portland, for mortar, a number of materials were used for coarse aggregate to make concrete.

Lime concrete for footings and underpinning was generally made with gravel or ballast, but the fire proof flooring systems variously specified coke breeze (obtained from gas works), cinders, broken brick, burnt clay, limestone and even clean sieved rubbish. Rubbish was the term used to describe the arisings from taking down old buildings and was therefore a mixture of broken brick, plaster and mortar.

There was a theory that concrete made with aggregates that had gone through fire performed better in a fire than that made with those which had not, hence the use of breeze, brick and burnt clay as aggregate for concrete for fireproof flooring. Crushed brick is a good aggregate for concrete but breeze, which was first introduced by Matthew Allen in 1862,[17] is much more commonly found. Breeze is liable to be incompletely burnt and to contain a proportion of coal, giving a combustible content to 'fireproof' floors. Breeze also tends to be acidic, because of the sulphates and sulphides originating in coal, and hence can actively promote rusting of embedded iron or steel in damp conditions.

A solution to foundation problems

The history of concrete foundations is covered in Chapter 7 by Chrimes, but it is necessary to mention its early use here, if only because of the use of cement mortar in the brickwork. There is little doubt that Smirke's was the first effective use since Roman times of concrete in foundations, when he used lime concrete in his remedial works for the sinking foundations at Millbank Penitentiary (1817–22)[18] and for the underpinning following the partial collapse of the London Custom House (1825–27)[19,20] and elsewhere.

At Custom House he excavated 12 ft wide and 12–15 ft deep, down to a bed of gravel, and filled the trenches with lime concrete, on top of which he bedded large York stone slabs or landings, and then pinned up to the old footings with slate above 12 courses of hard stock bricks laid in (Roman) cement mortar (Figure 3.2). The lack of any sign of subsequent movement in the superstructure is a testament to the success of his underpinning work.

In following years, Smirke continued to use lime concrete as an artificial foundation when building on bad ground, sometimes as a raft, as in the rebuilt centre of Custom House, and sometimes as a footing, doubtless in conjunction with brick footings laid in Roman cement mortar.

Occasionally a lime concrete footing is found in buildings on a smaller scale, but that must be regarded as a exception and would only have been considered in particularly soft ground; normally the brick footing courses were laid directly on the earth, perhaps incorporating chain bond timbers to help bridge over soft spots.

Portland cement concrete footings started to be use for large and important buildings in the 1860s but did not come into general use until the 1880s and even later for domestic buildings.

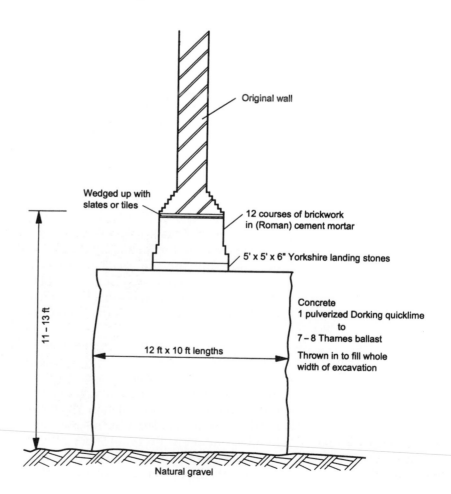

Figure 3.2   Section through Robert Smirke's underpinning of Custom House in 1825.

The early use of mortar and concrete in flooring

Plaster floors for upstairs rooms were used in the east Midlands from Elizabethan times up to the 19th century, with straw or more commonly reed laid over the joists and then covered with 2 in or more of plaster or mortar, made with gypsum or occasionally lime, mixed with burnt clay or broken brick or ashes[21–23] (Figure 3.3). Plaster floors were also to be found in the Cotswolds but survivals are more to be expected in the Nottingham area, where they continue to serve their purposes, but are known to be susceptible to point loads from bed legs, baths and pianos.

Recipes for plaster floors vary from neat coarse gypsum, to the use of the waste mixture of lime and ashes from the bottom of a lime kiln, to one-third lime, one-third coal ashes with one-third loamy clay and horse dung.

Lime ash concrete was shown by John Foulston as a finish to the stone slabs carried on cast iron inverted tees for the suspended floors of Bodmin Asylum (1818)[24] and it appears probable that this is the germ of the idea developed by Henry Hawes Fox for the fireproof flooring he invented for the private asylum he built at Northwoods, Nr. Bristol, in 1833.

Fox's Patent of 1844[25] for the floor, which was developed and marketed by James Barrett and hence became known as Fox and Barrett, and is found in very many buildings up to 1885 or even 1890, consists of cast iron inverted tee joists at about 1 ft 6 in centres with stout timber laths spanning between the outstanding flanges to provide formwork for 8:1:1 (sieved rubbish or road grit : coal ashes : lime) concrete or pugging (Figure 3.4). The surviving floors of Northwoods show that this is a far stronger material than would be expected from the weak mix specified. The concrete was laid on ¾ in of 1:2 lime mortar pushed through the gaps between the stout laths to provide a key for the plaster on the soffit which completed the encasing to the cast iron tees and provided the first truly fire resisting floor, since previously the iron joists supporting brick arches or stone slabs had been exposed on the soffit and thus vulnerable to fire. Barrett was not only one of the first to appreciate the need to provide protection against fire to the undersides of the iron

Figure 3.3   Lime ash on reed flooring (by courtesy of Philip Hartley of SPAB).

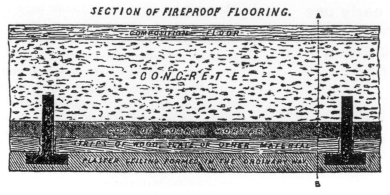

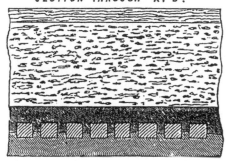

Figure 3.4　Fox & Barrett
flooring with cast iron joists,
before 1851 or 1852 (p. 163 in
Ref. 76).

joists, but he was also the first to appreciate the value of composite action, when he wrote, in his paper to the ICE in 1853:

> '… the force of compression acts upon the joists, only through the medium of the concrete, and this material is well known to be of the best for resisting that force.'

He also appreciated the '*considerable accession of strength*' by building the ends of the joists firmly into the walls.[26]

After the patent had expired in 1859, the Fox and Barrett system came into general use because its components — iron joists, timber laths and concrete — were readily available to anyone, and the above principles having not been clearly enunciated were not perpetuated. An ordinary lath and plaster ceiling was used, with battens interposed between the ordinary laths and the stout laths to accommodate the bottom flanges of the wrought iron joists, and the finish was timber boards on small joists resting on the top flanges, left exposed above the concrete, which was then little more than pugging. It is in this form that Fox & Barrett flooring is more likely to be found (Figures 3.5 and 3.6).

A builder from Nottingham, Charles Colton Dennett also developed a concrete fireproof flooring system using tried material when in 1857 he patented 'arches composed of sulphate of lime and an artificial puzzolana of burnt clay or Porous cinders' effectively gypsum concrete, as was already in use for plaster floors in his home town.[27] His arches however sprung off cast or wrought iron inverted tees with bottom flanges directly exposed to fire and hence were no more fire proof than a similar floor with brick arches, but they were marketed as being less costly (Figure 3.7).

As the 19th century progressed losses of buildings in fires produced greater and greater pressures to make buildings fireproof and indeed a series of Acts of Parliament sought to enforce the requirement, initially for parts of buildings needed as means of escape and latterly for certain types of buildings. However, there was considerable discussion and confusion about what constituted fireproof or fire resisting construction, with a school of thought which regarded incombustible as

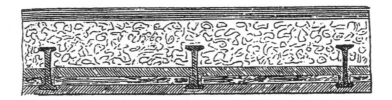

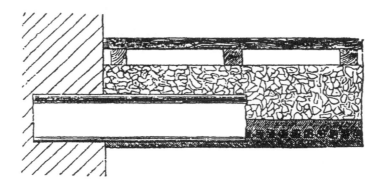

Figure 3.5   Fox & Barrett flooring with wrought iron joists, after 1851 or 1852 (p. 249 in Ref. 26).

Figure 3.6   Fox & Barrett flooring with battens and plaster on ordinary laths, at Finsbury Barracks (1857).

synonymous with 'fireproof', generally citing satisfactory behaviour probably in small fires which were insufficiently hot to cause failure of the exposed structural metal. This led to the Fire Brigades being unwilling to enter burning 'fireproof' buildings because of the risk of abrupt and catastrophic collapse when the exposed cast or wrought iron columns and beams succumbed to the heat.[28]

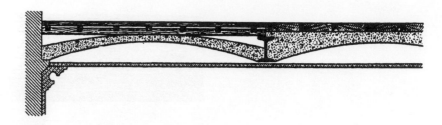

Figure 3.7   Dennett's arch floor.

There was nevertheless an increasing market for fireproof flooring as the number of patents for novel ways of solving the problem economically displays. Almost all of the patents used Portland cement for screeds and bedding and many incorporated Portland cement concrete. Further details of many of the systems can be found in Sutcliffe,[29] Potter,[30] Webster's paper to the ICE,[31] Farrow's BFPC Red Book,[32] and Lawford's paper to the Society of Engineers.[33]

Hoop iron in brickwork and concrete

Hoop iron — a thin strip of wrought iron from 1½ in × 15 gauge (1.8 mm) to ¾ in × 20 gauge (1.0 mm)[34] termed hoop because of its use by coopers for binding barrels — was in general use as bond in brickwork, from about 1830 to the end of the 19th century, as a replacement for bond timbers, particularly chain bond timbers, in the centre of walls, which suffered from rot whatever timber was used.[35] Advocates of *in-situ* concrete walling adopted it to control cracking and one recommended 'a prodigal expenditure of hoop-iron bond'.[36]

It was that material which Brunel and Francis adopted for their experimental brick beams in 1835–38. Francis and Sons at their Vauxhall (Roman) cement works and Brunel at the Thames Tunnel site both built beams of brickwork laid in neat Roman cement with hoop iron bond laid in the bed joints and tested them to failure. Pasley made his own tests at Chatham and also tested an unreinforced beam and a reinforced beam built with lime mortar for comparison. When his four course high brick and cement beam incorporating five strips of hoop iron was examined after gradual collapse the centre strip was found to be bent, the two lower strips had fractured and the two upper strips were buckled and the bricks crushed. The unreinforced beam failed abruptly at a small load and the iron strips in the lime mortar beam were 'drawn inwards', i.e. the bond with the mortar failed (Figure 3.8).

This clear understanding of the behaviour of reinforced masonry, as reported by Pasley in his 'Observations' published in 1838[37] was not adopted and applied practically by the designers of buildings and does not seem even to be mentioned in the Transactions of the RIBA, the forum for discussion about building construction.

This is perhaps surprising considering the publicity that surrounded Brunel's works at the Thames Tunnel and where he had also built 'two extraordinary semi-arches for experiment' of brickwork in neat (Roman) cement with hoop iron laid in the courses cantilevering either side of a central pier and rising 10 ft, one 60 ft and the other over 37 ft long and loaded with 62,700 lb[38] (Figure 3.9).

The hoop iron that engineers of today are likely to encounter will be as bond in brickwork, as could be seen in 1994 in quantity on the Mansion House Square site at No. 1 Poultry in the City of London, hanging down from the back of the Cheapside elevation where the party walls had been demolished (Figure 3.10). In Thomas & Frank Verity's specification for the new building to be erected at 96 Piccadilly in 1890, the contractor was to allow for providing five tons of hoop iron in the brickwork, which equates to something between 6 and 20 miles of the material, and incidentally he was required to use lime concrete footings for the walls but cement concrete foundations for the cast iron stanchions.[39]

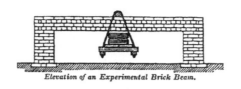

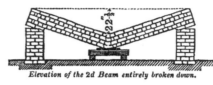

Elevation of an Experimental Brick Beam.

Elevation of the 2d Beam entirely broken down.

(a)

Section of an Experimental Brick Beam.

Figure 3.8   Pasley's 10ft span brick beams reinforced with hoop iron (pp. 234, 236, 238 in Ref. 11). (a) Reinforced Roman cement beam failed with 4532 lb. (b) Reinforcement lime mortar beam failed with 742 lb (unreinforced Roman cement beam failed with 498 lb).

View of the third Beam when broken down.

(b)

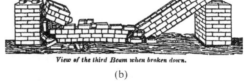

Figure 3.9   M.I. Brunel's reinforced brick cantilevers (p. 38 in Ref. 77).

Flat floors without iron — 'Bold; be not too bold'

This quotation is taken from the heading of the chapter in George Sutcliffe's book on *Concrete its nature and uses*[40] published in 1893 on flat concrete floors without iron, of the sort that engineers today would not contemplate building and, if they know of them, would dread encountering in a survey or appraisal of a Victorian building for fear of knowing what to say. Is over 100 years of not unsatisfactory use a reason to recommend or permit retention of a floor which cannot be shown by calculation to be structurally sound and which would collapse abruptly and catastrophically if anything occurred to disturb its abutments?

There were at that time a number of architects and engineers who held that steel or iron joists or reinforcement in floors constituted unnecessary expense and indeed some who held that they were a positive disadvantage.[41] There are reports of plain concrete floors 10 ft × 10 ft × 4 in thick and even 26 ft × 20 ft × 7 in thick which behaved satisfactorily and stood up to tests. Others built shallow unreinforced concrete arches and even cantilevers. Col. Seddon's reported tests included an unreinforced slab 14 ft 6 in × 13 ft 6 in clear span × 6 in thick which broke suddenly under a weight of 10 t or 120 lb/ft^2.[42]

Figure 3.10   Hoop iron bond in party wall at Mansion House Square site in 1994.

Most of these slabs had good bearings on thick solid walls and many were of concrete made with broken brick aggregate, thus ensuring a tenacious material. Sutcliffe goes on to say:

> 'They are, however, economical, and there is no reason why they should not be used in houses and other buildings where they will not be subjected to intense heat or heavy impingent loads, up to spans of 10, 12 or even (with care) 14 feet.'[43]

It is probable that some of these unreinforced floors are still extant, serving their purpose without revealing that they cannot be justified to modern standards — to engineers appraising Victorian buildings Sutcliffe's adage could be changed to 'Beware'! (examples are listed in Table 3.11).

Another form of unreinforced construction which does certainly exist in some quantity is tile creasing — flat floors and arches formed with up to three layers of ordinary flat clay roof tiles bedded in neat cement or strong cement mortar and laid to break bond (Figures 3.11 and 3.12).

Tile creasing is generally held to have been invented by Charles Fowler in about 1835 where it was used for the floors and roofs of Hungerford Market, spanning

Table 3.1   Examples of concrete floors without reinforcement[14]

No.	Length (ft)	Breadth (ft)	Thickness (in)	Where used	Authority	Remarks
Flat slabs						
1	20	20	14	Atrium Vestae, Rome	Prof. Middleton	Ancient Roman work
2	6	6	3	Footway of bridge	F. Caws	
3	14.5	7	6	Brigade depots	Col. Seddon	1 P.c. to 4 breeze, slag, brick or burnt clay, to pass ¾ in mesh
4	10	10	4	–	F. Caws	–
5	12.5	11.5	6	–	C.A. Adams	–
6	21	12.5	13	Warehouse	F. Caws	1 P.c. to 4 brick
7	26.5	19.5	7	–	F. Caws	Thicker for 9 in around margin, 1 P.c. to 4 brick
Arched slabs						
8	–	–	9	12 to 5	Stables	W.B. Wilkinson
9	–	9.5	11–4	Corridor	W.B. Wilkinson	Granite-concrete
10	50	12	11–3	Warehouse	Broughton	Carrying 'immense weight of machinery and men'
11	–	12	15–6	Pantechnicon	Lockwood	–
12	19	13	11–4½	House	J. Tall	–
13	40	16	11–4½	Drawing room	J. Tall	–
14	70	16.5	7–3½	Roof of barn	Trench	Intrados rising 1 in ft of span — 1 P.c. to 5 gravel
15	90	20	15	Roof of warehouse	W.C. Street	Segmental arch with 3ft rise — P.c. to 5 Thames ballast
Cantilevers						
16	–	4	8–3	Balcony	Potter	1 P.c. to 5 brick
17	50	4	11–3	Balcony	Broughton	1 P.c. to 4 ballast (part, 1 to 6)

P.c. = Portland cement.

Figure 3.11   Flat roof of plain tiles and cement known as tile creasing (p. 175 in Ref. 77).

Section of a Flat Roof of plain Tiles and Cement.

between iron beams 4½ ft apart to form the terrace,[44] but Pasley in 1838 says it was first proposed by 'the late ingenious Mr. Smart' (George Smart, d. 1834). Brunel used it for supporting the garden above his drawing office where cast iron beams at 5 ft centres satisfactorily supported tile creasing spanning 5 ft.[45]

That form of construction, flat and arched, can be found today in entrance hall floors, flat roofs, porch roofs, balconies and vaults of early and mid-Victorian houses built in London. It was also used and still exists as an arched non-combustible fire break in the House of Lords' roof, in ceilings to timber floors on cast iron beams in the Palace of Westminster and in the Waterloo Building at the Tower of London, and no doubt elsewhere.

Practical application of reinforcement to floors

William Boutland Wilkinson's patent of 1854[46] is generally agreed to be the beginning of reinforced concrete in England. His coffered concrete floor reinforced with old colliery ropes in the ribs in the reinforced concrete cottage he built in Newcastle in about 1865 (Figure 3.13) as described in detail in the careful records made during its demolition in 1954,[47,48] is an example of his work. Wilkinson's patent also cites the use of hoop iron as reinforcement. The papers and discussion at RIBA meetings in the 1860s and 1870s, when the use of concrete for building was discussed at length[49–56] indicate however that neither Wilkinson's patent, nor Lambot's nor Coignet's patents of the following year were adopted by the building industry.

Figure 3.12   Underside of
back addition roof formed in
tile creasing, in c.1840
terrace house in Paddington
area.

Figure 3.13   Wilkinson's
reinforced concrete cottage.

Reinforced concrete floors of the 1860s and 1870s were not distinguished from
the other forms of fireproof flooring which were being developed to satisfy the
demand arising out of public concern about the increasing cost of damage and
the loss of life caused by building fires and enable compliance with the resulting

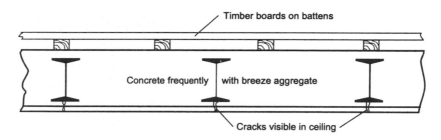

Timber boards on battens

Concrete frequently   with breeze aggregate

Cracks visible in ceiling

Figure 3.14   Ordinary filler joist floor.

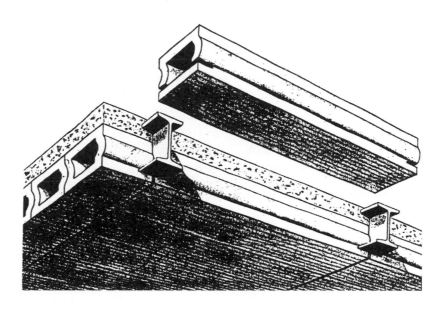

Figure 3.15   King floor, with fireclay tubes, usually 3 ft long × 4 to 8 in deep, but the tubes have thinner walls than indicated in this plate from King's 1935 brochure.

legislation. The majority of fireproof flooring systems incorporated rolled or fabricated iron joists at 1 ft 6 in to 3 ft centres supporting various forms of cellular clay tiles or concrete usually made with breeze aggregate, sometimes containing a particular form of reinforcement.

These floors lead in the 1890s to the ordinary filler joist floor — with iron or later steel joists at about 3 ft centres embedded in a slab of breeze concrete — which still survives in quantity and will be frequently encountered by engineers working on buildings up to the First World War (Figure 3.14).

The other type of fireproof floor, which continued in use during the 1920s and can indeed be found up to the Second World War, is formed with hollow clay tubes or lintels spanning between steel joists, with the ends notched to embrace the bottom flange and finished with concrete, usually with breeze aggregate, or with screed; examples of this were by J.A. King (Figure 3.15), Homan & Rogers (Figure 3.16), and Fawcett (Figure 3.17).

Thaddeus Hyatt, who crossed the Atlantic to England in the early 1870s and was responsible for over 40 patents concerned with pavement lights and concrete flooring, had a series of tests carried out by Kirkaldy, and published in 1877 a pamphlet setting out why reinforcement was only needed in the bottom of a Portland cement concrete slab and why all iron needed to be protected by concrete cover from fire.[57] The need to provide cover to protect iron was stated as early as 1838 by Pasley, who suggested a coating of tiles laid in cement,[58] and in 1853 by Barrett who referred to the plaster to protect the iron joists from the action of fire from below,[59] but it was not generally understood until the 1870s and 1880s that it was not sufficient just to provide a non-combustible structure to achieve fire resistance. Hyatt and the tests he had carried out certainly helped this understanding. The need for an appreciation of construction that could really be shown to

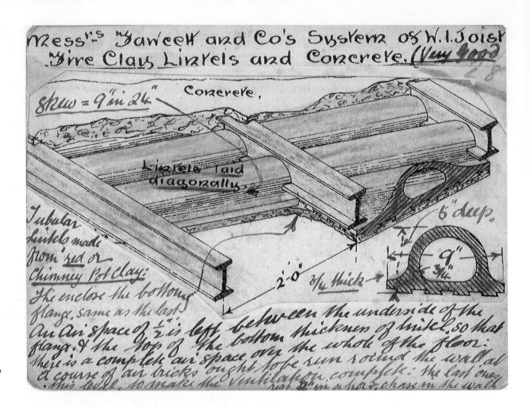

Figure 3.16   Homan & Rogers floor as sketched by the late B.L. Hurst in the late 1890s.

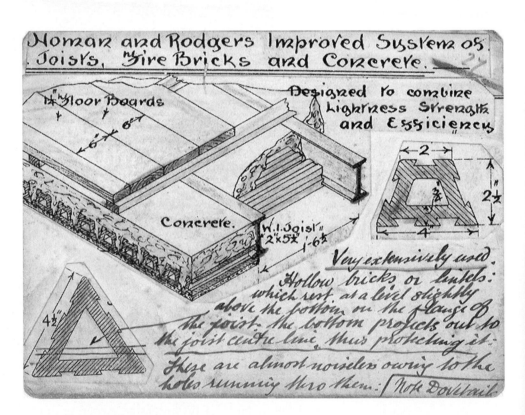

Figure 3.17   Fawcett's floor as sketched by the late B.L. Hurst in the late 1890s.

survive a fire was one of the reasons for the formation of the British Fire Prevention Committee in 1897 by Edwin Sachs. The BFPC's first testing station was in use less than a year later to carry out instrumented documented fire tests on a variety of building materials.[60] The development of this story leading to the formation of

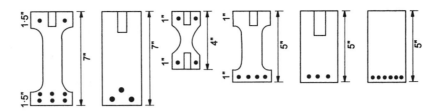

Figure 3.18   F.G. Edwards reinforced concrete joists with nailing strips as illustrated in Patent No. 2941, 1891.

the Concrete Institute which became the Institution of Structural Engineers is told by Anita Witten in Chapter 14.

Other patents were taken out for what was effectively reinforced concrete, by Gen. Scott,[61] of Royal Albert Hall fame, Matthew Allen,[17] who in 1862 was one of the first to advocate Portland cement and breeze concrete, F.G. Edwards,[62] for I section joists with a breeze concrete nailing strip in the top (Figure 3.18); and others; but none of them came into general use or can be expected to have survived in any quantity.

However, it would not be unexpected to find examples of the two floors mentioned earlier — Fox & Barrett and Dennett (or later Dennett & Ingle) in use in buildings constructed before 1890.

In-situ walls

The other type of patent in the lists for the 1860s, 1870s and 1880s for concrete work is for apparatus for forming concrete walls, taken out by Tall (7), Drake (3), Payne (2) and others. These were for formwork systems and sometimes included loose descriptions of reinforcement and recipes for concrete — there is even one system of climbing formwork! Most were more ingenious than useful and do not appear to have benefited their inventors or the industry. Their products do not appear to have survived in any quantity. They were generally advocated as a less costly way of providing 'workers' dwellings', a philanthropic tendency of that time, and hence it would not be totally unexpected to find mid and late Victorian small houses with *in-situ* concrete walls. Anerley New Church, now converted to housing, by W.J.E. Henley, manager of the Concrete Building Company (1883), a remarkable neo-Gothic *in-situ* concrete building shows that it is always wise to be alert for the unexpected.[63]

Discussions at the RIBA were a dialogue between the disciples of concrete and the sceptics, who thought that even if it remained standing when the formwork was struck it would crumble away within a year or two. There was probably right on both sides, because they were generally talking of different materials — the sceptics had in mind examples of concrete made with lime or defective cement and the disciples carried out tests and supervised the work closely to ensure a satisfactory product. It is probable that some examples of the latter survive, such as 63 Lincoln's Inn Fields designed by William Simmons, with *in-situ* mass concrete walls and Hyatt's floors[64] (Figures 3.19 and 3.20).

Official acceptance of *in-situ* concrete as a material for walls came in clause 2a of the Bye-Laws made in 1886 under the provisions of the Metropolis Management and Building Acts Amendment Act 1878, where it could be substituted for brickwork of the same thickness. The concrete had to be of Portland cement, clean sand, and clean ballast, gravel, broken bricks, or furnace clinker, passing a 2 in ring, in proportions 1 : 2 : 3, carefully mixed with clean water, and carried up regularly, in parallel frames of equal height.[65]

Figure 3.19   63 Lincoln's Inn Fields with in-situ mass concrete walls and Hyatt's floors.

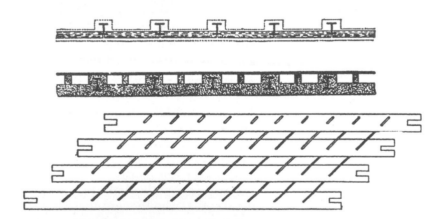

Figure 3.20   Hyatt's 'gridiron' floor (p. 62 in Ref. 28).

Precast walls and precast dressings to brickwork

It is not generally realized that a substantial proportion of the bay columns with ornate capitals and other painted decorative features on the elevations of late Victorian and Edwardian speculative housing are of precast concrete — an economical and more durable substitute for the soft easily carved stone alternative — nor is it realised that the use of precast components in buildings pre-dates most other uses of concrete.

From the early years of the 19th century, stucco decoration was frequently precast, using Roman cement,[66] rather than being formed in-situ, and, in 1832 and 1834 William Ranger patented[67–69] and used precast concrete blocks, made with lime slaked with boiling water, for building dock and sea walls and a number of buildings in Brighton and London and possibly elsewhere. It is probable

*Figure 3.21 Lascelles'
system at Central Buffet.*

that they might still be found behind stucco on terraces in Brighton of that date. Ranger's concrete also had military applications as described by John Weiler in Chapter 19. Precast concrete, or as it was generally termed 'artificial stone', continued to be used, as whole blocks, or probably more commonly facings to mass concrete walls, usually with dovetail strips or ribs to bond them to the *in-situ* work.

Castle House, Bridgewater, dating from the 1851, incorporates precast concrete facing panels, an interesting and possibly unique form of iron and tile floor, and a pitched roof with stretcher bond crucks probably laid in neat Roman cement.[70]

In 1995 there was an opportunity to inspect an ingenious precast system, patented by William Lascelles in 1875, used in the construction of the Central Buffet and Dock Managers Office at the Royal Albert Dock in London's Docklands whose elevations have been conserved and restored.[71,72]

Lascelles' system[73] employed 3 ft × 2 ft breeze concrete slabs, screwed to timber studs. They were 1–1½ in thick, reinforced with two diagonal wrought iron rods. Similar panels could be used on the inside in lieu of plaster, as they were in the ceilings of houses at Sydenham,[74] and on the floors in lieu of boards, but were not at Royal Albert Dock. Some of the wall slabs at Royal Albert Dock had a rough-cast finish but they could be cast to resemble brickwork or tile hanging and be coloured accordingly. The Central Buffet also incorporates a further layer of pre-cast components outside the slabs, fixed through them to the timber framing, of columns, with bases and ornamental capitals, friezes and decorative panels, and a precast concrete cornice incorporating the gutter (Figure 3.21).

Lascelles' concrete could be self-coloured — red, buff or grey. Red, obtained by lining the mould with a grout of 'Spanish brown' was used at the Royal Albert Dock buildings and the finish on surviving unweathered areas is good enough to be mistaken for terracotta. Even the fire surrounds and overmantels in the Central Buffet are Lascelles' precast concrete.

Lascelles' brochure lists the range or architectural dressings of all kinds supplied by the company, many no doubt from stock, and it would have been from workshops such as his that the speculative builder would purchase the bay columns and other decorative work for speculative housing.

Lascelles' system is likely to have been chosen for the buildings at Royal Albert Dock on the grounds of speed and economy. Moulds for precast work could be made quickly and cheaply in Lascelles' joinery works so that the precast concrete components could be delivered to site within weeks of being ordered. In his contribution to the discussion at the RIBA in 1876,[75] Lascelles said the slabs were removed from the mould in three or four days and were then ready for fixing to the walls.

To produce such decoration the only alternatives would be terracotta or stone; the former would take months for pattern and mould making, drying and firing, while stone would have to be sourced as well as carved, so precast concrete was undoubtedly both quicker and cheaper.

It is interesting to note that one of the latest concrete repair techniques — re-alkalization — has been used to conserve a substantial proportion of Lascelles' units at the Royal Albert Dock buildings.

One of the leading Victorian architects prepared to use this new material was Norman Shaw, who seems to have been retained by Lascelles to produce typical house designs, and it was he who designed the elaborate facade of Lascelles' award-winning pavilion at the Paris Exhibition of 1878.

Summary of structural uses of cement and concrete in the 19th century

The primary purpose of this chapter has been to introduce engineers of today to the development of the structural uses in buildings of cements during the 19th century and hence to describe forms of construction they are likely to encounter when working on Victorian buildings.

Until 1890 concrete was not used in Britain for building frames, only for slabs, lintels and occasionally walls. Most forms of fireproof flooring which had been developed since 1850 incorporated concrete or screed and by 1890 this would have been made with Portland cement, which had by then replaced the lime and occasionally gypsum used as the binder in the 1850s and 1860s.

Fireproof flooring, frequently finished with timber boards on battens, was used in commercial and institutional buildings but not exclusively except for the stairs, landings and corridors forming the escape routes. It was also used in the higher class of domestic buildings but only in the lower floors, or sometimes at ground and second floor levels — to separate the kitchens from the family living rooms and to separate the bedrooms from the rooms below.

Reinforced concrete beams and columns would be unusual, except for lintels although concrete casing to rolled wrought iron beams (steel would be unlikely before 1890) and occasionally to columns is not uncommon. The concrete would more than likely be made with coke breeze aggregate. Concrete in floors had either breeze or occasionally broken brick aggregate. Gravel was uncommon but some patent systems used crushed limestone.

Two other forms of construction have been mentioned. Hoop iron which can be expected to be encountered in the brickwork of substantial buildings, sometimes in quantity, sometimes only in a couple of courses in each storey height, and tile creasing which was used for fireproof flooring in houses and also in external construction where timber would be liable to rot.

During the second half of the 19th century, a number of people had set down and published and patented the principles of reinforced concrete, but the forms of construction in general use indicated that no real understanding of the behaviour of reinforcement in conjunction with concrete existed in general practice. The ingredients were all there waiting for Hennebique and others to compile the recipes.

References

1. LCC (General Powers) Act 1909.
2. Godwin, G., Prize essay upon the nature and properties of concrete, and its application to construction up to the present period. Trans. IBA, 1836, 1, 1–37.
3. Abraham, R., Concrete used at Westminster New Bridewell. Trans. IBA, 1836, 1, 38–39.
4. Taylor, G.L., An account of the methods used in underpinning the long storehouse at His Majesty's Dock Yard, Chatham, in the year 1834. Trans. IBA, 1836, 1, 40–43.
5. Brunel, M.I., Particulars of some experiments on the mode of binding brick construction. Trans. IBA, 1836, 1, 61–64.
6. Smeaton, J., Narrative of the Building of the Eddystone Lighthouse, London, 1791.
7. Parker, J., A certain cement or terras to be used in aquatic and other buildings and stucco work. Patent No. 2120, 1796.
8. Thurston, A.P., Parker's 'Roman' Cement. Trans. Newcomen Soc., 19, 1938–39, 193–206.
9. Bristow, I.C., Exterior renders designed to imitate stone. Trans. ASCHB, 22, 1997, 13–30.
10. Pasley, C.W., Outline of a course of practical architecture, Chatham, 1862, but first issued as lithographed notes in 1826. It is clearly from that earlier date and was not revised before being printed in 1862, p. 91. (Reprinted Donhead, 2001.)
11. Pasley., C.W., Observations on Limes, Calcareous Cements, Mortars, Stuccos and Concrete, etc., 1st edn., London, Weale, 1838, 38.
12. Pasley, C.W., 1838, 84.
13. Pasley, C.W., 1826, 184.
14. Skempton, A.W., Portland cements 1843–1887. Trans. Newcomen Soc., 1967, 35, 117–52.
15. Dobson, E., A Rudimentary Treatise on Foundations and Concrete Works, London, Weale, 1850. (Reprinted Bath, 1970.)
16. Smith, T.R., New materials and recent inventions connected with building. Trans. RIBA, 1874–75, 199–216, 221–30.
17. Allen, M., Stairs, etc. Patent No. 244, 1862.
18. Crook, J.M., Sir Robert Smirke: A pioneer of concrete construction. Trans. Newcomen Soc., 389, 1965, 5–22.
19. Pasley, C.W., 1838, 16, 265–70.
20. Pasley, C.W., 1826, 16, 25–26, 27.
21. Barley, M.W., The English Cottage and Farmhouse, London, 1961, 83–95, 103, 259, 263.
22. Stevens, H.J., Plaster floors in Derby district. In: discussion following Burnell's paper (Ref. 38). Trans. RIBA, 1854, 63.
23. Allen, C. Bruce, Cottage Building and Hints for Improved Dwellings for the Labouring Classes, 6th edn., London, 1867: 40.
24. Foulston, J., The Public Buildings Erected in the West of England, London, 1838.
25. Fox, H.H., Fire-proof roofs, floors and ceilings. Patent No. 10047, 1844.
26. Barrett, J., On the construction of fire-proof buildings. Proc. ICE 12, 1852–53, 251.
27. Dennett, C.C., Floors and ceilings. Patent No. 685, 1857.
28. Hyatt, T., Portland-Cement-Concrete Combined with Iron, as a Building Material. London, 1877 (reprinted by ACI, Detroit, 1976).
29. Sutcliffe, G.L., Concrete: Its Nature and Uses, London, 1893, Chapter 20.
30. Potter, T., Concrete: Its Use in Building and the Construction of Concrete Walls, Floors etc., 2nd edn., Winchester, 1891; 3rd edn., London, 1908.
31. Webster, J.J., Fire-proof construction. Proc. ICE, 105, 1891, 249–86.
32. Farrow, Frederick, Fire-resisting Floors Used in London. BFPC Red Book No. 7, 1898.
33. Lawford, G.M., Fireproof floors. Trans. Soc. Engrs, 1889, 43–70.
34. Pasley, C.W., 1838, 243.
35. Pasley, C.W., 1838, 169, 239.

36. Wonnacott, T.H., On the use of Portland cement concrete as a building material. Trans. RIBA, 1870–71, 175–80.
37. Pasley, C.W., 1838, 233–40.
38. Brunel, M.I., Trans. IBA 1836, 61–64.
39. Verity, T. & F. T., Specification of work required to be done in the erection and completion of club premises at Nos. 96 and 97 Piccadilly, 1890.
40. Sutcliffe, G.L., London, 1893, Chapter 20.
41. Tall, J., In: discussion to Payne's paper (Ref. 55). Trans. RIBA, 1875–76, 232.
42. Seddon, Major H.C., Experiments with Concrete Slabs. School of Military Engineering: Chatham, 1880.
43. Sutcliffe, G.L., 1893, 263.
44. Fowler, C., On terrace roofs. Trans. IBA, 1836, 1, 47–51.
45. Pasley, C.W., Observations etc., 2nd edn, Part 1, London 1847, 174–75.
46. Wilkinson, W.B., Construction of fireproof buildings etc. Patent No. 2293, 1854.
47. Cassie, W.F., Early reinforced concrete in Newcastle-upon-Tyne, Magazine of Concrete Research, 1955, 25–30.
48. Brown, J.M., W.B. Wilkinson (1819–1902) and his place in the History of Reinforced Concrete. Trans. Newcomen Soc., 39, 1967, 129–42.
49. Burnell, H.H., Description of the French method of constructing iron floors, and discussion. Trans. RIBA, 1853–54, 36–74.
50. Lewis, T.H., Fire-proof materials and construction. Trans. RIBA, 1864–65, 109–26.
51. Wonnacott, T.H., On the use of Portland cement concrete as a building material. Trans. RIBA, 1870–71, 175–80.
52. Blomfield, A.W., Remarks on concrete building. Trans. RIBA, 1870–71, 181–87.
53. Seddon, Captain, Our present knowledge of building materials and how to improve it. Trans. RIBA, 1871–72, 143–57, 177–84.
54. Smith, T.R., On new materials and recent inventions connected with building. Trans. RIBA, 1874–75, 199–216, 221–30.
55. Payne, A, Concrete as a building material. Trans. RIBA, 1875–76, 179–92, 225–54.
56. Cates, A., Concrete and fire-resisting constructions. Trans. RIBA, 1877–78, 296–312.
57. Hyatt, T., 1877, 4.
58. Pasley, C.W., 1838, 168.
59. Barrett, J., On the construction of fire-proof buildings. Proc. ICE, 1852–53, 12, 244–72.
60. Hurst, Lawrance, Edwin O. Sachs — Engineer & Fireman. In: David Wilmore (ed.) Edwin O. Sachs Architect, Stagehand, Engineer & Fireman, Braisty Wood, 1998: 126.
61. Scott, H.Y.D., Floors and roofs. Patent No. 452, 1867.
62. Edwards, F.G., Concrete building etc. Patent No. 2941, 1891.
63. Concrete Quarterly 88/90, April–September, 1971, 50–51.
64. The Builder, 8th December, 1888, 421, 422.
65. 41 and 42 Vic. cap. 32, 22 July 1878. Cited on p. 69 of Harper, R.H., Victorian building regulations, London, 1985.
66. Bristow, I.C., 1997, 19.
67. Pasley, C.W., 1838, 18, 141, 252.
68. Ranger, William, Artificial stone. Patent No. 6341, 1832.
69. Ranger, William, Artificial stone. Patent No. 6729, 1834.
70. New Civil Engineer, 30th October 1986, 18.
71. New Builder, 28th April, 1995, 32, 34.
72. McFarland, B., Woodhouse, J., Jeanes, C., Royal Albert Dock, 'the gentle touch', Construction Repair, 1995, 5, 20–23.
73. Lascelles, W.H., Constructing buildings. Patent No. 2151, 1875.
74. Stanley, C.C., Highlights in the History of Concrete. Cement and Concrete Association, 1979, 26.
75. Lascelles, W.H., In: Discussion on Payne's paper (Ref. 44). Trans. RIBA, 1875–76, 185.

Further Reading

Batty Langley, The London Prices of Bricklayers Materials and Works etc., London, 1749.

Braidwood, J., On fire-proof buildings. Proc. ICE, 1849, 8, 141–61.

Burn, R. Scott, The New Guide to Masonry, Bricklaying and Plastering. John G. Murdoch, London, c. 1872.

de Courcy, J.W., The emergence of reinforced concrete, 1750–1910. Struct. Engr., 1987, 65A, 315–22, 1988, 66, 128–30.

Francis, A.J., The Cement Industry 1796–1914: A History. Newton Abbot, 1977.

Halstead, P.E., The Early History of Portland Cement. Trans. Newcomen Soc., 1961–62, 34, 37–54.

Hamilton, S.B., A Note on the History of Reinforced Concrete in Buildings, National Building Studies Special Report No. 24, HMSO: London, 1956.

Hamilton, S.B., A Short History of the Structural Fire Protection of Buildings, National Building Studies Special Report No. 27, HMSO: London, 1958.

Newlon, H. (ed.), A Selection of Historic American Papers on Concrete 1876– 1926, American Concrete Institute, Detroit, 1976 (includes papers by Thaddeus Hyatt, W.E. Ward, A.N. Talbot, A.R. Lord, C.A.P. Turner, E.L. Ransome and D.A. Adams).

Newman, J., Notes on Concrete and Works in Concrete, London, 1887.

Reid, H., A Practical Treatise on Concrete and How to Make it, London, 1869.

Reid, H., The Science and Art of the Manufacture of Portland Cement, London, 1877.

Semple, G., A Treatise on Building in Water, Dublin, 1776.

4   The era of the proprietary systems

Michael Bussell

Synopsis

This chapter reviews developments in reinforced concrete construction, mainly in Britain, from the 1890s until the First World War. It focuses on the proprietary systems that offered structural design in this new composite material as an adjunct to the supply of reinforcement, or to the construction of the complete reinforced concrete structure. Of these the Hennebique system, through its British agent, L.G. Mouchel, was the most prolific, with a remarkable record of expansion following its arrival in 1897. Many other systems were patented, although only a few found wide use, notably those of Truscon (using the Kahn system), and Considère.

The Appendix contains further information to assist in identifying proprietary reinforcement, floors, and systems of this period in existing construction.

The early years of reinforced concrete in Britain

The 1854 patent by William Boutland Wilkinson of Newcastle upon Tyne was the first to propose the use of iron as reinforcement to concrete, recognizing the relative weakness of concrete in tension.[1] Like those of the Frenchmen Lambot and François Coignet in 1855, Wilkinson's patent attracted little interest in the British building industry. It was not until 1892 that François Hennebique, a French contractor, obtained a British patent for his system, which, with others later, would find wide use in Britain.[2]

The principal concern was to provide an economical and fire-resistant form of construction — something which was widely sought as a solution to the fatal and expensive fires that frequently consumed mills, warehouses and public buildings in particular. Attempts to deal with this problem had begun in the 19th Century; some of the most notable later 19th-century solutions have been reviewed in Chapter 3. Such solutions were often based on beams and columns of iron or steel. Floors used with such frames included brick vaulting, and flat and vaulted concrete slabs. Although themselves 'fireproof' (or at least incombustible), they did not always protect the vulnerable bottom flanges of the beams, nor the columns, from distortion or collapse when exposed to fire.

Hennebique's patent was based on plain round bars with fish-tailed ends, and stirrups of flat strips, all at that time being of mild steel. The concrete cover to the bars and stirrups afforded protection against fire, for a period depending on the thickness of the cover. The bars provided tensile resistance in beams and slabs, and supplemented the compressive capacity of the concrete in columns and walls. The stirrups provided shear resistance, although they were not mechanically anchored in the compression zone, as is the norm nowadays. Column bars were linked by strips of wire. A later development of the Hennebique system in 1897 was to provide bent-up bars in beams to provide hogging resistance and supplement shear capacity.

At this time, such construction was generally known as 'ferro-concrete' rather than as reinforced concrete. Significantly, the English version of the Hennebique house journal, first published in English in 1909 by Mouchel, was titled *Ferro-Concrete* (see below).

The question arises how such structures were designed. Hennebique carried out tests to establish the strength of beams, and had the technical assistance of a

Belgian engineer.[2] His approach was experience-based. Edmond Coignet, son of François, and Napoléon de Tédesco presented a paper to the French Society of Civil Engineers in 1894 which laid down the basis for calculations on the modular ratio method.[3] This assumed that plane sections remained plane, that concrete carried no tensile stress, and that capacity was dictated by limiting, or 'permissible stresses' under service load. This approach remained valid and in use for most of the 20th century. (Design methods of the time are discussed further in Chapter 5.)

Outside Britain, others were developing their ideas on the reinforcement of concrete. Indeed, it has to be said that, after Wilkinson, the initiative for the development of reinforced concrete came from the Continent (notably France and Germany) and from the USA. Britain lagged behind, and the early application of reinforced concrete construction in Britain was essentially due to the 'import' of Continental and American systems.

Joseph Monier obtained a French patent in 1867 for reinforcing concrete plant tubs with wire and rods. His ideas were taken up by the German G.A. Wayss, and the 'Monier system' — regarded on the Continent as a synonym for reinforced concrete — was widely used there, but found little application in Britain.[4] In the USA, Thaddeus Hyatt developed reinforcing systems using nuts,[4] or cross-bars placed through holes in square bars,[5] to anchor the steel to the concrete. E.L. Ransome obtained an American patent in 1884 for a square twisted bar which was stronger (being cold-worked), and had superior bond performance.[4]

Numerous other patents were obtained, and systems developed, using various reinforcing section profiles. Expanded metal was introduced as reinforcement to concrete around 1890, and was widely used.[4] The Kahn bar of 1902–1903 is of unusual profile, being a square section with two projecting strips on diagonally opposite corners. These are slit to be bent up diagonally in short lengths, forming shear reinforcement. Kahn saw these as forming a 'trussed' beam, and indeed the business which used his system in the USA, and later in Britain, was called the Trussed Concrete Steel Company, affectionately remembered by many as Truscon.

A detailed recital of the various patent reinforcement profiles and arrangements, and their related construction systems, would be unduly lengthy. It will, however, be helpful for today's engineer to be able to identify the commonest profiles and systems likely to be found when investigating existing construction of this period. These are illustrated in the Appendix. Further details of these (and many other) systems can be found in contemporary texts such as Cassell[4] or Marsh.[5] Table 4.1 lists the proprietary systems in common use in Great Britain in 1907.

Hennebique's 1892 patent was followed by the opening of an office in Brussels, and the construction of two early reinforced concrete structures in France — a refinery in Paris in 1894, and a framed mill in Tourcoing in 1895. In 1897 he obtained a further British patent, and appointed L.G. Mouchel as his agent in Britain, with offices in Victoria Street, Westminster (the traditional home of consulting engineers until recent years). Hennebique's first building in Britain was Weaver's Mill of 1897 in Swansea (Figure 4.1, demolished 1984). The working drawings for this were done in Nantes in France, from where too were supplied the cement, aggregate and reinforcement; the licensed contractor was also French![2] Apparently the only indigenous element of the construction was the water used in mixing the concrete.

The expansion of the Hennebique enterprise was remarkable. By 1899, 3061 projects had been undertaken.[2] A major reason for this success was Hennebique's insistence on the use of suitable materials and experienced labour, to ensure that the resulting construction was of a good standard. This explains the use of French materials and a French contractor for Weaver's Mill, although it was more common for the company to license contractors whose work would be, in part at least, undertaken and supervised by men who had gained experience with Hennebique. In modern jargon, one would say that the process was one of 'technology transfer', and certainly such was essential, bearing in mind the phenomenal workload

Table 4.1   The common proprietary systems in use in Great Britain in 1907

No.	Name of system	Form of tension bars	Form of compression bars	Form of shear reinforcement	Method of fixing shear reinforcement	Direction of shear reinforcement
1	Coignet	Round straight bars	Round straight bars	Round rods bent to U-shape	Looped under tension bars and twisted above compression bars	Vertical
1a	Coignet	Round straight bars and round bars bent up near supports	Round straight bars	Bent up ends of extra tension bars	Continuous with extra tension bars	Diagonal
2	Considère	Round straight bars and round bars bent up near supports	Round straight bars	(a) Bent up ends of extra tension bars and (b) round rods lapped around the main tension and compression bars	(a) Continuous with extra tension bars and (b) bent round tension and compression bars	(a) Diagonal (b) Vertical
3	Hennebique	Round straight bars and round bars bent up near supports	Round straight bars	Steel strip bent to U-shape and made with spring clip	Sprung on to tension bars and bent over for anchorage in concrete	Vertical
4	Improved Construction	Round straight bars and round bars bent up near supports	Round straight bars	Round rods wound around the main tension and compression bars	Bent round tension and compression bars	Spiral
5	Indented	Corrugated square bars, bent up near supports	Corrugated square straight bars	Bent-up ends of tension bars	Continuous with tension bars	Diagonal
6	Johnson	Round straight bars woven with wire lattice	– [sic]	Trough of wire lattice with rectangular or with diamond mesh	Woven with tension bars	Vertical or diagonal strands, according to mesh used
7	Kahn	Square bars generally straight, sometimes bent up towards supports	Square straight bars	Wings attached to main part of tension and compression bars	Continuous with tension and compression bars	Diagonal
8	Ridley-Cammell	Angle bars straight, and corrugated sheeting	Angle or other bars, straight	Trough of corrugated sheeting	Riveted or bolted	Continuous plate
9	Wells	Twin round bars connected by short web	Round straight bars	Steel strip hangers and bonders [sic]	Bent round tension bars	Vertical
10	Williams	Rolled steel sections, straight	Rolled steel sections, straight	(a) Round bars (b) rolled steel sections (c) spiral coils of steel wire sometimes used in addition	(a) Ends split for anchorage in concrete (b) riveted or bolted to tension and compression bars	(a) Vertical (b) Diagonal (c) Spiral

Source: *Concrete and Constructional Engineering*, 1907–1908, 2, 433.

in those early years and later. By 1909 the Hennebique system had been used for nearly 20,000 structures, and the company had 62 offices: 43 in Europe (including Britain), 12 in the USA, four in Asia and three in Africa. A list of British Hennebique projects issued in 1911 by what had by then become L.G. Mouchel & Partners totalled 1073.[2] (A second notable reason for Hennebique's success was his company's assiduous attention to promoting its achievements.)

A standard specification of 1917 by L.G. Mouchel & Partners[6] is of interest. It covers 'ferro-concrete', and cost 2s. 6d. (12½p). Steel for reinforcement was to

Figure 4.1 Weaver's Mill, Swansea (1897, demolished) (by courtesy of J.W. Figg).

comply with BS 15 of 1912,[7] to be Trisec bars (a patent high tensile steel), or to be shell discard of prescribed quality (the First World War still had a year to run). Concrete proportions by volume were approximately 1.1 : 2 : 4 of Portland cement : sand : coarse aggregate. Provision was made for load testing if required. (Such testing was common practice at that time.)

The deflection criterion for passing a load test carried out on the completed structure was onerous by today's standards. Under 1½ times the imposed load, the structure was not to deflect more than 1/600 times the span! It should, however, be borne in mind that such structures were designed using elastic modular ratio theory, with permissible stresses typically of $16{,}000\,lb/in^2$ ($110\,N/mm^2$) in the reinforcement and of $600\,lb/in^2$ ($4.1\,N/mm^2$) in the compressed concrete. Consequently the section sizes and the reinforcement were more generous than a present-day design would require. A further factor — certainly for the floor slabs — is that the framing plans usually provided beams in two directions (on the precedent of iron and steel frames). The typical floor slab was accordingly supported on all four sides and, when loaded, would tend to behave more as a shallow dome in compression than as a slab in flexure, with the beams acting both as supports and perimeter ties. Such behaviour would generate smaller deflections in the slabs.

A typical if early surviving Hennebique building structure is the Co-operative Wholesale Society (CWS) warehouse on Quayside, Newcastle-upon-Tyne[5,8] (Figure 4.2). The building generally dates from 1897 to 1900, although the ninth floor was added in 1908. The foundation is a raft in view of poor ground conditions. Actually, this 'raft' is a series of haunched slabs spanning between substantial ground beams. These slabs are 7 in (178 mm) thick at their centres, with a square column grid of 14 ft 6 in (4.42 m). Floors were designed for an imposed load of no less than $6\,cwt/ft^2$ ($32.3\,kN/m^2$). Typical suspended floor slabs — also 7 in (178 mm) thick — were satisfactorily test-loaded to 50% above this value.

Procurement of concrete construction

How were such buildings procured? Clearly Hennebique and his competitors were in effect providing a 'design and build' service, although the basic structural layouts and performance specification would usually be defined by the client and his architects.

Figure 4.2   Co-operative Wholesale Society Warehouse at Quayside, Newcastle-upon-Tyne before recent repairs (1897–1900).

There were very few independent consulting engineers practising at the turn of the century, so far as building design was concerned. Sven Bylander, for example, who had worked on steel framed structures in the USA, was employed by Waring White, its builder, for the structural design of the 1904 steel-framed Ritz Hotel in London. This building, incidentally, had concrete floor slabs reinforced using the 'Columbian' system with Bonna bars of cruciform or 'cross of Lorraine' section (see Appendix).

Architects would generally be competent to design most elements of the building, including foundations and structural masonry. These elements were largely sized by empirical rules relating foundation width to wall thickness, and wall thickness to height of building. Similarly, tables and standard textbooks were available to assist the architect to design and size timber floors and roofs.

When it came to steel or concrete structures, however, it was normal for their design — that is the calculations and drawing-up of details — to be carried out by 'specialists'. For steelwork this would be the steel fabricators, and for reinforced concrete it would be Hennebique, Truscon, Considère and others.

Nor was the appointment of a specialist just a matter of selecting one such company to implement the architect's scheme.

The example of the recently refurbished YMCA building in Manchester is illuminating.[9] Its structure was originally conceived in 1908 by the architects Woodhouse, Corbett & Dean as a steel skeleton with brick walls. Two particular features of the building led to a decision to consider reinforced concrete. One was the provision of a swimming pool on the top floor, the box structure for this logically being best done using reinforced concrete. The other was the need to have a column-free hall at lower level. The original plan was to span this by storey-height steel trusses built into masonry walls on the floor above, but clearly it would be a sensible alternative to design the walls in concrete as deep beams, with considerable savings in steel and altogether simpler construction.

The drawings were sent out to five specialists, of which the Trussed Concrete Steel Company was successful in its bid. Each was required to prepare a scheme in concrete and to submit it together with a tender prepared by a contractor, to be nominated by the specialist. So the 'main' contractor was, effectively, the choice of the concrete specialist. This apparently back-to-front approach makes sense when the importance of good materials and workmanship, already stressed, is taken into account. In their own interests, the concrete specialists might be expected to propose only contractors licensed by or approved by themselves to use their system.

Four of the five specialists submitted schemes with the tenders. It then appears that the architects spot-checked selected typical designed members from each scheme and calculated their strengths using the 1907 RIBA Joint Committee Report.[10] These calculations were compared with the required strengths, to give an indication of the soundness of the tenderers' designs.

Having accepted the Trussed Concrete tender using Kahn bars, the client required the final design by the specialists to follow the RIBA Report methods, with C.F. Marsh, author of the first British book on reinforced concrete,[5] acting as checking engineer. The specialists were required to pay his fee for this service! There were at this time no regulations for reinforced concrete construction in Manchester, and Mr Marsh's 'seal of approval' was noted as being most helpful in obtaining building control consent.

The picture one has, therefore, is of the design and construction of reinforced concrete at this time clearly being a specialist activity. Established concrete specialists such as Hennebique were anxious, of course, to increase turnover and profits, but were also no doubt concerned that use of unsuitable materials and inexperienced labour, without established standards or codes of practice, could lead to failures and consequent loss of confidence in reinforced concrete as a structural material. They were willing to engage in competitive bidding against each other for work — indeed, they had no choice — and that necessarily meant preparing designs and pricing schemes competitively, with no guarantee of payment. Against the previously quoted figure of 3061 projects realized by Hennebique in the period 1892–99 must be set 8078 not realized.[2]

The ethical position for a professional engineer employed by a specialist concrete firm was in those days ambiguous; at least one ICE member was expelled for over-zealous efforts to 'procure' work for his employer. The fiercely competitive commercial environment may explain this eagerness, although of course it could not (officially) be condoned by Edwardian professionals.

The systems develop

Hennebique was, with Mouchel, for several years the only major concrete specialist in Britain. The company began working in Britain in 1897. The Trussed Concrete Steel Company took out a British patent in 1903 to cover use of its American Kahn bar.[11] In 1904 Edmond Coignet built a group of tobacco warehouses in Bristol (recently demolished) while the British Reinforced Concrete Co. (BRC) was established in 1905. Armand Considère took out a British patent in 1902 for the use of helical binding in columns, based on tests he had conducted. These showed — as is now well appreciated — that adequate restraint to the compression bars is essential, both to restrain them from premature buckling and to enhance the strength of the concrete core. But this form of reinforcing was not used until 1907. Considère later developed the U-hook to anchor main bars. Other forms of reinforcement included expanded metal, as illustrated in the Appendix.

Anchorage and bond were recognized as important design matters, and in this respect the relatively inferior performance of the plain round or square bar was one factor in the proliferation of patents for the use of deformed or profiled bars. Of these, the Ransome and Kahn bars have already been noted. Others included Johnson's indented bar, patented unsurprisingly by the Patent Indented Steel Bar Company, and much used by the British Concrete Steel Company. An advertisement in a contemporary textbook[12] by the former company notes that this bar profile has a continuous mechanical bond with the concrete, without slip. Significantly, it also points out that the bar is offered without fee or royalty, and may be used for any concrete system. In other words, the company was offering a product — the bar — to anyone who would buy it for use in their own reinforced concrete construction. This is in strong contrast to the 'concrete specialists', mainly from the Continent, who were offering a complete system.

The clients for early reinforced concrete structures included, perhaps paradoxically, some of the more 'conservative' bodies such as the Government's Office of Works, dock and harbour companies, railways, water boards, and agricultural and co-operative societies. Their interest in the new construction material was, surely, commercially motivated. Concrete was clearly economically competitive with steel and masonry from the outset, for if this had not been the case then its use would never have been considered. At the same time, it was seen to offer other advantages. The reinforcement appeared to be inherently protected against corrosion by its embedment in the concrete. Only later did it become embarrassingly clear that the actual thickness of concrete cover, the concrete mix design, its compaction, and (particularly) water content were all vitally important in determining durability. The concrete cover was also acknowledged to protect the reinforcement against fire. Concrete was pourable, so that structures could be built with non-orthogonal forms at only slightly greater cost. The Manchester YMCA swimming pool has already been cited as a case where the mouldable nature of concrete, its structural strength, and its potentially water-resistant nature combined to offer the 'ideal' material. Water tanks, silos, bunkers, and other essentially functional structures were obviously similarly suited to concrete construction.

The Government's initial attitude towards reinforced concrete was ambivalent. On the one hand, the Local Government Board was unwilling to provide loan finance for local authorities to build in reinforced concrete on the same terms as were applied to traditional construction and even to the relatively recent newcomer, steel framing. But many public architects (then as now) were keenly interested in new structural developments. Sir Henry Tanner became Chief Architect in the Office of Works in 1898 and encouraged the use of reinforced concrete in its projects.[13] Crown buildings were exempt from building regulations, and so there was no procedural objection to using reinforced concrete for their construction at a time when building bye-laws made no provision for this new material. Consequently, the Public Office and the larger Sorting Office of the General Post Office in King Edward Street in the City of London[12] were built on the Hennebique system in 1907–10. The demolition of the Sorting Office in 1998 regrettably deprived London of one of its earliest and boldest reinforced concrete structures. However, the demolition exposed briefly the ambitious scale of the construction (Figure 4.3). It also showed the simple butt-connection of column bars in the monolithic frame, a detail that might surprise today's engineer accustomed to providing lapped connections (Figure 4.4).

Figure 4.3   A cross section through the Sorting Office of the General Post Office, King Edward Street, City of London (1907–1910), exposed during demolition in 1998 (by courtesy of Lawrance Hurst).

It is no coincidence that Sir Henry took the chair of the RIBA Concrete Committee when it was established in 1905. He was also an active member of the Concrete Institute. (His role in promoting the wider use of concrete is discussed further in Chapters 5 and 14.)

The ready embrace of reinforced concrete by public utilities (including railway companies and port authorities), and by commercial bodies, was undoubtedly spurred by practical considerations of economy, including what was perceived as minimal maintenance cost. Today we may question this strictly utilitarian approach on aesthetic grounds. Unfortunately, too, durability was inadequately understood, as will be considered in more detail in Chapter 5. Weaver's Mill in Swansea (Figure 4.1) was not a visual delight as it stood abandoned in the early 1980s, and it was demolished despite being the oldest surviving reinforced concrete structure in Britain. It is not certain which structure now holds that title, although the 1897–1900 CWS Quayside warehouse in Newcastle (Figure 4.2) is a contender. (It is shown here before recent repairs.)

Many concrete structures of this period were bold in concept. One of the most spectacular was the 1909 Royal Liver Building at Pier Head, Liverpool, architect W. Aubrey Thomas[4] and once more engineered by Mouchel using the Hennebique system (Figure 4.5). One hundred and sixty-seven feet (51 m) high at roof level,

Figure 4.4   Column in the Sorting Office of the General Post Office during demolition, showing butt-jointed main bars with small tubes used as location aids (1907–1910) (by courtesy of Lawrance Hurst).

and crowned by the Liver bird with an overall height of 310 ft (94 m), it offered arriving transatlantic voyagers the sight of what was then the nearest resemblance in Europe to a New York 'skyscraper'.

The tallest concrete building in the world before this structure was erected had been the 18-storey Ingalls building in Cincinnati, Ohio, of 1902, at 210 ft (64 m). This was reinforced on the Ransome system of twisted bars, mixed with plain round bars. Column main bars were spliced using early 'couplers' — sleeves of wrought iron tubing filled with grout (similar to the details seen in the GPO Sorting Office, shown in Figure 4.4). Writing of the Ingalls building, Twelvetrees notes wryly that 'the architectural design is not remarkable for novelty',[12] and indeed it can be said that many early concrete buildings, as opposed to structures, did not advertise their concrete construction. From the outside, and often even from the inside, the structure could be of steel or of concrete — it was hard to tell. The expressionist use of concrete in habitable buildings had to wait for the birth of the modern movement (see Chapter 2). On the other hand, functional structures were usually clear in displaying their concrete form and finish.

This review suggests that, until the First World War, the concrete specialists 'had it all their own way' so far as design was concerned. While this is true in terms of the preparation of designs, drawings and schedules, it was often required of the

Figure 4.5   Royal Liver Building, Pier Head, Liverpool (1909) (by courtesy of British Cement Association).

specialists that they also satisfy an independent consultant. The example of Marsh and the Manchester YMCA building has already been described above. As for construction, the specialists either did the work themselves, or they licensed contractors who were guided by the specialists in sound construction practice, usually in return for paying a royalty for using the system.

Construction practice

Actual construction practice was inevitably more basic than today, although the principles of sound practice were little different. Materials were to be clean and properly stored; formwork was to be clean, rigid and secure. It was in the mixing, transporting, placing and treatment of concrete that the main differences were to be found.

All concrete was site-batched — there were no ready-mix trucks, no skips manoeuvred by tower cranes, no pumps. Hand-mixing was common, although mixing machines were already being used. Concrete was mixed by volume proportions (a practice that is still to be found today on smaller sites), or in 'recipe' form. A good example is the 1917 Mouchel specification.[6] This typically called for 6 cwt (305 kg) of cement to be batched with $13\frac{1}{2}$ ft^3 (0.38 m^3) of sand and 27 ft^3 (0.76 m^3) of coarse aggregate to give a 1.1 : 2 : 4 mix, with a probable cube strength of around 15–20 N/mm^2.

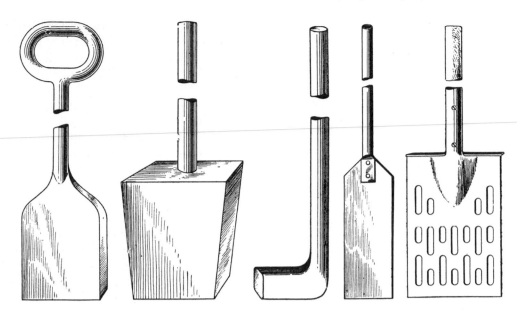

Figure 4.6 Typical concrete punning tools from the early 20th century.

Transporting was likely to be by tipper trucks running on a narrow-gauge railway, by chute, or by wheelbarrow once the fresh concrete had been raised to the required level by a hoist or a derrick. Placing would be by spade or shovel, working the mix as far as possible into its final location. It would then be compacted, up to a point, by ramming, tamping, or 'punning' by hand. The tools for this work came in a variety of shapes (Figure 4.6), including a rod with an enlarged box-end and another shaped like a hockey stick.[4] Not until 1917 did Eugène Freyssinet establish the essential importance of mechanically compacting the concrete to expel air and ensure thorough filling of the profiles, completely enclosing the steel reinforcement with concrete. Even then, it was 1924 before he himself used vibrating tools in practice. Before then, the importance of filling the shutters and fully surrounding the reinforcement with concrete was understood; but, without mechanical compaction, and with (often) congested reinforcement, the all-too-obvious answer was to make the concrete wetter, so that it flowed more readily and was 'self-compacting'. Some specialists, including Ransome, argued in the 1900s that a wet mix would indeed be superior to a dry mix.[11] It is perhaps just a coincidence that Duff Abrams completed his research on concrete mixes in 1918, only a year after Freyssinet's work on the need for mechanical compaction. Abrams's law declares that the water–cement ratio is a fundamental determinant of concrete strength. In simple terms, a 'wetter' concrete will be weaker when hardened than a 'drier' concrete.

Variations of technique and form

Numerous construction techniques and structural forms were introduced during the period under review which, at first glance, might be thought to be later in origin. Some more notable examples are briefly described below.

Precasting

Examples of early precast cladding have been given in Chapter 3. Precasting of structural elements was the subject of Hennebique patents as early as 1897, both as self-contained units and as permanent shuttering with projecting reinforcement to provide composite action with *in-situ* concrete. The British Precast Concrete Federation was formed as early as 1918.[14]

A simple precast floor system was revealed in the 1904–1905 Joshua Hoyle building in Manchester prior to refurbishment. This building has a steel frame and terracotta cladding, and — clearly an innovation for this time — an unattended automatic lift.[15] However, it is the floor units that are of most interest. They are precast unreinforced half arches that are laid in pairs to sit on the lower flange of the I-beams, butting at the crown. The units thus form simple three-pinned mass concrete arches. The coarse aggregate is clinker (burnt coal), and timber shavings as are nowadays found in woodwool slabs. Re-use proposals had to take account of the potential combustibility of such concrete.

Filler joist floors and clinker concrete

The introduction of 'filler joist' floors in the second half of the 19th century was described in Chapter 3. They continued to be used in the early part of the 20th century. Unreinforced concrete slabs, spanning effectively as flat arches, were supported by I-beams or joists at typically 0.6–1.0 m spacing (Figure 4.7). Originally, the joists were of wrought iron, while the concrete would contain either lime or Portland cement as binder, with coarse aggregate of broken brick, slag, clinker or 'breeze' from coal fires, or stone. Incomplete combustion of the coal could leave a combustible fraction in the floor. Steel, as it became more widely available from the 1890s, replaced wrought iron in the joists, while the concrete was increasingly made with Portland cement, sand and stone aggregate, typically as a 1:2:4 mix.

This filler joist floor construction was widely used in offices and blocks of flats, notably the 'mansion blocks' of Edwardian times and later, where its high mass gave good resistance to air-borne sound. Timber flooring laid on battens (particularly when carpeted) also gave good structure-borne sound resistance, resulting in excellent acoustic insulation between dwellings.

Tests to assess the enhancement of strength by what is nowadays termed composite action between the filler joists and the concrete were carried out at the National Physical Laboratory as late as 1922–23.[16] The results were incorporated into BS 449[17] in 1932, and indeed remain in the 1969 metric edition still, with later amendments, in use today. However, the wide adoption of orthodox reinforced concrete floors, both *in situ* and precast, resulted in the demise of filler joist construction, with its requirement of formwork for the invariably *in-situ* infill concrete.

A problem often found with clinker concrete in such construction arises when the floor or roof becomes wet, as in neglected bathrooms or kitchens, or on flat roofs with degraded waterproofing. The clinker contains compounds of sulphur, and also of nitrogen and chlorine, which in wet conditions can severely aggravate corrosion of the iron or steel joists, leading also to spalling of adjacent concrete. Repair can be expensive and disruptive.

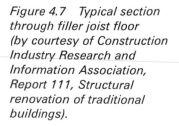

Figure 4.7   Typical section through filler joist floor (by courtesy of Construction Industry Research and Information Association, Report 111, Structural renovation of traditional buildings).

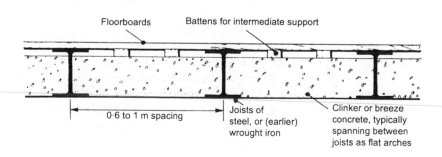

Floorboards

Battens for intermediate support

0·6 to 1 m spacing

Joists of steel, or (earlier) wrought iron

Clinker or breeze concrete, typically spanning between joists as flat arches

Concrete blocks

Concrete was quickly recognized as a potentially cheaper but adequate substitute for fired-clay materials including bricks and tiles. A patent for hollow concrete blocks had been taken out as early as 1850,[18] and block-making machines were in use by c. 1860.[14] Steam curing was in use by the early 1900s[18] to accelerate curing and increase productive use of the block moulds.

Asbestos–cement and woodwool

The first asbestos–cement sheets are believed to have been made in Austria in 1900. Within a decade the material was being widely used in its basic flat sheet form as a durable, non-corroding building board and roof sheeting. Corrugated sheeting followed soon after in the early 1910s, achieving longer spans from its deeper profile.[14] This is a relatively early example of fibre-reinforced cement, a predecessor of grc or glass-reinforced cement, used for architectural cladding in recent decades.

Another cement-based composite with a lengthy history is woodwool which, like clinker concrete, makes use of what would otherwise be a waste product. Wood shavings from the planing of timber and cement paste are pressed together to form slabs, which are lightweight and provide good thermal insulation.

Unreinforced concrete

The emphasis in this chapter is mainly on reinforced concrete, but the use of 'plain' or unreinforced concrete was — and remains — still widespread. This followed the precedent of using lime-based concrete (described in Chapter 3) for foundations pads, strips, and rafts, and other structural elements subject to essentially compressive loads. Plain concrete was an obvious choice where bulk or mass was needed, as in gravity dams for reservoirs, retaining walls to uphold the sides of railway cuttings, and marine works. It was also used in the walls of housing and other buildings, where its presence may be deduced from the thicker wall section, the presence of visible horizontal daywork joints, and often the erosion of the cement paste to leave a stony surface.

The term 'mass concrete' should perhaps have been confined to the use of the material in bulk, but in practice it has always been synonymous with plain or unreinforced concrete.

Shotcrete

Shotcrete, also known as gunite and nowadays as sprayed concrete, was developed almost a century ago in 1907 by one Carl Ethan Akeley, an American naturalist and taxidermist.[19] Seeking a means of creating animal models and mounts, he devised the idea of 'shooting' a pressurized blend of cement and sand onto a metal armature, adding water as the mixture left the 'cement gun'. Built up in thin dense layers, this produced realistic free-form profiles with a hard durable surface.

The Cement-Gun Company was formed in Allentown, Pennsylvania, in 1910. It developed Akeley's invention and applied it on a larger scale to building and civil engineering uses. The company registered the trade name of 'Gunite' for the product, which was quickly adopted for hydraulic work and particularly for tunnel linings (see Chapter 15). It also found wide use as sprayed fire protection to steel frames, and for repair and strengthening works, when the ability to add a reinforcement cage and then form the concrete profile without formwork proved very effective.

The process was introduced into the UK in the 1920s. The original mixing method came to be known as 'dry-mix'. Later, the 'wet-mix' method was introduced, in which the water was mixed with the cement and sand before it left the gun. This in effect uses the same principle as present-day concrete pumping, although the discharge speed was much higher. In both methods 'rebound' of some of the concrete was inevitable, although an experienced operator could limit this, and would also control the amount of water added.

Publications of the time

For those faced with appraising or refurbishing such structures today, it may be helpful to review the technical information and design guidance available to clients, architects and structural practitioners during the years 1897–1915. The libraries of the Institutions of Civil and Structural Engineers, and also of the Royal Institute of British Architects, have good collections of useful material and can between them provide all the references cited here.

This pioneering period began with construction of the first British Hennebique Mouchel designs, and closed with publication of the London County Council Reinforced Concrete Regulations[20] which offered the first 'code of practice' for reinforced concrete design. Regulatory practice is discussed further in the next chapter; here, attention is briefly drawn to contemporary texts and specialist literature.

As already noted, *Reinforced Concrete* by C.F. Marsh appeared in 1904, being the first British textbook on reinforced concrete design and construction.[5] It is a substantial text, well illustrated, and can be fairly called a 'state-of-the-art' review. Later editions followed in collaboration with W. Dunn. These gave increasing space to calculation methods as the use of reinforced concrete widened.[18]

A more enduring record of concrete practice began for the general reader with the publication in 1906 of the first issue of *Concrete and Constructional Engineering*, which continued until 1966. This journal is a rich source of information on both individual buildings and practice of the time.

Essentially a house journal, but none the less useful for that, was Hennebique's *Le Béton Armé*, produced from 1898. An English sister journal, *Ferro-Concrete*, began publication in 1909. Today this is a valuable if scarce source of detailed information for British Hennebique buildings such as those mentioned here. The early editor was W. Noble Twelvetrees, also the author of early guides to theory and practice in reinforced concrete.[12,21]

The Concrete Institute was set up in 1908 for architects, engineers and contractors, publishing its first *Transactions* in 1909. Its development and metamorphosis into the Institution of Structural Engineers is traced in Chapter 14.

Other concrete specialists published literature — both technically supportive and openly promotional — during this period, while various 'independent' textbooks started to appear. Their growth into a wide range of design guides is outlined in Chapter 5.

The Appendix gives more comprehensive guidance on sources of contemporary information.

The era comes to an end: The First World War and later

The outbreak of the First World War in 1914 brought many things to an end. From the viewpoint of structural history and that of today's structural engineer, it can be said to mark the end of the 'golden age' of the proprietary systems. The publication of the London County Council Reinforced Concrete Regulations in 1915 put an acceptable method of designing reinforced concrete structures into the public domain.[20] Using plain mild steel bars, the engineer or numerate architect or builder could now do the calculations and design concrete structures. That is not to say that the specialists disappeared; but an element of 'free market' theory had arrived. Henceforward, it would not need a specialist to design concrete structures; and the designs could be assessed against published codes, and approved under regulatory bye-laws. Nevertheless, companies such as Hennebique continued to design and build structures. L.G. Mouchel, for example, formed a consulting engineering practice that continues to this day. Companies making reinforcement continued to offer design services — indeed, a few still do.

Another aspect of the First World War was, of course, the application of reinforced concrete to military structures, most commonly as protection against shelling. The defensive 'pill-box' has become for some an archaeological novelty, but it should be recalled that its design, however empirical, involved the first serious attempts to assess the effects of blast loading on reinforced concrete structures.[22]

Additional general sources of information on this period

Additional general sources for the period covered in this chapter include studies by de Courcy,[23] Stanley,[24] and Mainstone.[25] Some sources of guidance for those called upon to assess or alter such structures is cited at the end of Chapter 5, which traces the continuing development of concrete in the 20th century, and is supplemented by more comprehensive guidance in the Appendix.

Acknowledgements

The author wishes to thank his fellow co-authors for their encouragement and for helpful criticism of the text. The source of Table 4.1 was drawn to the author's attention by Mr B.N. Sharp. Sir Alan Muir Wood kindly provided advice on the development of shotcrete.

References

1. Brown, J.M., W.B. Wilkinson (1819–1902) and his place in the history of reinforced concrete. Trans. Newcomen Soc., 1966–67, 39, 29–142.
2. Cusack, P., François Hennebique: the specialist organisation and the success of ferro-concrete: 1892–1909. Trans. Newcomen Soc., 1984–85, 56, 71–86.
3. Coignet, E., Tédesco, N. de., Du calcul des ouvrages en ciment avec ossature metallique. La Société des Ingénieurs Civils de France: Paris, 1894.
4. Jones, B.E. (ed.), Cassell's Reinforced Concrete, 2nd edn. Waverley Book Company: London, 1920.
5. Marsh, C.F., Reinforced Concrete. Constable: London, 1904.
6. L.G. Mouchel & Partners, Standard Specification for Ferro-Concrete. L.G. Mouchel & Partners: London, 1917.
7. Engineering Standards Committee. Standard Specification for Structural Steel for Bridges and General Building Construction. Engineering Standards Committee: London, 1912, BS 15.
8. Anon., Ferro-concrete warehouse at Newcastle-on-Tyne. Engineering, 1903, April 17, 514–15.
9. Lakeman, A., The YMCA building, Manchester. Concrete & Constructional Engineering, 1911, 6, 368–77 (and discussion, 501–15).
10. Joint Committee on Reinforced Concrete. Report of the Joint Committee on Reinforced Concrete. J. Roy. Inst. Br. Archit. (3rd ser.), 1907, 14(15), 513–41.
11. Hamilton, S.B., A Note on the History of Reinforced Concrete in Buildings. National Building Studies Special Report No. 24. HMSO: London, 1956.
12. Twelvetrees, W.N., Concrete-Steel Buildings. Whittaker & Co.: London, 1907.
13. Gray, A.S., Edwardian Architecture: A Biographical Dictionary, 2nd edn. Wordsworth Editions: Ware, 1988.
14. Hudson, K., Building Materials. Longman: London, 1972.
15. Anon., A Piccadilly improvement. Manchester City News, 1905, January 21.
16. National Physical Laboratory Reports, 1922, p. 8; 1923, p. 169. Etchells, E.F. (ed.), Modern Steelwork. Nash & Alexander: London, 1927.
17. British Standards Institution. Specification for the Use of Structural Steel in Building. BSI: London, 1932, BS 449.

18. Marsh, C.F., Dunn, W., Manual of Reinforced Concrete and Concrete Block Construction. Constable: London, 1908 (and various other editions).
19. King, E.H., Shotcrete. In: Bickel, J.O. *et al.* (eds), Tunnel Engineering Handbook, 2nd edn. Chapman & Hall: New York and London, 1996, 220–30.
20. London County Council. Reinforced Concrete Regulations. LCC: London, 1915.
21. Twelvetrees, W.N., Concrete-Steel. Whittaker & Co.: London, 1906.
22. Mallory, K., Ottar, A., Architecture of Aggression: A History of Military Architecture in North West Europe 1900–1945. Architectural Press: London, 1973.
23. De Courcy, J.W., The emergence of reinforced concrete 1750–1910. Struct. Engr, 1987, 65A(9), 315–22 (and discussion of paper in Struct. Engr, 1988, 66(8), 128–30).
24. Stanley, C.C., Highlights in the History of Concrete. Cement & Concrete Association: Slough, 1979.
25. Mainstone, R.J., Developments in Structural Form, 2nd edn. Architectural Press: Oxford, 1998.

5 The development of reinforced concrete design and practice

Michael Bussell

Synopsis

The previous chapter described the introduction of reinforced concrete into the UK via proprietary systems, which flourished from the end of the 19th century until the First World War. This was in the absence of an agreed — or rather a codified — design method, which did not appear until 1915. This chapter reviews developments in the understanding of reinforced concrete behaviour, and charts the gradual standardization of structural design, materials, codes, standards and textbooks up to 1948 when CP 114 first appeared. It continues this review briefly from 1948 to the present day. Some significant structures built between the First World War and 1948 are noted. Landmarks in architectural concrete use and other notable developments in shell roofs, prestressed concrete, bridges, and maritime structures are covered in more detail in other chapters.

Appendix B contains information to assist in identifying proprietary concrete floors of this period in existing construction.

Early understanding of reinforced concrete behaviour

Early studies of the behaviour of reinforced concrete were led by American, French and German workers. The first book on reinforced concrete (as opposed to articles) was published in 1877 by Thaddeus Hyatt, an American living in London.[1] He used data from tests carried out at David Kirkaldy's testing laboratory in Southwark. His work recognized that composite behaviour between iron or steel and concrete is dependent on bond between the two materials, and is aided by the near equivalence of their coefficients of thermal expansion. Hyatt's work acknowledged that, at least under working loads, strain compatibility renders stresses in materials proportional to their relative elastic moduli.[2]

G.A. Wayss published in 1887 a book on the Monier system,[3] which his company was using in Germany and elsewhere. Working with K. Koenen, he concluded that the steel should be designed to take all the tensile stresses, and also that bond between steel and concrete was essential to transfer the internal forces to maintain equilibrium. A similar study was published in 1894 in France by Edmond Coignet and Napoléon de Tédesco,[4] and this — together with a paper by P. Christophe of 1899, published as a book in 1902[5] — laid the foundations for elastic modular ratio theory. This was practised until well after the Second World War, and indeed is still used for serviceability calculations in today's limit-state codes.

Early design guidance

In Britain the first textbook, by C.F. Marsh,[6] appeared in 1904; but it was 1906 before the Royal Institute of British Architects appointed a Reinforced Concrete Committee to review the use of structural reinforced concrete. Its first report was published in 1907.[7]

Events in Britain now gained momentum. BS 12, covering Portland cement, appeared in 1904, reflecting concern about the quality of cement. Similar concern led to the establishment of a Special Commission on Concrete Aggregates by the British Fire Prevention Committee in 1906. The previous chapter drew attention to the use of clinker aggregates. These often contained unburnt coal; in the event of a fire the nominally 'fireproof' floor could then actually support combustion! The Committee's founder and chairman was E.O. Sachs, who was also editor of *Concrete and Constructional Engineering*, which began publication in the same year.

The year 1908 saw the formation of the Concrete Institute, an open forum for all interested in the use of reinforced concrete, not just 'the specialists'. This, in due course, became the Institution of Structural Engineers. Chapter 14 gives a fuller account of its birth and development.

The year 1909 was notable for the passing of the London County Council (General Powers) Act.[8] This is well known for its schedule covering — for the first time in British building regulations — the design of steel, wrought iron and cast iron structures, although by the time of its issue steel was almost exclusively the only ferrous structural metal in use. The Act also made provision for the introduction of regulations controlling the use of reinforced concrete in inner London, although these would not appear until six years later. (It also, for the first time, specified floor loadings to be taken into account in design.)

The Institution of Civil Engineers set up its own Committee on Reinforced Concrete, which reported in 1910.[9] The committee reviewed the concrete specifications of the specialist firms (Hennebique, etc.), overseas rules on concrete, and available test data. It also heard statements from practising consulting engineers, who were generally reluctant to take responsibility for designing larger concrete projects. Of particular relevance for today's appraising engineer is the evidence for wide variation in cover requirements (or none at all) and a preference for using wetter mixes to achieve 'good' compaction, when mechanical compacting methods were neither yet available nor recognized to be beneficial. Low cover and a weaker 'wet' concrete have obvious implications for durability. This is now understood, but was not recognized at the time.

In 1911 the RIBA Committee was restructured as a Joint Committee, with added representation from the Concrete Institute and the London County Council (LCC). It produced a revised report in the same year.[10] This again did not make specific recommendations for cover or care in use of wetter concrete mixes, so that durability was not well addressed. Fire resistance needs were taken into account, with cover of ½ in (13 mm) being specified for slabs and 1 in (25 mm) for beams. A 1 : 2 : 4 concrete mix was assumed, with a cube strength of 1800 lb/in^2 (12.4 N/mm^2) at 28 days. The permissible concrete compressive stress was to be one-third of this value. The standard notation for symbols, introduced in 1909 by the Concrete Institute, was adopted. Shear reinforcement was to be provided when concrete shear stress exceeded 60 lb/in^2 (0.41 N/mm^2) — one-tenth of the permissible compressive stress — and was to comprise either bent-up bars or stirrups spaced no further apart than the beam depth. Column axial compressive capacity was recognized to be enhanced when helical binding was present, acknowledging the containing effect of this steel. Design for bending was based on the by then generally accepted elastic modular ratio theory, using a ratio of 15.[2] Concern about reinforcement anchorage to the concrete was reflected in guidance that bar ends should be split, bent, or otherwise mechanically secured.

The 1915 Reinforced Concrete Regulations

These recommendations were largely embodied in the LCC's Reinforced Concrete Regulations of 1915.[11] The symbol Q (with units of stress) was defined as the ratio of applied moment to (member width times the square of effective depth); it was a useful and long-lived aid to engineers making calculations (with a slide rule, of course), as it was a measure of how hard the section was working. Tensile and shear steel was to be effectively anchored, by hooking or otherwise, at its ends. Hook internal diameter was to be four bar diameters.

Cover was again dictated by fire considerations. Columns (known then as 'pillars' if vertical, otherwise as 'struts') were to have cover of 1½ in (38 mm), or the bar diameter if this were greater.

From 1918 to 1934

The LCC regulations offered the first codified British design and construction guidance for reinforced concrete, but they were little used until the end of the First World War. The next decade, the 1920s, was a period of often (but not always) poor quality construction, carried out by building contractors who were frequently unaware of the need for adequate cover to steel and the need for proper treatment of concrete. This contrasted sharply with the pre-war period, when the specialist concrete firms might admittedly charge higher prices, but would take care to build a sound structure. They had both the understanding of what was needed, and the motive of maintaining sound reputations.

A dramatic example of a good 1920s reinforced concrete structure is the 1926 New Royal Horticultural Hall in Westminster, London, by architects Easton and Robertson (Figure 5.1). Design of the structure was by British Reinforced Concrete Ltd., one of the longer-lasting specialist firms providing a design service. Such a structure shows a clear use of concrete arches that could never be mistaken in form or appearance for either steel or masonry.

Another notable structure of this period, facing imminent demolition at the time of writing, is Wembley Stadium (1921–24, Figure 5.2). It was designed by the architect Maxwell Ayrton and the engineer Owen Williams for the British Empire Exhibition, to accommodate 125,000 people. The contractor was Sir Robert McAlpine. This was one of the several large structures on the exhibition site, including the Empire Pool. The stadium was presented more in the style of a masonry structure both in form and in the false jointing applied to concrete surfaces to give the appearance of ashlar masonry. This was on the 'public' faces of the building. The engineer Oscar Faber was disappointed with the more utilitarian concrete work elsewhere.[12]

Figure 5.1   The New Royal Horticultural Hall (1926) (by courtesy of the British Architectural Library).

*Figure 5.2   Wembley
Stadium (1921–24)*

Williams himself went on to engineer some rugged Highland bridges (see Chapter 11), before becoming architect–engineer for the 1930 Dorchester Hotel, London (in collaboration with architect W. Curtis Green and consulting engineers Considère & Partners) and many other structures. Of these perhaps the most notable are the 1931 Daily Express building in London, its 1939 successor in Manchester, and the 1932 Boots 'wets' factory at Beeston, Nottingham (Figure 5.3).[12,13]

The two last-mentioned buildings are particularly significant in their use of flat slabs. These had been developed independently in the USA and in Switzerland, as an alternative to the traditional drop beam solution. Beams were essential in a frame of iron or steel because these materials are used as one-dimensional 'sticks', whereas concrete can be readily formed to any required shape. In addition, the omission of downstanding beams simplifies the formwork and eases construction.

Nevertheless, early concrete floors and roofs were usually structured with one-way or, more commonly, two-way grids of beams supporting slabs. Indeed, in some buildings the concrete beams are haunched near the column supports to look more as if they are steel beams cased in concrete and supported on stiffened gussets (Figure 5.4)! Flat slab column heads, in contrast, are clearly not derived from steelwork practice (Figure 5.5). O.W. Norcross took out a patent for flat slabs in the USA in 1902. His fellow-countryman, C.A.P. Turner, was building flat slabs by 1906, and the first American load tests on flat slabs were performed by A.R. Lord in 1910.

American practice favoured four layers of slab reinforcement, rather than two as is now customary. The additional reinforcement layers at 45° to the column grid were provided in pragmatic recognition that the principal moments were not

Figure 5.3   Boots 'wets' factory (1932). Under grey skies before recent restoration and cleaning.

Figure 5.4   Early concrete structure resembling steel-framed construction (Co-operative Wholesale Society warehouse on Quayside, Newcastle-upon-Tyne, a Hennebique structure of 1897–1900).

always parallel to the column lines. Nevertheless, the simpler arrangement of two layers at right angles can resist the bending moment components adequately if designed for the appropriate values.

Robert Maillart in Switzerland advocated this simpler approach, and based his designs on tests carried out in 1908 and 1913–14.[14] 'Mushroom' column heads of more or less elegance were developed independently in the USA and mainland Europe in this period, with the Americans leading in the adoption also of 'drop' panels as another way to cope with high shear stresses around columns.

Some use were made of flat slabs in the UK before the early 1930s by innovative engineers such as Sven Bylander, but wider application was probably inhibited because design guidance was not readily available.

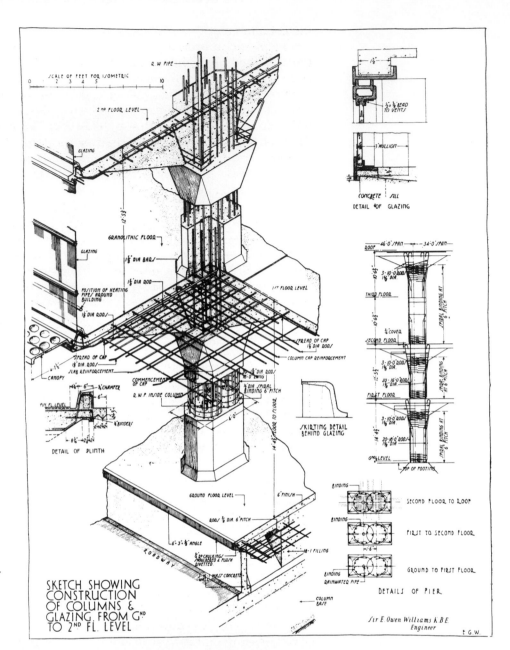

Figure 5.5 *Flat slab reinforced concrete construction with 'mushroom' heads (Boots 'wets' factory, 1932, by Sir Owen Williams) (by courtesy of the Builder Group).*

The first British concrete code

Design of flat slabs was one of the significant additions to guidance provided in the first British code of practice (as opposed to regulations) in 1934.[15]

Another was improved guidance on column design. Oscar Faber, a practising engineer with a strong interest in structural behaviour, questioned the validity of the modular ratio design method in columns. He carried out tests in 1927 to assess the effects of shrinkage and creep on the current assumptions for reinforced concrete design, and concluded that the modular ratio method was illogical for columns. It obliged the designer to assume that the column steel was elastically stressed at, typically, 15 times the compressive stress in the adjacent concrete. The reality was quite different. Creep of the compressed concrete would increase compression in the steel, and shrinkage would shed further load from the concrete into the steel. Work at the Building Research Station confirmed this.[16,17] So it was logical on that evidence to abandon the modular ratio method for columns, and to use instead the arithmetic sum of allowable loads on concrete and steel to give the column's capacity. (This was, of course, still some time away from the more recent 'plastic' approach of working out the column's ultimate strength, based on the combined contribution of both concrete and steel, and taking its load capacity as a suitably factored proportion of that.)

The 1934 code was the result of a review of current practice by the Reinforced Concrete Structures Committee appointed by the Building Board of the Department of Scientific and Industrial Research in 1931. The 1915 regulations, which formed their starting point, had been based on the Joint Committee's report of 1911 — so there were two decades of experience and development to be considered.

Attention was given to the strength requirements for concrete mixes. Volume batching was still the norm (as it was to be until the 1950s). There were four classes, I–IV, these being designated respectively $1:1:2$, $1:1.2:2.4$, $1:1.5:3$, and $1:2:4$; these were the proportions of cement:fine aggregate:coarse aggregate by volume. But each class could be of three grades, use of which depended on the level of control and supervision. Ordinary Grade concrete was to have cube tests taken only when directed by the designer. Higher Grade concrete was to have prescribed preliminary and works tests (including daily works consistence or 'slump' tests), and the work was to be carried out under the direction of a foreman and a supervisor or clerk of works, both to be experienced in such work. For Special Grade concrete, further controls on materials consistency were called for, while the designer was to design the structure taking account of continuity in all members. In return for this, greater stresses were permitted on the concrete. For a $1:2:4$ Ordinary Grade concrete, the permissible compressive stress of $600\,\mathrm{lb/in^2}$ ($4.1\,\mathrm{N/mm^2}$) in the 1915 regulations was increased now to $750\,\mathrm{lb/in^2}$ ($5.2\,\mathrm{N/mm^2}$). The corresponding Higher Grade figure was $950\,\mathrm{lb/in^2}$ ($6.5\,\mathrm{N/mm^2}$). Special Grade compressive stresses were to be calculated from preliminary test cube results divided by 5, but not to be more than 25% above the Higher Grade figures.

The modular ratio for design was now linked to the concrete strength (recognizing that Young's modulus for concrete is, of course, related to its strength). The modular ratio in Imperial units was 40,000/(3 times the permissible bending stress, i.e. the specified 28-day works cube strength); in SI units the numerator would be 276. For an Ordinary Grade $1:2:4$ mix, with a 28-day works cube strength of $2250\,\mathrm{lb/in^2}$ ($15.5\,\mathrm{N/mm^2}$), the modular ratio would now be about 18.

For reinforcement, permissible stresses in bending and shear were given as 0.45 times the steel yield point stress. For ordinary mild steel this was $18,000\,\mathrm{lb/in^2}$ ($124\,\mathrm{N/mm^2}$). In compression the permissible stresses were three-quarters of those for bending. Reference was also made to BS 15 for a higher-grade steel and to BS 165 for hard-drawn steel wire,[18,19] for which higher permissible stresses were given, acknowledging the developing use of high-tensile steels for rod and fabric.

The handbook to the code[20] comments on the benefits of mechanical vibration as an effective means of compacting concrete using a lower water–cement ratio, with benefits to strength and hardening rate. Again durability was not explicitly identified as a consideration.

The factor of safety for reinforcement was declared in the handbook as 1/0.45, i.e. 2.2, on yield. That for concrete was 3 on 28-day works cube strength. The handbook warned that use of higher-strength steel could increase the risk of cracking in tension zones. Also, the modular ratio design method meant that the compressive steel stress in flexure was limited to, typically, 11–18 times the adjacent concrete stress; so there was no great benefit, in general, in using high-tensile steel for strength unless cracking was either acceptable or investigated. Neither did the code recognize any higher permissible bond stress for deformed (cold-worked) bars, although it had long been argued, rightly, that deformed or ribbed bars had improved anchorage and bond properties.

Guidance was given on the calculation of moments in flat slabs and continuous members, and redistribution of hogging moments by up to 15% was permitted in beams and slabs. This acknowledged the relaxation of hogging moments due to creep in the compressed concrete at monolithic supports. It was also of practical benefit in reducing steel congestion and the associated risk of poor concrete compaction over supports.

Appendices to the code included loadings, as a British standard on loadings had yet to appear.

The code was based on working-load design, a philosophy that was unchallenged in the UK until the issue of CP 110 in 1972, as discussed below.

Structural concrete in the 1930s

The code arrived in time to be of use to the new generation of structural engineers and contractors, mostly from outside Britain, who were keen to collaborate with architects of the Modern Movement. Ove Arup, Oscar Faber, and Felix Samuely relished the chance to design unusual structures with architects such as Berthold Lubetkin and the Tecton group. Chapter 2 has considered this in more detail; here it is of interest to note two buildings which were structured with only limited assistance from the code.

The 1933 Penguin Pool at London Zoo (Arup and Samuely with Tecton, Figures 5.6 and 5.7) was an exercise in structural gymnastics in which the spiral ramps are clearly subject to torsion. British concrete codes before 1972 were coy about torsion, although spiral staircases and other structures reliant on torsional strength had been built as early as 1900. So the calculations were made from first (if rather substantial) engineering mathematical principles.

Similarly, the pioneering 1935 Highpoint I flats at Highgate, London (Arup and Tecton) had a 'box-frame' structure above ground-floor level, with slabs carried on walls with minimal downstanding beams (Figures 5.8 and 5.9). The code of 1934 required reinforced concrete walls to be designed and built to an equivalent standard of strength to that for other members, but gave no further guidance. This hinted that the orthodox steel-oriented view of structures being made up of columns and beams was not easily shaken off.

In 1938 the Ministry of Health promulgated new model building bye-laws which barely mentioned concrete.[21] However, it was by now easier to obtain building

Figure 5.6   Spiral ramp reinforcement for The Penguin Pool, London Zoo (1933) (by courtesy of Ove Arup & Partners).

regulations approval with the aid of the established 1915 London regulations, and subsequently the 1934 code, than it had been in the early 1930s. Further encouragement came from the LCC's 1938 guide to construction of buildings in inner London[22] and from a code issued by the Building Industries National Council in 1939,[23] although both were somewhat pushed aside by the events of the Second World War.

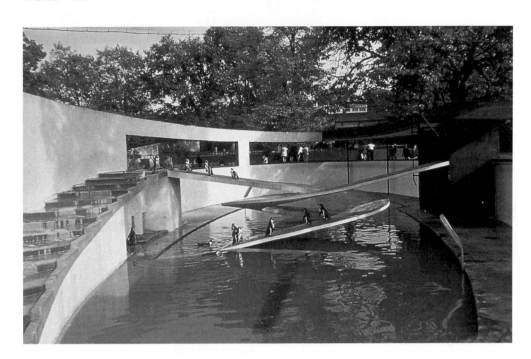

Figure 5.7   The Penguin Pool, London Zoo completed (by courtesy of Ove Arup & Partners).

Figure 5.8   Highpoint I flats at Highgate, London under construction (1935) (by courtesy of Ove Arup & Partners).

Figure 5.9   Highpoint I flats completed (by courtesy of Ove Arup & Partners).

Materials developments proceeded. The 1934 code countenanced use of rapid-hardening Portland cement, Portland-blastfurnace cement, and — significantly — high-alumina cement, even though not all were covered by British standards. Chapter 6 gives fuller attention to changes in cement and concrete.

The long-awaited BS 882 for coarse and fine aggregates appeared in 1940.[24] Reinforcement standards were developing. BS 785 appeared in 1938 for hot-rolled bars and hard-drawn wire,[25] BS 1144 in 1943 for cold-worked bars,[26] and BS 1221 in 1945 for steel fabric, i.e. mesh.[27] Developments in reinforcement strengths and permissible tension stresses in bending are reviewed in Table 5.1.

The Second World War

But this is to run ahead of the Second World War, which at once both halted most civilian projects and generated huge demand for military works, both offensive and defensive. With awareness of the surely inevitable war, and the equal certainty of bombing raids on town and cities, many local authorities had been already considering air raid precautions. Arup and Lubetkin argued for massive concrete shelters, in view of their greater safety. Circular shelters were conceived with spiral floor slabs that could be used as car parks after the war. Construction of these as proposed would be on the 'top–down' principle, as was later widely used on commercial sites in the 1980s for speed of building and minimising damage to adjacent buildings from ground movements.[28] For various reasons (mainly political and economic) their proposals were not pursued.

Equally original in concept, but this time realized, were the concrete structures of the Mulberry Harbours, designed and built for the Allied invasion of occupied France.[29] One can imagine the design of these being carried out with an eye to rapid construction and 'buildability' (see Chapter 13).

One effect of the war was inevitably that many structures were subject to blast and fire.[2] Study of blast-damaged concrete structures led to conclusions that will surprise few engineers who have, of sad necessity, been involved in more recent investigations of structures damaged by terrorism or gas explosions. Structural redundancy was desirable, so that a damaged structure might still be able to stand even though suffering serious local damage. Generous reinforcement lap lengths should be provided. Reinforcement should be provided where reversals of stress,

Table 5.1 Developments in specified reinforcement strength and permissible tensile stresses in bending from 1915 (Part 1 and Part 2)

Code and year	British Standard	Steel type	Ultimate tensile strength lb/in^2 (N/mm^2)	Yield stress (f_y) lb/in^2 (N/mm^2)	Permissible tensile stress due to bending lb/in^2 (N/mm^2)
Part 1					
London Reinforced Concrete Regulations, 1915	–	Mild	–	–	16,000 (110)
Building Research Board (DSIR)	BS 15	Mild	62,720–71,680 (432–494)	–	18,000 (124)
Code, 1933	BS 15	Defined yield point	–	44,000 (minimum) (303)	0.45 f_y, i.e. 20,000 (138)
London Byelaws, 1938	–	All	–	–	18,000 (124)
CP 114: 1948	BS 785	Mild	62,720–71,680 (432–494)	–	18,000 (124)
		Medium tensile*	73,920–85,120 (510–587)	36,960–43,680 (255–301)	0.5 f_y, maximum 27,000 (186)
		High tensile*	82,880–96,320 (571–664)	42,560–51,520 (293–355)	0.5 f_y, maximum 27,000 (186)
	BS 1144	Cold worked single twisted:			0.5 f_y, maximum 27000 (186)
		ø < 3/8 in	80,000 (551)	70,000 (482)	
		ø ≥ 3/8 in	70,000 (482)	60,000 (414)	
		Cold worked twin twisted	63,000 (434)	54,000 (372)	
	Fabric to BS 1221	To BS 785	As above for BS 785		
		To BS 1144	As above for BS 1144		
		Expanded metal	75,000 (517)	50,000 (345)	20,000 (138) ø ≤ 1½ in
CP 114: 1957	BS 785	Mild, no defined yield point	62,720–71,680 (432–494)	–	20,000 (138); ø > 1½ in 18,000 (124)
	BS 785, BS 1144, BS 1221	Guaranteed yield stress, high-bond or mesh	–	f_y	0.5 f_y, maximum 30,000 (207)
CP 114: 1965	BS 785, BS 1144, BS 1221	Guaranteed yield stress, high-bond or mesh	N/A	f_y	0.55 f_y maximum (33,000) 227 for ø ≤ 7/8 in, (30,000) 207 for ø > 7/8 in
CP 114: Part 2: 1969 (Metric)	BS 785	Mild, no defined yield point	432–494	–	140 for ø ≤ 40 mm, 125 for ø > 40 mm
	BS 785, BS 1144, BS 1221	Guaranteed yield stress, high-bond or mesh	–	f_y	0.55 f_y maximum 230 for ø ≤ 20 mm, 210 for ø > 20 mm
CP 110: 1972	BS 4449	Hot rolled mild	–	250[†]	–
	BS 4449	Hot rolled high yield	–	410[†]	–
	BS 4461	Cold worked high yield	–	Maximum 460 for ø ≤ 16 mm, 425 for ø > 16 mm[†]	–
	BS 4482	Hard drawn wire (ø ≤ 12 mm)	–	485[†]	–
CP 110: 1972 (amended 1980)	BS 4449, BS 4461	Grade 460/425	–	Maximum 460 for ø ≤ 20 mm, 425 for ø > 20 mm[†]	–
	BS 4482, BS 4483	Hard drawn wire (ø ≤ 12 mm)	–	485[†]	–

Table 5.1   Continued

Code and year	British Standard	Steel type	Ultimate tensile strength lb/in^2 (N/mm^2)	Yield stress (f$_y$) lb/in^2 (N/mm^2)	Permissible tensile stress due to bending lb/in^2 (N/mm^2)
CP 110: 1972 (amended 1983)	BS 4449	Grade 460	–	460[†]	–
BS 8110: 1985	BS 4449	Hot rolled mild	–	250[†]	–
	BS 4449, BS 4482, BS 4483	High yield (hot rolled or cold worked)	–	460[†]	–

*Ultimate tensile strength and yield stress varied depending on bar diameter, smaller bars having higher strengths.
[†]Redefined as characteristic strength in and after CP 110: 1972.
Note: during the mid-1960s, the Lancashire Steel Manufacturing Co. produced a hot rolled ribbed bar (Lancs 80) with f$_y$ of 82,000 lb/in^2 (565 N/mm^2) which was used with a permissible tension stress due to bending of 0.55 f$_y$, i.e. 44,000 lb/in^2 (303 N/mm^2).

as from uplift pressure, could otherwise cause supports and slabs to fail in tension or shear. Similarly, fire is most likely to severely damage structures with highly stressed steel having shallow cover, particularly cold-worked steel which loses its strength enhancement and reverts to mild steel at around 400°C.

A committee convened by the Institution of Structural Engineers and chaired by Oscar Faber sat in 1942–43 with a brief to propose post-war building methods which would produce rapid, economic construction.[30] The particular circumstances were, of course, related to the severe destruction of large areas of towns and cities by bombing, allied to the generally accepted need to improve housing standards.

The committee's technical conclusions were generally sound, prudent and undramatic, such as that the reinforcement standard BS 785[25] should include yield stress. Previously it gave only ultimate tensile strength figures. Reinforcement could then be stressed at one-half of this yield stress, effectively reducing the steel factor of safety from 2.2 to 2. Permissible concrete compressive stresses should be increased by 10%; increased stresses due to wind load only should be permitted; prestressed concrete should be encouraged, with a code to be provided when experience allowed (it came out as CP 115 in 1959).[31] The mechanical vibration of concrete should be encouraged, although — as before — the argument for this was stated to be on grounds of improved strength, without reference to durability.

Of some interest were the report's recommendations relating to improved design and construction methods. Design and construction in reinforced concrete should be carried out only by those with suitable experience. More effort should be made to train concretors and to give them the status of craftsmen. Formworking and steel fixing should also be recognized as specific trade activities, and supervisors should be suitably trained. Standard specifications should be adopted. These calls echo across more than half a century with a somewhat depressing familiarity.

One particular recommendation was for the engineering designer to be involved at the outset of a project. This is, or should be, now taken for granted; but many architects were still keen to take their design to an advanced stage, so that beam and column layouts and dimensions were basically fixed, before a consulting engineer or design-and-build contractor was appointed (who would then complete the detailed design). One reason for this could be that an architect was understandably loath to relinquish control of the structural planning. Another consideration was that the architect's fee might be reduced by the client if a consulting engineer were appointed, or indeed the architect might be expected to pay the consultant's fee himself. This problem did not arise if a concrete specialist were employed.

In contrast, there were some who favoured the closely involved approach. Owen Williams as architect-engineer straddled both disciplines. Ove Arup — having left the contracting firm of Kier for which he designed the Penguin Pool and

Highpoint — collaborated as an independent consulting engineer with like-minded architects, founding Ove Arup & Partners in 1946. Oscar Faber was an early practitioner of multi-disciplinary engineering consultancy.

CP 114: 1948

Soon after the Faber report appeared, the war ended and work began on updating the 1934 code. Conditions were difficult. Structural steelwork and reinforcement were in short supply, which gave an incentive to prestressed concrete, as prestressing steel was not rationed (see Chapter 10). In the absence of orthodox reinforcement, many concrete elements of this period were reinforced with old tram rails, scrap electrical conduit, and the like. (This approach, of necessity rather than choice, is an interesting precursor of today's environmentally-favoured 'recycling'.)

It was decided that the new code should fall within the structural series being prepared by the British Standards Institution (originally numbers 111–118 covering the common structural materials), and accordingly CP 114 first appeared in 1948.[32] Water–cement ratio was prescribed, with a 20% reduction in water for vibrated concrete. Such vibrated concrete could be stressed 10% higher in compression. Cover on external faces was to be increased, as also for internally corrosive conditions, by ½ in (13 mm) above the nominal internal value of ½ in (13 mm), or the bar diameter if this were greater.

A maintenance clause was introduced, which recommended inspections at suggested intervals of 3–5 years to identify cracking or corrosion of reinforcement. It pointed out that a little work done when the problem was small would be preferable to remedying a major problem if it were neglected. Unfortunately, the cost of access for external inspections was often seen as too high (especially for high-rise buildings) and, as now, maintenance inspections were all too often deferred.

A supplementary code, CP 114.100–114.105,[33] was introduced in 1950. This covered suspended concrete floors and roofs (including stairs) and comprised six 'sub-codes'. Sub-code 114.100 gave general recommendations for these elements, while the remaining five sub-codes added specific recommendations for solid slabs supported by beams; flat slabs; ribbed slabs; pre-cast elements; and filler joist construction.

Post-war reconstruction

The design of concrete structures in the late 1940s and early 1950s was restricted by shortages of materials (particularly reinforcement) and planning controls. Nevertheless, some notable structures were built. These included the Brynmawr rubber factory (described in Chapter 8) and some early multi-storey public housing projects such as the 1949 Rosebery Avenue project in Finsbury, London, by Tecton and Ove Arup & Partners[34] (Figures 5.10 and 5.11). This developed the Highpoint principles, and used a novel jacking system to fix and strip the reusable formwork.

In Bristol, Felix Samuely designed an unusual factory structure (Colodense, 1951, Figures 5.12 and 5.13). This combined precast concrete arch frames with prestressed ties, and used composite floors with prestressed planks placed, as if they were reinforcement, in tension zones.[35] The high strength of prestressing steel allowed more efficient use of the limited quantities of steel available in periods of shortage.

Much rebuilding was needed after the war, and various solutions were employed using concrete. Prefabricated housing and bungalows were developed to a myriad of designs. Soon would come the large panel precast concrete systems for tower blocks of flats (and, often overlooked, many lower-rise buildings, too).[36,37] Some systems were purpose-designed for building types, such as CLASP (Consortium of Local Authority Schools Projects). Although conceived around a steel frame, this system included precast concrete cladding units, the use of architectural

Figure 5.10 Box-frame flats under construction at Rosebery Avenue, Finsbury, London (1949) (by courtesy of Ove Arup & Partners).

Figure 5.11 Detail of shuttering system used at Rosebery Avenue (by courtesy of Ove Arup & Partners).

Figure 5.12   The Colodense factory under construction at Malago, Bristol (1951) (by courtesy of F.J. Samuely & Partners).

Figure 5.13   Detail of prestressed planks used as reinforcement in the Colodense factory (by courtesy of F.J. Samuely & Partners).

cladding in concrete[38] having grown vigorously since the war. (A subsequent period when exposed concrete fell out of favour, largely due to poor appearance after weathering when not thoughtfully detailed, has recently been succeeded by another upsurge in the use of 'architectural concrete'.)

Another form of concrete construction widely used for housing in the early post-war period was 'no-fines', which as its name indicates was based on a concrete made with cement and coarse aggregate only, omitting the sand. The resulting 'honey-comb' consistency was thermally efficient and resistant to water penetration if built of reasonable thickness. Originally developed in Holland,[36] the technique was employed almost exclusively in the UK by the contracting firm Wimpey, who combined it with reusable shuttering to produce economical housing using relatively unskilled labour. Reportedly some 300,000 low-rise dwellings were built;[39] the relatively modest strength of the material led to construction that resembled traditional masonry wall construction, without large windows or narrow piers between them. No-fines was also used in high-rise concrete frames as infill walling.

Proprietary concrete floor systems proliferated, mostly incorporating precast components. A selection in use in the middle part of the 20th century is illustrated in Appendix B.

Some floors, such as those by Siegwart and Armocrete, dated their origins back to early this century. Others, of more recent origin, used prestressed wires in long-line moulds to produce standard concrete joists that could literally be cut to the required length after the concrete had hardened. For these, high-alumina cement (HAC — developed in France and originally known as ciment fondu) was widely if not always used in the concrete, as its very rapid hardening and gain of strength allowed swift turnaround and efficient use of the moulds. Several localized failures of purpose-designed HAC units in the mid-1970s drew attention to some loss of strength due to chemical changes in the HAC.[40] However, since then, no further failures have been reported in what remain as several million square metres of such construction, most of which is different in the details from the structures that suffered failure, and contains mass-produced quality-controlled units.

Other precast floor systems rely on composite action between precast units, or ribs, and an *in-situ* infill. The precast sections can often span in the temporary condition between supports and carry the wet concrete weight without propping. With joists and ribs a variety of infill blocks, often hollow, can be laid to reduce floor weight and cost. Still other floor systems include reinforced or prestressed hollow-core and plank units. All such flooring systems generally, of course, eliminate the need for formwork and minimize or eliminate use of propping.

Codes and standards since 1948

It may be useful to link briefly the code of 1948, the first CP 114, forward through its successors to today's BS 8110.[41]

The 1957 edition of CP 114,[42] although still based on working-load design philosophy, introduced the load-factor method for slab and beam design as an alternative to modular ratio design. It also allowed mixes to be specified by strength — typically 3000, 3750, or 4500 lb/in^2 (21, 26, or 31 N/mm^2) — and to be designed by the contractor to achieve this specified figure. This gave scope for economy in the mix design. A 10% permissible stress increase was allowed for a designed-mix concrete. The load-factor approach was based on ultimate load theory, but loads considered in design remained service loads, and the concrete cube strength and steel yield stress were factored down to give permissible stresses. The factor of safety for steel was now 1.8, a further 10% reduction, while that for concrete was 3 for nominal (by volume) mixes and 2.73 for designed mixes, a 10% reduction over nominal mixes.

The sub-codes CP 114.100–114.104 issued in 1950 were incorporated into the 1957 edition of CP114, while CP114.105 for filler joist construction was withdrawn from the concrete code as its recommendations were, rather more logically, to be found in the structural code for steel, BS 449.

The amended 1965 CP 114,[43] a reprinting, recommended that all beams (except for lintels and other such minor items) should be provided with nominal shear reinforcement.

Separate codes for prestressed concrete design, CP 115,[31] and for precast concrete design, CP 116,[44] appeared in 1959 and 1965 respectively.

Metric editions of all three codes were introduced in 1969, with CP 116 being amended in 1970[45] to give enhanced recommendations for tying and robustness, following the Ronan Point collapse in 1968.[46]

In 1972 the three codes were unified into one document, CP 110.[47] This adopted limit-state design exclusively, with design generally being based on strength and stability, and with detailing rules being used to satisfy stiffness and serviceability once basic member sizes were known.

This in turn was replaced by the present BS 8110, first issued in 1985.[41] The European pre-standard DD ENV 1992-1-1[48] was issued in 1992.

Some structural engineers were not persuaded that longer, more complex codes based on limit-state principles were necessary, or indeed desirable, for the design of routine building structures. In 1987 a referendum of its members was held by the Institution of Structural Engineers, in which a large majority of those voting were in favour of retaining the 'permissible stress' design methods embodied in CP 114 and its predecessors. In response, the Institution commissioned preparation of a design manual incorporating permissible stress recommendations updated from CP 114, together with supplementary design guidance.[49] This appeared in 1991, six years after a similar design manual based on BS 8110.[50]

Standards for reinforcement and concrete have also evolved, reflecting changes in practice and materials.

High yield reinforcement has gradually displaced mild steel. Initially the higher strength was achieved by cold-working, typically by stretching and/or twisting the bars (either singly or in pairs) to produce a distinctive 'barley sugar' spiral profile. This was reflected in the brand names of bars such as Twisteel and Twin Twisted (also known as Isteg). Tentor bars were probably named in recognition of their being subject to both *ten*sion and *tor*sion during the cold-working process. Fabric or mesh has usually been made from hard-drawn wire, spot-welded at the rod intersections with care taken to avoid annealing the steel back to a weaker mild state. More recently, hot-rolled high yield reinforcing bars have been widely used.

For concrete, durability considerations have at last received their necessary recognition, with minimum concrete grades, cement contents, and maximum water–cement ratio being now defined in BS 8110 for various conditions of exposure. Typical grades of concrete are consequently now likely to give $35–40 \, N/mm^2$ cube crushing strength — well above the 21–31.5 range typically quoted in CP 114.

Changes in reinforcement strengths and permissible stresses are given in Table 5.1. Typical concrete grades and permissible flexural compressive stresses have changed too, as shown in Table 5.2.

Changes in design practice

This chapter has attempted to chart the evolution of design theory and practice from before the turn of the century to 1948, and briefly beyond there to the present day. This chapter's title could equally be 'the development of standards for the use of reinforced concrete'. Certainly, in more recent years, the tendency has been for more and more design issues and details to be codified and prescribed. There is thus less incentive for the designer — particularly of 'routine' structures — to take other routes, even though current building regulations for structure (e.g. for England and Wales[51]) make it clear that the designer is not obliged to use an available code if another sound method can be justified.

Increasingly, too, the whole process of concrete design is being automated. Today, engineers and technicians can model, view, analyse, design and detail structures sitting in front of a PC. This is somewhat different from the days — still within the experience of engineers working today — when the tools of the trade were a drawing board and T-square, pencil, pad, slide-rule, and one or more design aids — books, and 'ready reckoner' charts and tables.

Older practitioners will remember, too, the use of moment distribution as devised by Professor Hardy Cross,[52,53] itself an advance on the laborious hand solutions of slope-deflection equations or the slightly less tedious 'theorem of three moments'. Concrete codes from the outset recognized the monolithic nature of reinforced

Table 5.2 Developments in specified concrete compressive strength and permissible flexural compressive stresses from 1915

Code and year	Concrete grade or mix	Batch proportions by volume	Works cube strength at 28 days lb/in² (N/mm²)	Permissible flexural compressive stress lb/in² (N/mm²)
London Reinforced Concrete Regulations, 1915	–	1 : 1 : 2	2200 (15.2)	750 (5.2)
		1.5 : 2 : 4	2000 (13.8)	700 (4.8)
		1.2 : 2 : 4	1800 (12.4)	650 (4.5)
		1 : 2 : 4	1600 (11.0)	600 (4.1)
Building Research Board (DSIR) Code, 1933	Ordinary grade	1 : 1 : 2	2925 (20.2)	975 (6.7)
		1 : 1.2 : 2.4	2775 (19.1)	925 (6.4)
		1 : 1.5 : 3	2550 (17.6)	850 (5.9)
		1 : 2 : 4	2250 (15.5)	750 (5.2)
	Higher grade	1 : 1 : 2	3750 (25.8)	1250 (8.6)
		1 : 1.2 : 2.4	3600 (24.8)	1200 (8.3)
		1 : 1.5 : 3	3300 (22.7)	1100 (7.6)
		1 : 2 : 4	2850 (19.6)	950 (6.5)
	Special grade	1 : 1 : 2	Not more than 25% above values for higher grade concrete	
		1 : 1.2 : 2.4		
		1 : 1.5 : 3		
		1 : 2 : 4		
London Byelaws, 1938	Ordinary and higher grades	1 : 1 : 2	As 1933 Code (1 : 1–2 : 2.4 mix not included)	
		1 : 1.5 : 3		
		1 : 2 : 4		
	Special grade	–	Not included from 1933 Code	
CP 114: 1948	Aggregate to BS 882	1 : 1 : 2	4500 (31.0)	1500 (10.3)
		1 : 1.5 : 3	3750 (25.8)	1250 (8.6)
		1 : 2 : 4	3000 (20.7)	1000 (6.9)
CP 114: 1957	Ordinary	1 : 1 : 2	4500 (31.0)	1500 (10.3)
		1 : 1.5 : 3	3750 (25.8)	1250 (8.6)
		1 : 2 : 4	3000 (20.7)	1000 (6.9)
	Special designed mix	N/A	u_w	$u_w/3$
CP 114: 1965	Ordinary	All	As CP 114: 1957	
	Designed mix	N/A	u_w	$u_w/2.73$
CP 110: 1972	Prescribed mix	Prescribed in Table 50 (by weight)	(15)	N/A
			(20)	
			(25)	
			(30)	
	Designed mix	N/A	f_{cu}	N/A
BS 8110: 1985	Ordinary prescribed mix	See BS 5328 (by weight)	Concrete grade must be chosen with regard to exposure conditions and durability (see 33.3 and Tables 3.2–3.4)	
	Designed mix	N/A		
	Special prescribed mix	See BS 5328 (by weight)		

concrete, and formulae were provided to assist in the calculation of fixity moments. Nevertheless, continuous beams and slabs of varying spans and frames of asymmetrical dimensions and member sizes could oblige the designer to undertake time-consuming and tedious calculations.

Technical information and design guidance

The following briefly highlights aspects of the much more comprehensive advice on information sources to be found in the Appendix.

Of the standard textbooks, that by Reynolds[54] has been, over six decades, one of the most used guides, together with the explanatory handbooks to the various design codes from 1934.[20,55,56]

In addition to textbooks already noted, special mention should be made of the magazine *Concrete and Constructional Engineering*, published from 1906 until 1966.

Much useful guidance for designers and constructors was contained in the numerous books from Concrete Publications Limited, which are listed in the Appendix.

The Cement and Concrete Association (founded in 1935, and now the British Cement Association), and the Concrete Society (founded in 1966), have both published many useful design and practice guides and other reference works, as have other more specialist sources. Such guidance, whether in books, on CD-ROM, or in some yet-to-be-imagined future medium, emphasises that design and construction in reinforced concrete is not entirely circumscribed by codes, but is indeed a *practice*.

For today's engineer called upon to assess or alter such structures, these contemporary publications are invaluable. The library of the Institution of Civil Engineers and that of the Institution of Structural Engineers both hold excellent collections of such material. More recent works include a review of UK reinforcement standards[57] and a paper on British concrete codes and regulations prior to the issue of CP110 in 1972.[58] Numerous reports and pamphlets on the identification, investigation, and assessment of proprietary concrete systems for housing and other building types are available from the Building Research Establishment. The Institution of Structural Engineers has published a general guide to the appraisal of existing structures.[59]

Guidance on the repair and maintenance of concrete structures is also widely available. A BRE report[60] describes general principles, while the assessment and repair of corrosion-damaged concrete is discussed by Pullar-Strecker.[61] BS 6089 gives guidance on the assessment of concrete strength in existing structures.[62] A growing number of BRE and Concrete Society publications deal with specific aspects of investigation, assessment and repair.

Acknowledgements

The author wishes to thank his fellow co-authors for their encouragement and for helpful criticism of the text. He also thanks Mr D.K. Doran for information on the use of 'no-fines' concrete, and Dr L.G. Booth for greatly enlarging on the author's passing reference to the works produced by Concrete Publications Limited. Dr Booth's bibliography of their works is included in the Appendix with his kind permission.

References

1. Hyatt, T., An Account of Some Experiments with Portland-Cement Concrete, Combined with Iron as a Building Material, with Reference to Economy of Metal in Construction, and for Security Against Fire in the Making of Roofs, Floors and Walking Surfaces. Chiswick Press: London, 1877.
2. Hamilton, S.B., A Note on the History of Reinforced Concrete in Buildings. HMSO: London, National Building Studies Special Report No. 24, 1956.
3. Wayss, G.A., Das System Monier. G.A. Wayss: Berlin, 1887.
4. Coignet, E., Tédesco, N. de., Du calcul des ouvrages en ciment avec ossature metallique. La Société des Ingénieurs Civils de France: Paris, 1894.
5. Christophe, P., Le beton armé et ses applications, 2nd edn. Béranger: Paris, 1902.
6. Marsh, C.F., Reinforced Concrete. Constable: London, 1904.
7. Joint Committee on Reinforced Concrete, Report of the Joint Committee on Reinforced Concrete. J. Roy. Inst. Br. Archit. (3rd ser.), 1907, 14(15), 513–41 (discussion 497–505).
8. London County Council. London County Council (General Powers) Act. HMSO: London, 1909.
9. Committee on Reinforced Concrete, Interim Report on Reinforced Concrete. Institution of Civil Engineers: London, 1910.
10. Joint Committee on Reinforced Concrete, Second Report of the Joint Committee on Reinforced Concrete. Royal Institute of British Architects: London, 1911.
11. London County Council, Reinforced Concrete Regulations. LCC: London, 1915.
12. Cottam, D., Sir Owen Williams 1890–1969. Architectural Association: London, 1986.

13. Architects' Journal, Messrs Boots Factory, Beeston, Nottingham. Arch. J., 1932, 76, 3 August, 125–36.
14. Billington, D.P., Robert Maillart's Bridges: The Art of Engineering. Princeton University Press: Princeton, NJ, 1979.
15. Reinforced Concrete Structures Committee, Report of the Reinforced Concrete Structures Committee of the Building Research Board. HMSO: London, 1933.
16. Glanville, W.H., Studies in Reinforced Concrete II: Shrinkage Stresses. HMSO: London, Building Research Technical Paper No. 11, 1930.
17. Glanville, W.H., Studies in Reinforced Concrete III: The Creep or Flow of Concrete Under Load. HMSO: London, Building Research Technical Paper No. 12, 1930.
18. Engineering Standards Committee, Standard Specification for Structural Steel Bridges and General Building Construction. Engineering Standards Committee: London, 1912, BS 15.
19. British Standards Institution, Hard Drawn Steel Wire for Concrete Reinforcement. BSI: London, 1929, BS 165.
20. Scott, W.L., Glanville, W.H., Explanatory Handbook on the Code of Practice for Reinforced Concrete. Concrete Publications Limited: London, 1934 (with later editions accompanying and explaining the editions of CP 114 of 1948, 1957 and 1965).
21. Ministry of Health, Model Byelaws Series IV Buildings. HMSO: London, 1938.
22. London County Council, Construction of Buildings in London. LCC: London, 1938.
23. Building Industries National Council, Code of Practice for the Use of Reinforced Concrete in the Construction of Buildings. BINC: London, 1939.
24. British Standards Institution, Coarse and Fine Aggregates from Natural Sources for Concrete. BSI: London, 1940, BS 882.
25. British Standards Institution, Rolled Steel Bars and Hard-Drawn Steel Wire for Concrete Reinforcement. BSI: London, 1938, BS 785.
26. British Standards Institution, Cold Twisted Steel Bars for Concrete Reinforcement. BSI: London, 1943, BS 1144.
27. British Standards Institution, Steel Fabric for Concrete Reinforcement. BSI: London, 1945, BS 1221 (covered expanded metal, as well as hard-drawn steel wire and twisted steel fabric).
28. Arup, O.N., Design, Cost, Construction and Relative Safety of Trench, Surface, Bomb-proof and Other Air-raid Shelters. Concrete Publications Limited: London, 1939.
29. Institution of Civil Engineers, The Civil Engineer in War, Vol. 2. ICE: London, 1948.
30. Reinforced Concrete Structures Committee, Reinforced Concrete Structures. HMSO: London, Post-war Building Studies No. 8, 1944.
31. British Standards Institution, The Structural Use of Prestressed Concrete in Buildings. BSI: London, 1959, CP 115.
32. Codes of Practice Committee for Civil Engineering, Public Works and Building, The Structural Use of Normal Reinforced Concrete in Buildings. British Standards Institution: London, 1948, CP 114.
33. The Council for Codes of Practice for Buildings, Suspended Concrete Floors and Roofs (Including Stairs). British Standards Institution: London, 1950, CP114.100–114.005.
34. Collins, A.R. (ed.), Structural Engineering — Two Centuries of British Achievement. Tarot Print: Chislehurst, 1983.
35. Higgs, M. (ed.), Felix James Samuely. Archit. Assoc. J., 1960, 76, June, 2–31.
36. Finnimore, B., Houses from the Factory: System Building and the Welfare State, 1942–74. Rivers Oram Press: London, 1989.
37. White, R.B., Prefabrication: A History of its Development in Great Britain. HMSO: London, National Building Studies Special Report No. 36, 1965.
38. Morris, A.E.J., Precast Concrete Cladding. Fountain Press: London, 1966.
39. Reeves, B.R., Martin G.R., The Structural Condition of Wimpey No-fines Low-rise Dwellings. BRE: Garston, 1989, Report BR153.
40. Bate, S.C.C., High Alumina Cement Concrete in Existing Building Structures. HMSO: London, 1984, Report BR235.
41. British Standards Institution, Structural Use of Concrete. BSI: London, 1985, BS 8110.
42. British Standards Institution, The Structural Use of Reinforced Concrete in Buildings. BSI: London, 1957, CP 114.
43. British Standards Institution, The Structural Use of Reinforced Concrete in Buildings. BSI: London, 1965, CP 114 (reset and reprinted).

44. British Standards Institution, The Structural Use of Precast Concrete. BSI: London, 1965, CP 116.
45. British Standards Institution, Large Panel Structures and Structural Connections in Precast Concrete. BSI: London, 1970, Addendum No. 1 (1970) to CP 116: 1965 and CP 116: Part 2: 1969.
46. Ministry of Housing and Local Government, Report of the Inquiry into the Collapse of Flats at Ronan Point, Canning Town. HMSO: London, 1968.
47. British Standards Institution, The Structural Use of Concrete. BSI, London, 1972, CP 110.
48. British Standards Institution, Eurocode 2: Design of Concrete Structures. General Rules for Buildings (Together with United Kingdom National Application Document). BSI: London, 1992, DD ENV 1992-1-1.
49. Institution of Structural Engineers, Recommendations for the Permissible Stress Design of Reinforced Concrete Building Structures. ISE: London, 1991.
50. Institution of Structural Engineers and Institution of Civil Engineers, Manual for the Design of Reinforced Concrete Building Structures. ISE: London, 1985.
51. Department of the Environment and the Welsh Office, The Building Regulations 1991: Approved Document A — Structure. HMSO: London, 1991.
52. Cross, H., Continuity as a factor in reinforced concrete design. J. Am. Concrete Inst., 1929, 25, December, 669–711.
53. Cross, H., Analysis of continuous frames by distributing fixed-end moments. Trans. Am. Soc. Civil Eng., 1932, 96, 1–10 (discussion 11–156).
54. Reynolds, C.E., Reinforced Concrete Designer's Handbook. Concrete Publications Limited: London, 1932 (later editions in 1939, 1946, 1948, 1957, 1961, 1971, 1974, 1981 and 1988).
55. Bate, S.C.C. et al., Handbook on the Unified Code for Structural Concrete (CP 110: 1972). Cement and Concrete Association, London, 1972.
56. Rowe, R. et al., Handbook to British Standard BS8110: 1985, Structural Use of Concrete. Palladian Publications: London, 1987.
57. Steel Reinforcement Commission, UK reinforcement standards 1938 to 1990. Concrete, 1990, 24(3), 40–41.
58. Matthews, D.D., The background to CP110: 1972: The structural use of concrete — notes on earlier British concrete codes and regulations. Unified Code Symposium, Institution of Structural Engineers: London, 1973, 5–18.
59. Institution of Structural Engineers, Appraisal of Existing Structures, 2nd edn. ISE: London, 1996.
60. Currie, R.J., Robery, P.C., Repair and Maintenance of Reinforced Concrete. BRE: Garston, 1994, Report BR254.
61. Pullar-Strecker, P., Corrosion Damaged Concrete: Assessment and Repair. CIRIA/Butterworth: London, 1987.
62. British Standards Institution, Guide to Assessment of Concrete Strength in Existing Structures. BSI: London, 1981, BS 6089.

6 Cement and concrete as materials: changes in properties, production and performance

George Somerville

Synopsis

This chapter records the changes in cement and concrete as materials since 1900, in terms of their properties, production and performance. While largely material orientated, the influence of changes in design standards and construction practices is also covered. A major factor is increase in strength; as this has occurred, the emphasis in design has switched to durability, although an integrated approach — design, materials and construction quality — has not yet fully evolved. It is shown that change has been triggered by a range of factors, including production methods, customer demand for improved performance, construction methods and changes in fashion.

Introduction

The starting point for this chapter is the early years of the 20th century, when the first attempts were made in the UK to produce authoritative specifications for both cement and concrete. Other contributions in this series cover the earlier developments for both materials — mainly the inventive and entrepreneurial periods in the 19th century. A whole range of cements were then available and reinforced concrete was very much promoted via proprietary systems, each having its own design method. Without wishing to duplicate that coverage, the author did find it necessary to delve into previous practice in order to establish clearly his starting point; useful references in this respect are the works by Gooding and Halstead,[1] Halstead,[2] Francis,[3] Davis[4] and Hamilton.[5] The first four of these relate to cement; the last is almost a standard work on reinforced concrete.

There were three significant events in the first decade of the century, which were to shape the future in the UK. The year 1900 saw the formation of The Associated Portland Cement Manufacturers (1900) Ltd, an amalgamation of 24 companies. Then in 1904 the first edition of BS 12,[6] a specification for Portland cement, was published by the Engineering Standards Committee of the ICE. Three years later in 1907, what was effectively the first Code of Practice for reinforced concrete was published by the RIBA.[7] Before 1904, there were very many different cements on the market, more than matched by the number of client specifications. The agreement of a common BS 12 was therefore a major step, subsequently underpinning the design code,[7] which had a strong emphasis on materials as well as on design calculations.

Using these events as a base, one objective of this chapter is to record the changes that have since occurred in the characteristics of cement and concrete, while trying to understand the underlying reasons for these. The second and final objective is to outline the significance of these changes, in terms of the required and actual performance of concrete structures in service. The emphasis is therefore on the significance of change, and not on the basic characteristics of the materials — for which reference should be made to standard works, such as the books by Lea[8] and Neville.[9]

Cement standards

Since BS 12[6] was first published, there have been 14 editions in total, the most recent one being in 1991.[10] In 1904, the characteristics that were specified included fineness, specific gravity, chemical composition, tensile strength (from briquettes), setting time and soundness. The basic constituents then, as now, were Portland cement clinker and calcium sulphate (to regulate setting time). Requirements for fineness and specific gravity have now gone, and the limits for other characteristics have changed, as indeed have methods of sampling and testing, of defining chemical composition, and of testing for strength. It is therefore virtually impossible to make comparisons on pure specification terms.

Cement standards have changed and developed in other ways. Still operating from a base of Portland cement clinker and calcium sulphate, there is a range of manufactured cements available, created by introducing additional main or minor constituents and additives. In producing specifications for these, there is now a major European influence. A recent publication[11] indicates what cements are available, while giving guidance on selection. Table 6.1, taken from that publication, provides a summary; it is interesting to note that the dominant factor in the selection process is durability, representing a substantial change in performance requirements.

Table 6.1 indicates how the different manufactured cements have evolved, in their own right. The British Standard for Portland slag cement (BS 146) first appeared in 1923, while that for Portland fly ash cement (BS 6588) first appeared in 1991. Much of the additional development recorded in Table 6.1 has been triggered

Table 6.1   Comparison of British and European cements

Cement designation to DD ENV 197-1	British Standard cement	Cement type to DD ENV 197-1	Notation in DD ENV 197-1 CEM	Clinker content (%)	Content of other main constituents (%)
Portland cement	BS 12:1991 BS 4027*	I	I	95–100	–
Portland slag cement	BS 146:1991	II	II/A-S	80–94	6–20
			II/B-S	65–79	21–35
Portland silica fume cement	None		II/A-D	90–94	6–10
Portland pozzolana cement	None		II/A-P	80–94	6–20
			II/B-P	65–79	21–35
			II/A-Q	80–94	6–20
			II/B-Q	65–79	21–35
Portland fly ash cement	BS 6588: 1991		II/A-V	80–94	6–20
	BS 6588: 1991		II/B-V	65–79	21–35
	None		II/A-W	80–94	6–20
			II/B-W	65–79	21–35
Portland burnt shale cement	None		II/A/T	80–94	6–20
			II/B-T	65–79	21–35
Portland limestone cement	BS 7583: 1992		II/A-L	80–94	6–20
	None		II/B-L	65–79	21–35
Portland composite cement	None		II/A-M	80–94	6–20
			II/B-M	65–79	21–35
Blast furnace cement	BS 146: 1991[†]	III	III/A	35–64	36–65
	None[†]		III/B	20–34	66–80
	None[†]		III/C	5–19	81–95
Pozzolanic cement	None	IV	IV/A	65–89	11–35
	BS 6610: 1991		IV/B	45–64	36–55
Composite cement	None	V	V/A	40–64	36–60
			V/B	20–39	61–80

*Cement to BS 4027, sulphate-resisting Portland cement, is included here since it complies with DD ENV 197-1, CEM I, although it will eventually be covered specifically in a future part of EN 197.
[†]BS 4246: 1991 covers a blast furnace slag content of 50–85%.

by the move towards European standards. Work on European cement standards started perhaps 25 years ago and was initially very slow; however, it all accelerated in the 1980s, and what we are looking at in Table 6.1 is essentially the first editions. However, both slag and pulverized fuel ash (PFA) also have their own British standard,[12,13] and a significant part of UK practice is to combine these with Portland cement as mixer blends. This practice is comprehensively reviewed in a Concrete Society report;[14] again, durability features strongly in the recommendations made.

Changes in the properties of Portland cement: the effect on concrete

General

In the early 1980s, a growing awareness of increases in the strength of ordinary Portland cement led to several published papers giving details of changes in a number of characteristics, as well as reviewing the effects of these on the properties of concrete. This process was begun by Corish and Jackson,[15] followed by Nixon,[16] the Concrete Society[17] and, most recently, by Corish.[18] The data presented are comprehensive and will not be repeated here, except to illustrate some of the key points.

Changes in strength

Table 6.2 is an abridged version of Corish's[18] data on cement strength from 1960 to 1992. Also in this table, an attempt has been made to extrapolate back to the postwar period, using BRE data from the Concrete Society;[17] this is for the weighted mean values only, and has to be treated with some caution, although the general trend is approximately correct. Extrapolation back from that period is difficult, because of the lack of reliable data; there is a strong testing influence here — prewar testing showed considerable variability, and the methods used may well have underestimated the real strength of the cement. It is interesting to note that in introducing the then new CP 114, Faber,[19] in 1949, commented that the strength of cement and rebars had not increased in the previous 10 years; it maybe deduced therefore that the top figures in Table 6.2 also applied in the late 1930s. Going back beyond that period, the author found no reliable test data; the only clue was perhaps the cube strength required on the common $1:2:4$ concrete, being $16.55\,N/mm^2$ at 28 days in 1907,[7] and $20.69\,N/mm^2$ in 1933.[20] Having made these qualifications, Figure 6.1 gives an estimate of estimated average cement strength up to 1980,

Table 6.2   Changes in cement strength, according to Corish[18]

Year	Concrete strength (N/mm²)						Ratio	
	3-day		7-day		28-day		3-day : 28-day	7-day : 28-day
	Weighted mean	Range	Weighted mean	Range	Weighted mean	Range		
Postwar	(13)	–	(20)	–	(32)	–	0.406	0.625
1952	(14)	–	(22)	–	(33)	–	0.424	0.667
1957	(15)	–	(23)	–	(34)	–	0.441	0.676
1960	16	14–20	24	20–27	35	27–42	0.457	0.686
1965	17	14–24	25	21–31	37	32–42	0.459	0.676
1970	20	13–26	28	22–35	40	35–46	0.500	0.700
1975	21	15–26	30	22–35	42	36–45	0.500	0.714
1980	24	18–28	33	24–38	44	36–46	0.545	0.750
1985	25	22–29	35	30–38	46	42–48	0.554	0.761
1990	26	22–29	36	31–38	47	42–49	0.553	0.766
1992	25	21–28	34	30–38	46	42–49	0.554	0.739

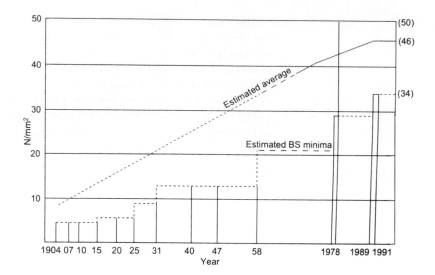

Figure 6.1 Estimates of actual and BS minimum 28-day cement strengths, in terms of concrete testing at 0.6 water/cement ratio.

Table 6.3   Range of C_3S and C_2S in Portland cement, 1914–1990[8,17,18]

	1914–1922	1928–1930	1944	1960	1980	1990
C_3S (%)	15–48	19–58	30–50	36–55	45–64	46–60
C_2S (%)	15–26	53–14	45–20	37–12	30–11	28–13

relative to the minimum strength in BS 12. Comparisons with Table 6.2 would indicate that the average curve levels off, in the 1980s and 1990s at about 46–47 N/mm^2.

Table 6.2 shows an increase in 28-day strength of just over 30% in the period 1960–1992, but with a tendency to level off from 1985 onwards. It has been noted that fineness has not changed over this period,[17] as determined by the air permeability test for surface area. However, this test gives no real measure of particle distribution, and there is evidence that modern cements have become more uniform, with less coarse and less very fine material. Conceivably, this is a contributor to increased strength. A further factor is the increased efficiency in kiln operation, particularly in the cooling phase, affecting the formation of alite (C_3S) crystals.

Equally significant in Table 6.2 is the ratio of early strength to 28-day strength. This is attributed to changes in the proportions of tricalcium silicate (C_3S) to dicalcium silicate (C_2S).[15,17,18] Lea[8] has indicated that these changes have been going on for most of this century (see Table 6.3), and therefore it is inferred that the ratio of early strength to 28-day strength has been increasing steadily over that time, but with a tendency to level off from 1980 onwards.

A review of the literature suggests that these changes have been brought about by a combination of factors. Improvements in the manufacture of cement are one such factor, both in terms of the process itself and in quality control and testing. Changes in customer demand are a second factor, which shows itself in two ways: in increases in general strength requirements in sequential design codes (as knowledge of behaviour increased); and in changes in the way in which concrete is produced, e.g. via ready-mixed concrete or precast concrete (where a high early strength is important). A third factor is the introduction of new technologies (e.g. prestressed concrete) and construction methods.

So what are the effects of these strength changes for concrete and for performance of structures in service? This has been discussed by Nixon and Spooner.[21] The increase in the early strength of concrete is clearly an advantage in precast concrete and in modern construction methods, with the emphasis on speed of construction and buildability. However, the higher early strength implies less strength gain up to 28 days. Nixon and Spooner[21] demonstrate that there is still significant strength gain beyond 28 days, although this is no longer taken into account in UK structural codes.

Possibly the major concern is with durability and with the resistance to corrosion provided by the quality of the cover concrete. With the increase in cement strength, concrete specifications based only on strength could be satisfied with lower cement contents and higher water/cement (w/c) ratios — both of these are trends which increase permeability and decrease durability. In fact, this problem was recognized in the early 1980s[22] between the issuing of CP 110 in 1972 and the appearance of its successor BS 8110[23] in 1985. Without going into detail, the importance of w/c ratio and cement content was recognized for durability, as was the practicality of testing for strength in compliance terms: the net effect is durability grades of concrete in BS 8110, and, since cement characteristics have now substantially stabilized (Tables 6.2 and 6.3) these can be used with confidence. It is interesting to note that action on these matters has been taken only relatively recently — although the importance of w/c ratio was first established in 1918.[24]

Alkalis

The alkali content of cement became an issue in the 1970s in relation to alkali–silica reaction (ASR), an expansive reaction in concrete brought about by critical combinations of reactive aggregates, a high reactive alkali concrete, and moisture. ASR first came to light in the USA in the 1930s, with the earliest-built structure dating back to 1914. In the UK, the American experience was known about in the late 1940s, but conclusions from investigative work in the 1950s and 1960s indicated that there were no known deposits of reactive aggregates in the UK; this was still the situation in 1971, when BRE Digest 126 was published.

In 1976, diagnosis of the reaction in foundation blocks for three electricity substations in south-west England and in a dam in Jersey changed the situation, and around 200 cases have since been identified. Much development work has been done in the past 20 years, first to produce specifications which minimize the risk of the reaction, and second to assess the effects on those structures already diagnosed.

Guidance to minimize the risk of ASR is now incorporated into British Standards,[25] based on work by BRE (Digest 330) and the Concrete Society (Technical Report 30). Control is generally achieved by limiting the reactive alkali content of the concrete mix to $3.0\,kg/m^3$ of Na_2O equivalent; this can be done by limiting the cement content or by selecting Portland cement, ground-granulated blast-furnace slag (GGBS) or PFA with low reactive alkali contents. For cement contents less than $500\,kg/m^3$, a Portland cement with a certified average alkali content of less than 0.60% is deemed to satisfy.

Changes in the levels of alkalis in cement are therefore of interest. Table 6.4, an abridged version of data by Corish,[18] shows the trend since 1960. After peaking in the early 1970s, levels have come down following positive efforts by the cement industry; however, the most significant factor has been the elimination of UK-produced high-alkali cement. Current research is aimed at reviewing the present level of $3\,kg/m^3$ in concrete.

Other factors

The review of cement properties[15–18] covered many other factors, where change is perhaps less significant. These include:

- *Setting times*: Initial setting time has fallen since 1960 by about 20%.
- *Soundness*: Of major concern 100 years ago, this is no longer an issue.

Table 6.4   *Alkalis in Portland cement since 1960*

Year	Alkalis as Na_2O equivalent (%)	
	Weighted mean	Range
1960	(0.64)	(0.4)–(1.2)
1965	(0.66)	(0.5)–(1.2)
1970	0.69	0.5–1.2
1975	0.69	0.4–1.1
1980	0.61	0.4–1.1
1985	0.65	0.5–1.0
1990	0.62	0.41–0.83
1992	0.62	0.50–0.83

- *Reactivity*: Improvements in manufacturing, in the preparation of the raw materials and in the burning and cooling of the clinker have changed the mineralogy of cements. Improvements in finished grinding have changed the particle size distributions of cements. As a result, the activity of the cement has increased.
- *Heat of hydration*: The early rate of heat development has increased. Heat of hydration is generally only an issue in the adiabatic conditions prevailing in large sections, with the possibility of thermal cracking. Special low-heat cements are available, if then considered necessary.

A relatively recent concern is long-term expansion due to delayed ettringite formation (DEF) in concretes subjected to early heat curing.[26] While research is continuing, the procedure for avoiding DEF is well understood, and recommendations are contained in the Department of Transport Specification for Highway Works Part 5, and in the draft European concrete Standard pr-ENV 206. In essence, limits are set on the temperature of the concrete, on the curing temperature and on the rate of temperature increase.

The use of PFA and GGBS in concrete

Table 6.1 gives details of relevant standards for manufactured cements. Both PFA and slag have their own standards[12,13] and the use of these materials, when combined with Portland cement at the concrete mixer, has increased substantially since the 1970s. In mix design terms, a prime concern for all concrete is durability, and equivalence to a factory-made cement of the same composition is deemed satisfactory if the mixer blends comply with the same grades achieved by concrete made with the manufactured cement.[23,25] Effective curing is more significant for composite cements and mixer blends.

There is little doubt that the use of composite cements or mixer blends is advantageous in many durability situations, and guidance is available.[11,14] However, durability considerations have dominated developments to such an extent that they are considered almost the only issue in deciding what concrete to use; there has to be a balance with other performance requirements, such as speed of construction, and a perspective on this will emerge in due course.

In the context of this chapter — properties, production and performance of concrete — the use of PFA and slag is significant. In relation to documented changes, time is still short in a historical context. There is a great deal of laboratory evidence of the benefits, and it is known that the constituent materials can be reliably produced to a common standard. There are also well-documented procedures for demonstrating the equivalence of mixer blends, in terms of sampling (testing for strength and calculation of chemical composition). As yet, there is little documented evidence in the public domain of a change in performance in time in the manner recorded for Portland cement.[15–18]

Table 6.5   The main general codes for structural concrete since 1907

Group	Year	Brief listing
1	1907	Report of the RIBA Joint Committee[7]
	1911	Second report of the same RIBA Committee
	1916	LCC Regulations
2	1933	The DSIR Code[20]
	1938	LCC Regulations
	1939	Building Industries National Council Code
3	1948	CP 114 Reinforced concrete
	1957	CP 114 (with amendments in 1965, 1967 and 1973)
	1959	CP 115 Prestressed concrete (with an amendment in 1973)
	1965	CP 116 Precast concrete (with Addendum 1 in 1970)
4	1972	CP 110 Structural concrete
	1985	BS 8110 replacing and updating CP 110

Concrete: impact of changes in design standards and construction practices

Design standards

There have been four more or less distinct sets of rules for the general design and construction of structural concrete in the UK this century. These are listed in Table 6.5; this ignores other important codes for particular application (e.g. BS 5400 for bridges), and the recent developments with Eurocodes, but is sufficient for illustrating the historical interaction between design rules and the development of concrete as a material. A useful summary of this interaction is provided by Matthews.[27]

In the early part of this development — certainly up to 1948 and, arguably, as far as 1972 — there was a close relationship between material requirements, design rules and quality of construction. This is not surprising in 1907, since all concrete was designed by specialist contractors, having their own patented systems, and one of the main tasks of the RIBA Committee was to reconcile the different approaches while taking due account of the laws of mechanics! However, this integration was maintained and indeed emphasized in subsequent codes. Both the DSIR Code and CP 114 (1948) rewarded extra care in supervision and testing by the use of higher permissible stresses as a fraction of the required works cube strength. The subsequent segregation of material, design and construction matters really began in the late 1960s, and first showed itself in 1973, when a leaflet on 'specification of concrete for durability' was included in a paper by Matthews;[27] this has now developed to the point where design requirements[23] are separate from concrete specifications — effectively mirroring changes in the operating mode of the industry.

The next trend to note is the increases in strength over the years. The century began with $1:2:4$ concrete and 28-day cube strengths of 16–20 N/mm^2. With CP 114 (1948) the specified works cube strength was just over 20 N/mm^2 for $1:2:4$ concrete, but the existence of richer mixes was recognized (up to $1:1:2$) and the concept of designed mixes foresaw specified works cube strengths to 50 N/mm^2 or more. Both CP 115 (1959) for prestressed concrete, and CP 116 (1965) for precast, raised the level higher. By the time of the unified code — CP 110 in 1972 (Table 6.5) — there had also been a change in the method of specification, with designed mixes taking over firmly from what was now called prescribed mixes. This aspect of change is summarized by Beeby and Hawes.[28] In the last decade, there has been growing interest in high-strength or high-performance concrete, with strengths in excess of 100 N/mm^2 being achieved; in part, this is in response to durability concerns.

A close analysis of these changes shows a chicken-and-egg effect. The early specified concrete strengths were achievable with the cements available at that time; permissible stresses were between 20% and 25% of the cube strength. Permissible stresses increased as knowledge increased (up 20% by 1933) and new applications demanded higher-strength concrete and richer mixes. Broadly, cement strengths responded to these new needs; the one exception to this was in 1948, when a special clause was inserted into CP 114, indicating what to do if the specified strengths could not be achieved!

Concrete technology played a significant part in these developments. In 1948, a 10% increase in working stresses was permitted when the concrete was vibrated. The 1950s saw a sharp increase in research on mix design and quality control. It is fascinating to compare the proceedings of two conferences on mix design and quality control in 1954[29] and 1964.[30] Here, it is easy to detect the death knell of the works cube strength and its associated test regime; statistical control was the future.

The influence of research — or sometimes the lack of influence — on design standards and material development is also of interest. Two examples will suffice to make the point. The discovery of the importance of water/cement ratio on all concrete properties is generally attributed to Abrams[24] as long ago as 1918. This factor was slow to gain recognition, but it is now usually quoted in most codes in relation to durability requirements; in terms of control of concrete quality, however, it has led to little more than the development of the slump test for workability. On a more positive note, Faber's research on time-dependent effects in the 1920s followed by Glanville's work at BRE (BRE Technical Papers 11 and 12, 1930) led to a new treatment for column design in the DSIR Code of 1933, which negated the previous modular ratio approach, and accepted that the failure load was the sum of the yield of the steel and the capacity of the concrete; this in fact is ultimate load theory, not recognized formally in code terms until 1957. Time-dependent effects were, of course, integrated into the calculation of losses for prestressed concrete, but, even now, in general design, their consideration is relegated effectively to special structures where movement is considered to be significant. Regrettably, design is largely about strength, stiffness and stability, with insufficient attention to deformation and strain; concrete structures in service have often suffered as a result.

The net effect of these interactive changes in concrete properties and design standards is concrete structures which have become progressively lighter over the years,[28] with the ability to deform more, and yet constructed with materials which in themselves are stronger and stiffer. There is therefore less margin for error, in terms of effect on in-service performance.

Construction practice

Until the 1960s, responsibility for concrete making and placing rested with the contractor. All codes and manuals emphasized the need for good workmanship and skilled operatives. To quote Faber in 1949:[19]

> 'It is, I think, difficult to exaggerate the importance of this Clause [on supervision and workmanship], since in determining what is a reasonable factor of safety the excellence of the detailing and of the actual execution, plays an extremely important part'.

In the 1990s, procurement procedures are different, with the emphasis on speed of construction, construction management and specialist subcontracting; most concrete arrives on site in a ready-mix truck or in the form of precast concrete components. There are, therefore, more links in the chain which transposes basic materials into concrete in the final structure. While each link has undoubtedly become more efficient in itself, responsibility for the chain has become more diffuse, and care is essential to ensure that nothing is lost in the process. In particular, the actual placing, compacting and curing of concrete (having the right cover) is an installation sensitive operation, and an understanding by site operatives as to the significance of these activities is just as important as expertise on the site itself. Feedback from performance in service has demonstrated variable end results, and the good advice readily available[31–33] is clearly not always followed. The human element, especially in communications, is critical.

Changes in structural form (and fashions) and demands for new types of structure have had a major influence on construction practice, and a knock-on effect on the required properties of fresh and hardened concrete. The era of shell roofs and concrete trusses has been and gone. Dams were replaced by nuclear plants and offshore platforms. Novel construction methods, bigger spans, thinner sections, etc., have all contributed to the introduction of pumping and large area pours.

In the precasting arena, the century has seen the passing of several significant phases, not without trauma. In response to demand for housing after the Second World War, prefabricated reinforced concrete housing was introduced. A decade or so later, there was the industrialized building period of the 1960s. Neither was a technical or social success; both contributed to a growing awareness of the risks of corrosion. Another contributed to the corrosion issue — again mainly in the precast field — was the widespread use of calcium chloride as an accelerator which began in the postwar period; its use was effectively banned in 1977, in concrete containing embedded metal (it had been banned earlier for prestressed concrete).

Calcium chloride, as an admixture, is perhaps responsible for the general wariness among specifiers for the use of admixtures for other purposes. There has been a general increase in the use of admixtures over this century, but perhaps not as much as there should have been. Admixtures are available both as accelerators and retarders, for air entrainment and waterproofing and, especially, as plasticizers and super-plasticizers.[34]

Durability and whole-life costing

Lack of durability is perceived as a relatively new phenomenon, conceivably with changes in the basic materials — and the introduction of new ones — being a contributing factor. Concern over durability is not new; many technical papers dating back to the 1920s and 1930s drew attention to it, and this concern is reflected in the codes of that period. However, there are now very many more structures which have been in service for significant periods, thus emphasizing both the scale and nature of the problem.

With hindsight, what has been missing is the positive consideration of durability as a major performance criterion, to be compared with the provision of strength, stiffness and stability.[35] The history of the industry has seen change by evolution, punctuated by an occasional leap forward triggered either by a major technical innovation (e.g. prestressed concrete) or changes in society's needs (e.g. more housing in the middle part of the century, or the need to keep the motorway system clear by using deicing salts). Great leaps forward all too often mean that practice gets ahead of technology, but even the significance of slow evolutionary changes is not always picked up — there is no machinery to do that properly.

Durability is certainly a materials issue, but feedback from performance in service shows that design and construction are equally important. Changes in materials cannot be considered in isolation; the issue of design standards (including environmental loads) and of construction methods and quality are also significant — the reason why they have been addressed in this chapter.

The plethora of papers in the technical literature over the last 20 years[35] means that we now have a better understanding of durability, often quantified in physical or performance terms. Application of that understanding in practice still lags behind, however, particularly in terms of an integrated design, construction and materials approach. This is slowly changing. The emphasis is still too much on materials, but individual aggressive actions are being identified and addressed; the basis for doing that is illustrated in Table 6.6.

Table 6.6   Types of aggressive action for which material specifications have been developed

Aggressive action	General approach	Comments
Sulphate attack	Quantify the action	Specific material and mix proportions are recommended in most codes for defined ranges of sulphate concentration.
Alkali–silica reaction	Define ranges of intensity for it	The basic reaction and its possible effects are now well understood. Recommendations to minimize the risk of damage are published.
Freezing and thawing	Produce a specification for each range	Dealt with by choice of materials, mix proportions and concrete grade. Air entrainment for lower grades. Detail to exposure to moisture.
Abrasion		Specifications to cover aggregate properties, concrete grade and mix proportions, compaction and curing, methods of finishing, etc.

The recipe approach in Table 6.6 is perhaps most appropriate for aggressive actions which directly affect the concrete. Possibly the most serious durability issue is corrosion of reinforcement, due either to carbonation of the concrete or to the ingress of chloride ions from sources such as deicing salts or seawater. In the laboratory, carbonation of concrete has been known about for over 60 years; in practice, it became an issue only in the 1970s when the first wave of postwar construction had been in service for 25–30 years. Real awareness of corrosion due to chlorides was heightened by the use of deicing salts on motorways which began around 1960. This is a major subject, with the historical aspect being summarized by Somerville.[35] It is in this area that the interplay between design, material and construction aspects is most significant in influencing performance in service.

On a broader front, the technical performance of structures with time was first addressed formally with a code of practice in 1950;[36] this gave guidance on design lives for different types of building and building component. It was largely ignored. Its successor[37] has received more attention, since the passage of time has highlighted the importance of durability in relation to function. Until the last decade, the time factor was not recognized as part of the design process — levels of strength, stiffness, stability and serviceability were provided by calculation and tacitly assumed to obtain for the entire useful life (itself not normally given in the design brief). Whole-life costing[38] is beginning to change that. While still relatively new, this has the merit of recognizing that adequate performance is required over a significant period of time, and, in achieving that, should once again lead to the integration of design, construction and material factors.

High-alumina cement (HAC) concrete

HAC was first developed in France at the beginning of the century as a chemically resistant cement — from a fusion of limestone and bauxite in a reverberatory furnace. Early uses were in foundations and piling. However, its high early strength subsequently led to widespread use in precast concrete elements for buildings, particularly after the Second World War. At its peak, some 25 manufacturers produced a wide range of X- and I-section beam and slab elements; it has been estimated that nearly 90% of HAC use in the UK was for this application, with roughly $17 \times 106\,m^2$ of floors and roofs in some 30,000 buildings.

Major collapses of school roofs in Camden in 1973 and Stepney in 1974[40] — together with other local failures — led to major investigations by the Building Research Station and to detailed guidance on the assessment of buildings containing HAC concrete.[41] A recent BRE Digest[42] confirmed the validity of that earlier guidance, but expressed concern over the danger of chemical attack and drew attention to the risk of reinforcement corrosion due to carbonation.

The technical concern with HAC concrete was an inevitable change in mineralogical composition which led to a reduction in strength (especially in warm, damp conditions) of up to 50%. Even with design based on the reduced (converted) strength, there was still a risk. In effect, HAC concrete is no longer used for structural purposes in buildings and foundations, although there has been recent interest in using its high early strength for repair purposes, especially in roads and pavements — with concrete having a w/c ratio less than 0.40 and a cement content greater than $400 \, kg/m^3$.

Concluding remarks

As its title implies, the main objective of this chapter has been to record change — with the emphasis on the properties of cement and concrete in this century. Hopefully, that has been done.

Recording the significance of that change on in-service performance was a second objective. This has proven more difficult. What has emerged in this regard is the interactive influence of changes in design and detailing, materials and technology, construction and procurement methods. Coupled with that is the influence of fashion on structural form and, perhaps more important, the impact of customer demand. This has shown itself in three ways: volume of demand, which comes in waves, leading, for example, to the industrialized building phase of the 1960s; increases in design standards, e.g. in terms of loads or quality; and changes in use or function of 'normal' structures such as bridges or building, or for new types of structure, e.g. offshore platforms or superstores. Overriding all these changes is the shift in attitude to structural performance. Durability is now probably as significant as strength and stiffness, as we grapple with environmental issues and whole-life costing. We now know much more about that, but we still require the wisdom to apply that knowledge in practice.

References

1. Gooding, P., Halstead, P.E., The early history of cement in England. Proc. Third Int. Symp. Chem. Cement, London, 1952.
2. Halstead, P.E., The early history of Portland cement. Paper read to the Newcomen Society, Science Museum, London, 6 December 1961.
3. Francis, A.J., The Cement Industry 1796–1914: A History. David and Charles: London, 1977.
4. Davis, A.C., A Hundred Years of Portland Cement. Concrete Publications Ltd.: London, 1927.
5. Hamilton, S.B., A Note on the History of Reinforced Concrete in Buildings. National Building Studies Special Report No. 24. HMSO: London, 1956.
6. The Engineering Standards Committee, British Standard Specification for Portland Cement. Crosby Lockwood & Son: London, 1904.
7. Royal Institute of British Architects, Report of the Joint Committee on Reinforced Concrete. J. R. Inst. Br. Archit., 1907, 3rd series, 14, No. 15.
8. Lea, F.M., The Chemistry of Cement and Concrete, 3rd edn. Edward Arnold: London, 1983.
9. Neville, A.M., Properties of Concrete, 3rd edn. Pitman: London, 1982.
10. British Standards Institution, Specification for Portland Cement. BSI: Milton Keynes, 1991, BS 12.
11. Spooner, D.C., A Guide to the Properties and Selection of Cements Conforming to British and European Standards. Interim Technical Note 13. British Cement Association: Crowthorne, 1995.
12. British Standards Institution, Pulverised-Fuel Ash. Part 1. Specification for Pulverised-Fuel Ash for Use with Portland Cement. BSI: Milton Keynes, 1993, BS 3892: Part 1.

13. British Standards Institution, Specification for Ground Granulated Blast Furnace Slag for Use with Portland Cement. BSI: Milton Keynes, 1992, BS 6699.

14. Concrete Society, The Use of ggbs and pfs in Concrete. Technical Report No. 40. Concrete Society: Slough, 1991.

15. Corish, A.T., Jackson, P.J., Portland cement properties — past and present. Concrete, 1982, 16(7), 16–18.

16. Nixon, P.J., Changes in Portland Cement Properties and Their Effects on Concrete. Information Paper IP3/86. Building Research Establishment: Garston, 1986.

17. Concrete Society, Changes in the Properties of Ordinary Portland Cement and Their Effects on Concrete. Technical Report No. 29. Concrete Society: Slough, 1987.

18. Corish, P.T., Portland cement properties — updated. Concrete, 1994, 28(1), 25–28.

19. Faber, O., The structural use of normal reinforced concrete in buildings. Struct. Engr, 1949, April, 193–208.

20. Report of the Reinforced Concrete Structures Committee of the Building Research Board. HMSO: London, 1933.

21. Nixon, P.J., Spooner, D.C., Concrete proof for British cement. Concrete, 1993, 27(5), 41–44.

22. Deacon, R.C., Dewar, J.D., Concrete durability — specifying more simply and surely by strength. Concrete, 1982, 16(2), 19–21.

23. British Standards Institution, Structural Use of Concrete. Part 1. Code of Practice for Design and Construction. BSI: Milton Keynes, 1985, BS 8110.

24. Abrams, D., Bulletin No. 1. Lewis Institute: Chicago, 1918.

25. British Standards Institution, Concrete. Part 1. Guide to Specifying Concrete. BSI: Milton Keynes, 1991, BS 5328.

26. Lawrence, C.D., Dalziel, J.A., Hobbs, D.W., Sulphate Attack Arising from Delayed Ettringite Formation. Interim Technical Note 12. British Cement Association: Crowthorne, 1990.

27. Matthews, D.D., The background to CP110: 1972 'The structural use of concrete'. Notes on earlier British codes and regulations. Proc. Unified Code Symp., 27 September 1973. The Institution of Structural Engineers and the Concrete Society: London, 1973.

28. Beeby, A.W., Hawes, F.L., Action and Reaction in Concrete Design, 1935–1985. C&CA Reprint 3/86. British Cement Association: Crowthorne, 1986.

29. Cement and Concrete Association, Mix Design and Quality Control of Concrete. Proc. Symp., 11–13 May 1954. British Cement Association: Crowthorne, 1954.

30. Cement and Concrete Association, Proc. Symp. Concrete Quality, November 1964. British Cement Association: Crowthorne, 1964.

31. British Cement Association, Concrete on Site (formerly the 'Man on the job' series). Publication 45.201. British Cement Association: Crowthorne, 1993.

32. Dewar, J.D., Anderson, R., Manual of Ready-mixed Concrete. Blackie & Son: Glasgow, 1988.

33. Murdock, L.J., Brook, K.M., Dewar, J.D., Concrete Materials and Practice, 6th edn. Edward Arnold: London, 1991.

34. Hewlett, P.C. (ed.), Cement Admixtures — Uses and Application, 2nd edn. Longman Scientific and Technical: London, 1988.

35. Somerville, G., The design life of concrete structures. Struct. Engr, 1986, 64A(2).

36. British Standards Institution, Code of Functional Requirements of Buildings. CP3: Chapter IX: Durability. BSI: London, 1950.

37. British Standards Institution, Guide to: Durability of Buildings and Building Elements, Products and Components. BSI: London, 1992, BS 7543.

38. Concrete Bridge Development Group, Whole life Costing of Concrete Bridges. Proc. Sem., 25 April 1995. CBDG: Crowthorne, 1995.

39. Department for Education and Science. Report on the collapse of the roof of the assembly hall of the Camden School for Girls. HMSO: London, 1973.

40. S.C.C. Bate. Report on the failure of roof beams at St John Cass's School, stepney. BRE CP 68/74.

41. Building Regulations Advisory Committee, Report by Sub-committee. (High Alumina Cement Concrete) Document BRAC (75). HMSO: London, 1975.

42. Building Research Establishment, Assessment of Existing High Alumina Cement Concrete Construction in the UK. BRE Digest 392. BRE: Garston, 1994.

7 Concrete foundations and substructures: a historical review

Mike Chrimes

Synopsis

Concrete has been used extensively in foundations since the revival of interest in its use two centuries ago. This paper outlines changes in its applications and use for foundations and substructures over this period, paying particular attention to early 20th century developments with reinforced concrete, and some post-Second World War innovations such as diaphragm walls and the use of large diameter bored piles in London Clay.

Introduction

Mankind early recognised the value of sound foundations: the biblical references are well known.[1] Despite archaeological and written evidence of past foundation techniques the history of foundation engineering, as with much of the more practical side of civil engineering, has yet to be written. Professor Skempton has provided useful reviews of the development of soil mechanics,[2,3] Heyman has described the evolution of earth pressure theory and retaining wall design,[4] and Glossop described the development of specialist techniques,[5,6] but little has been written on the foundations themselves.[7-9]

For centuries timber was the chief means of providing adequate foundations for structures where ground conditions were unlikely to sustain the superstructure, being used for piles and also timber platforms and grillages. By the early 19th century many other expedients had been employed to improve the ground and spread the load of a structure — using fascine work (bundles of branches, etc.), crushed chalk, and sand. Empirical rules were developed for masonry footings, increasing the bearing surface with depth by using successively wider courses of masonry. No doubt many mistakes were made, but there was a considerable body of practical knowledge about the problems that could be encountered in ground engineering, the need for adequate site investigation, and the expedients that would be employed to secure good foundations.[10-13]

From the end of the 18th century, the requirement for good foundation practice was of increasing importance. The large factory buildings of the Industrial Revolution required foundations designed to sustain not only the static load of a multi-storey building, but also the static and dynamic effects of heavy machinery. Foundations were also needed for new warehouses in docks and alongside railways and canals, and for larger and more heavily loaded bridges.

It is believed that the Albion Mill, designed by Samuel Wyatt and erected 1783–86 close to the south end of Blackfriars Bridge, was the earliest example of a raft foundation, of masonry, covering the entire area of ground under a large building; it can be taken as an example of best foundation practice of the time.[14]

Early use of concrete in foundations

The construction of Albion Mill took place at a time when British engineers were taking increasing interest in the properties of mortars, most notably John Smeaton whose researches into pozzolannas were published in 1791.[15] Shortly afterwards Telford investigated the properties of Parker's 'Roman cement' on behalf of the

British Fisheries Society (Paxton 15a). These investigations were particularly concerned with identifying mortars that would set and be durable under water. In this context English engineers had been using foreign natural pozzolanas, particularly so called 'Dutch' trass for over a century. Italian pozzolana had been used by Charles II's engineers in their attempts to build a breakwater at Tangiers. Eighteenth century lock and sluice work by engineers such as John Grundy specified the use of trass for the mortar, and Colonel Henry Watson is known to have transported large quantitites to Calcutta for use in constructing a dockyard there around 1780.

However, although concrete was used in the medieval period in Europe, in Britain there is no evidence of its use being specified as a foundation material in the modern sense before the early 19th century.[16] Engineers' knowledge of its potential was probably limited to the description of the use of 'beton' for the foundations at Toulon (1748) in Belidor's *Architecture hydraulique*,[17] which all the leading engineers of the period: Smeaton, Rennie and Telford, are known to have possessed (Figure 7.1a). French interest in the use of concrete in the early 19th century, culminating in Vicat's work at Souillac, had no immediate impact on British practice.[18–21]

George Semple was the first British engineer who suggested the use of concrete for bridge foundations.[22] Lime 'concrete', described initially as 'grouted gravel' was first used by Sir Robert Smirke in 1817 for remedial work on the foundations of the Penitentiary at Millbank, then under construction. Claims for its earlier use at East India Docks appear unsubstantiated.[23,24] Smirke used it again at Lancaster Place (1820–23), Sir Robert Peel's House, 4 Whitehall Gardens (1822–23), and, most famously, to underpin the walls of the new Customs House, using a mix of one measure of Dorking (quick) lime to seven or eight of Thames ballast.[25,26]

His success at Custom House was widely reported. The contractors for many of his works were Samuel Baker and Sons, one of the leading contractors of the early 19th century, and this must have helped spread knowledge of the technique. Smirke used concrete again for the foundation of the British Museum (1833), and the Oxford and Cambridge Club (1835). The contractors for the latter were Grissell and Peto, who completed the foundations contracts for the Palace of Westminster, where lime concrete was again used.[27,28]

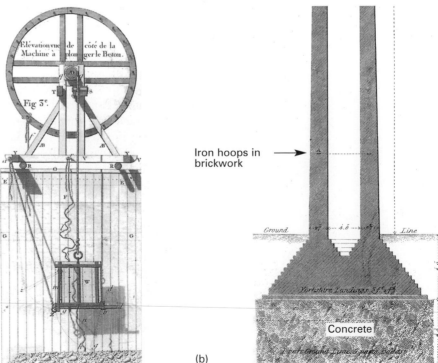

Iron hoops in brickwork

Figure 7.1   (a) Machine for placing concrete underwater at Toulon Harbour. (b) Chimney built at Fulham Gasworks on quicksand.   (a)   (b)

The earliest published specification for the use of concrete was the work of another architect, Robert Abraham, responsible for the new Bridewell prison (1830) in Tothill Fields Westminster (23). This talks about 1 ft layers of concrete being pitched from a height of 9 ft and 'immediately puddled and trodden down by men constantly employed in the works.' Less information is available about the early use of concrete foundations for industrial structures, but, for Fulham gasworks, on ground described as quicksand, Samuel Clegg used concrete for the foundations of a gasworks chimney in 1829 (Figure 7.1b). The concrete, of a 1:5 lime:ballast mix was placed to form a block 8 ft deep and 20 ft wide, surmounted by York stone landings and a brick footing 8 ft deep, offset. The chimney settled at 16½ in, but there were no other problems. In the second half of the century concrete was used extensively in gasworks, particularly for gasholder tanks (see Gould), and beneath retort houses as an alternative to rafts of brickwork and inverted arches, although brickwork arches were used in combination with concrete rafts at Beckton (1868) to reduce the weight.

Concrete was used by the contractor Hugh McIntosh for the foundations of the London–Greenwich Railway viaduct (1835).[23,24,29] In one area he had to excavate to a depth of 18 ft through a peat layer. Concrete was laid by filling the foundations with a layer of water perhaps 1 ft deep (305 mm) and tipping in it a mixture of 1:3:6 lime:sand:gravel which was then turned by shovels in the water. This was apparently developed from the methods used at East India Docks where Ralph Walker had suggested gravel tipped into water for the foundations for the Wall. It proved difficult in practice to mix the concrete adequately.

George Ledwell Taylor, architect for the government dockyards, attempted to underpin the foundations of a storehouse in Chatham Dockyard in 1834[30,31] (Figure 7.2). The contractor, Ranger, had developed his own patent for concrete using a 1:6 lime:gravel combination mixed with hot water, and compressed between timber forms. Whereas at the Custom House the whole of the excavated area was filled with concrete, at Chatham the concrete was contained within formwork affixed at the width of the original brickwork (7 ft) (2.13 m).

For the stores at the Clarence Victualling Yard, Gosport, where the ground was 'ooze or mud' Taylor sank boxes 4 ft square (1.22 m) into the mud, excavated the enclosed space, which was pumped dry to a sound base, and filled with concrete to act as a foundation for piers carrying iron columns at 15 ft (4.57 m) centres. He also used Ranger's patents in Chatham and Woolwich dockyard for dock and wharf walls.

Figure 7.2   Underpinning the Store House, Chatham Dockyard, 1834.

Sections shewing the Method of underpinning with Concrete, the Store Houses, Chatham Dock.

These proved vulnerable to water and frost action, and Pasley's observations to this effect may have delayed the use of concrete for retaining walls without any masonry facing.[24]

As the 19th century progressed lime concrete gained increased acceptance as a foundation material. Under 1879 bye-laws passed by the provisions of the Metropolis Management and Buildings Act, 1878, Section 16, the site of all buildings in the London area, unless founded in gravel, sand, or 'natural virgin soil' had to be covered in a layer of concrete at least 6 in thick (152 mm), with the foundations of the walls at least 9 in thick (229 mm) projecting at least 4 in (101.6 mm) off each side of the footings of the walls. The mix for the concrete was either 1 : 6 lime aggregate or 1 : 8 of Portland cement aggregate.[32,33]

Experiments with Roman cement, with its flash set, were unsatisfactory.[34–35] Even when it was successfully applied, as with the Great North Road at Archway, it was a very tedious process, but as Portland cement gained acceptance the possibility of stronger concrete for foundations grew.[36] In London's docks one can observe engineers' increasing confidence in concrete — from its use as a foundation, to a backing material for retaining walls, and finally mass concrete walls.[37–39] James Walker described the use of a very weak concrete at West India Docks — essentially a bed of gravel covered in a solution of lime, with a mix proportion of 1 : 20. At St Katharine's in 1826, Telford used a 12 in bed of concrete for the foundations of the dock walls.[26,40] West India Junction Dock (1850–53) had brick walls on mass concrete foundations, Commercial Docks' South Dock (1851–55) and Millwall Docks (1865–68) had combinations of brick facing and concrete backing. Royal Albert Dock (1876–80) had mass concrete walls (Figure 7.3).

Concrete was also successfully introduced as a cut-off for dam foundations.[41–44] At Woodhead the initial dam design was a failure due to excessive leakage through the fissured rock, and for his second design (1870–71) Bateman used concrete for the cut-off, which, unlike the original clay puddle, was not susceptible to erosion. This was followed by a concrete cut-off (1876) for the Upper Barden reservoir. By then possibly the first modern concrete dam had been designed by British engineers — the dam for Geelong waterworks, erected 1873–74.[45–46] It was originally intended to build the dam from local stone, but in the absence of suitable masonry a concrete mix of 1 part Portland cement to 6½ parts of aggregate, placed in rammed layers of 7 in (177.8 mm) in thickness, was used.

Raft foundations and concrete footings

A plain concrete slab became a typical foundation in the last quarter of the 19th century. In one spectacular example in 1895 the 3 ft thick concrete raft or pad for a 185 ft diameter steel gasholder at Middlesborough began to subside shortly after construction. The site was underlain with soft clay, and it had been hoped the overall load spread by the raft of $1.2\,t/ft^2$ would be sustainable. The problem of differential settlement may have been aggravated by the ground on one side being compacted by nearby railway lines. The solution was to drive boreholes on the higher side; the weight of the tank forced out soft clay through the boreholes (James IGSE).[46a]

Concrete in combination with iron, and later steel, was regularly used for the foundations of load bearing columns and other heavy loads by civil engineers by this time (Figure 7.4).[47] In the 1890s such practices would have been encouraged by knowledge of the work of engineers and architects in Chicago, where steel beams were used in combination with concrete for the first generation of skyscrapers.[48] The Blackpool Tower was founded on four concrete blocks 34 ft square (10.36 m) and 12 ft thick (3.66 m), in which 12 in × 6 in (304 mm × 152 mm) steel girders were embedded. The High Court at Calcutta, built in the late 1860s, was founded on a raft of concrete which contained two layers of hoop-iron interlaced to form an 18 in square (457 mm) mesh, an expedient which is likely to have been used elsewhere in the following decades.[49]

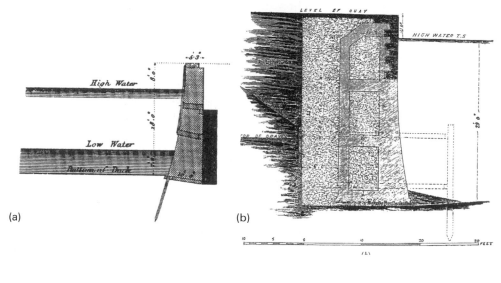

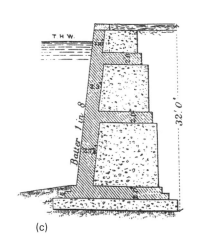

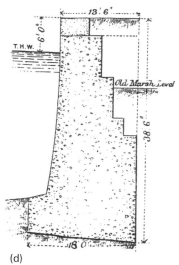

Figure 7.3   Dock wall profiles at (a) St Katharine's Dock, 1826; (b) West India Docks: South Dock; (c) Millwall Docks, 1865; (d) Royal Albert Docks, 1876.

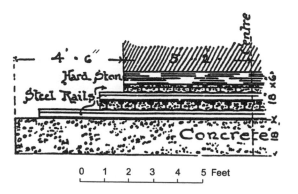

Figure 7.4   Foundation using concrete and steel rails, c. 1893.

According to a textbook of 1893 a raft would normally be at least 9 in (228 mm) thick, but 12 in (304 mm) was more usual, and 18 in (457 mm) for larger structures of the time.[50] A modification would be to have the main slab 6 in (152.4 mm) thick, but design the footings separately continued up above the raft, using the concrete slab merely as a basement/ground floor level. In practice walls always had footings courses. Many of the early Hennebique structures followed a similar pattern, but with the reinforced concrete columns carried on footings below the level of the basement/ground floor slab (Figure 7.5a). Lion Chambers in Glasgow has foundations of this type.[51]

When a reinforced concrete slab was used, 4 in (101.6 mm) notched boards would first be placed to permit the accurate positioning of the (round bar) reinforcement, with the cross bars being placed next, and wired to the main reinforcement at appropriate intervals, and with running boards supported above the reinforcement so it was not disturbed or deformed (Figures 7.5b and 7.5c). Normally the concrete would be placed in a relatively thin layer of about 1 in (25 mm) with the reinforcement being lifted clear, then placed on this layer and the remainder of the concrete would be placed, well tamped all the time. The floor would then be screeded over with the exception of those areas where the beam and column supports would be taken up into the rest of the building. When the slab incorporated pile caps, once the piles had been driven a box would be made to surround the pile group, the reinforcement positioned, and wooden forms placed for the beams connecting the pile caps, with the beam reinforcement connected up at its intersection with the piles. These would all then be concreted up to the underside of the main slab.[52]

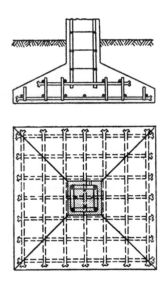

Figure 7.5a Typical Mouchel-Hennebique column foundation, c. 1907.

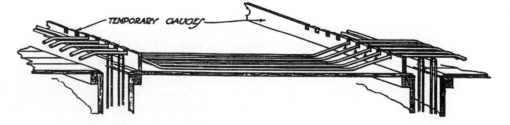

Figure 7.5b Foundation slab reinforcement held in notched templates, c. 1910.

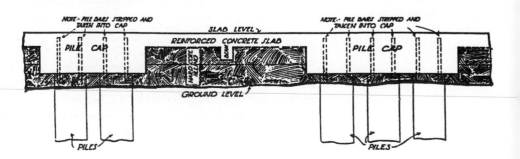

Figure 7.5c Section showing pile caps and reinforced concrete slab, c. 1910.

It should be noted that even with Hennebique buildings the foundations were not exclusively of concrete. The concrete columns of the transit sheds for the Number 9 Dock, Manchester erected 1903–1905 were founded on brick footings,[53] and masonry was again used for the column foundations of Hudson & Kearns printing works in south east London (1904) where the columns were supporting loads of 2400 lb/in^2 (16,547 kN/m^2). The reinforced concrete bases to the columns were 5 ft 3 in^2 (1.6 m^2) and 15 in (381 mm) to 9 in (229 mm) in depth distributing a load of 22,400 lb/ft^2 (1072.5 kN/m^2) on the masonry footings.[54]

The British Reinforced Concrete Engineering Company Limited (BRC), formed in 1905, expanded rapidly, exploiting the patents of the 'Paragon' system of reinforcement, which made use of special forms of hoops, stirrups and sectionalized helical wrapping to produce more rigid reinforcement.[52] The firm was taken over in 1908 and reorganized in 1911, when electrically cross-webbed steel wire fabric was introduced on the Clinton systems. This proved a great success. Column footings were pyramidical in form, similar in appearance to those of Mouchel, but with the base of the pyramid reinforced with BRC fabric, and a similar light fabric reinforcement was used for ground floor slabs. Heavier reinforcement was available for road pavements.[206]

An interesting variation of reinforced concrete foundations was used in the Cottancin system. Here reinforced brickwork walls were used in combination with reinforced concrete slabs. Footings were built up of parallel reinforced brick walls at appropriate centres supporting a thin reinforced concrete slab, to form a narrow caisson. In other cases a series of reinforced brickwork cells were built to support the basement/ground floor slabs. Knowledge of Cottancin's work came from reports to the 1900 Paris exhibition, although only a few Cottancin buildings were erected in Britain, the best known being Sidwell Street Methodist Church, Exeter.[55–57]

Pile foundations

The use of concrete pile foundations followed rapidly on the development of the Hennebique system of reinforced concrete in the 1890s. Many ideas used by the concrete systems had their origins in techniques developed over the previous century, or earlier. Most obviously timber was replaced by precast reinforced concrete piles. Even *in-situ* techniques can be traced to earlier ideas, such as sand piles[58] — a technique whereby timber piles were driven and withdrawn, and the holes filled with compacted sand. The use of a screw was well established for cast iron piles,[59] and boring techniques were used to sink cast iron cylinders, which were regularly filled with concrete for bridge foundations. Concrete filled cast iron columns had been used since the first half of the 19th century.[60,61]

Precast concrete piles

The earliest systems used in Britain were the Hennebique, Coignet and Considère systems.[55,62–67] The patent (No. 2703) taken out by Philip Brannon in 1871 may never have been used. Hennebique patented his reinforced concrete piles in 1896 (British patent 10203 1897) (Figure 7.7a). The piles were of a square section, and a similar reinforcement detailing to his columns, although wire cross ties were placed close together at the head and toe. The tops of the rods were about 2 in (50 mm) below the head of the pile, and the toe was similar to a timber pile's cast iron shoe. Rods were bent in at the foot to bear against the shoe. L.G. Mouchel developed a screw and a hollow precast pile in 1900 (British patent 4548, amended 1907) (Figure 7.7b), which was obviously lighter to transport and Mouchel believed had the same bearing capacity.[55,67] It was made with diaphragms containing forked spacers and wire ties connecting the longitudinal rods. The diaphragms supported tubular moulds forming the hollow core. Piles were initially cast in vertical forms

Figure 7.6a   Driving 43 ft
long Hennebique piles at
Plymouth, 1900.

Figure 7.6b   Vertical forms at
Southampton Cold Stores,
1904.

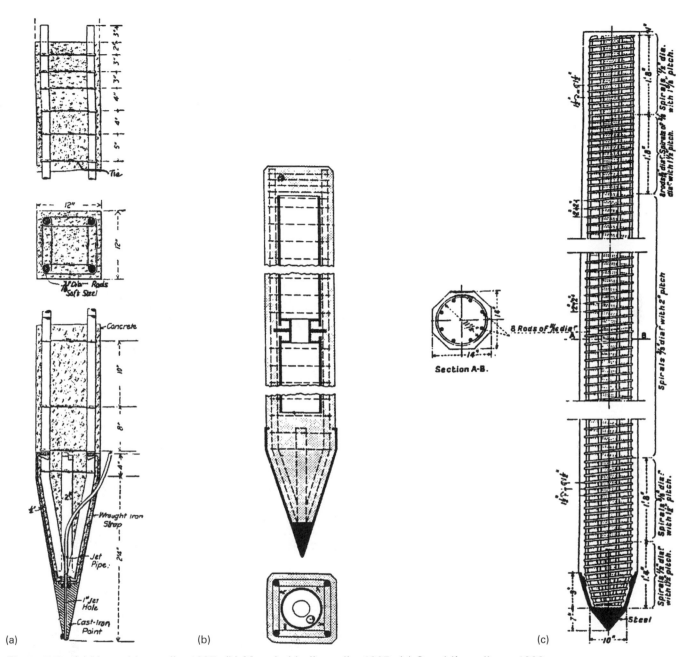

Figure 7.7   (a) Hennebique pile, 1897; (b) Mouchel hollow pile, 1907; (c) Considère pile, c. 1908.

within large racks. After about a week's curing they were stored on the ground, and normally not driven for a month or more (Figure 7.6b). Horizontal moulds soon became more common, and other methods of curing such as steam curing were developed before the First World War.[65,68,69]

An early British example of the use of Hennebique piles was the GWR grain warehouse at Plymouth (1900), where 30 ft long (9.144 m) piles 14 in × 14 in (355.6 mm × 355.6 mm) supported 12 in × 12 in (304.8 mm × 304.8 mm) columns, the ground floor being 6 in (152.4 mm) of mass concrete on rubble fill (Figure 7.6a).[67]

At Dagenham Docks (1902) A.E. Williams developed a 14 in × 14 in (355.6 mm × 355.6 mm) pile with rolled joist reinforcement, bent flat bars to strengthen the sides parallel to the webs, and loops in the concrete at 'frequent' intervals. The point was formed by cutting away the top and forging the flange to a point.[69a]

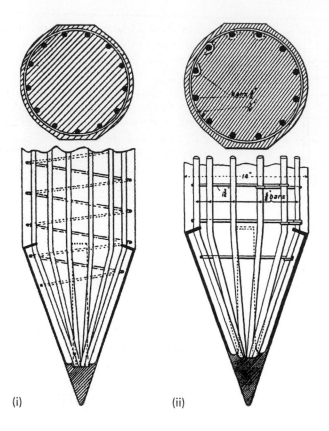

Figure 7.7d   Coignet pile:
(i) Early, 1906; (ii) Final
version, 1910.

(i)   (ii)

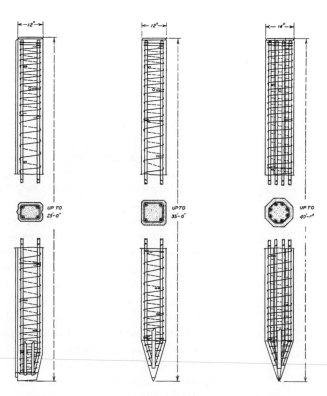

Figure 7.7e   Standard BRC
piles.

SHEET PILE   SQUARE PILE   OCTAGONAL PILE
                     TO TAKE 40 TONS   TO TAKE 60 TONS

The Armco (Armoured Concrete Construction Company) pile used angled irons at each corner connected with frequent straps, and a special diaphragm of wire bent in the form of a spring.[62] The Considère system (Patent 14871, 1902) (Figure 7.7c) used octagonal piles with longitudinal bars surrounded by spirals, similar to their columns, and was first used for a jetty at Thames Haven in 1907.[47,70] Coignet piles were the earliest to be widely used in Britain after the Hennebique system. Originally patented in France (1894) where experimental piles were driven at Levallois-Peret in 1894, and then Asnières on a trial basis, they were not used in Britain until 1906.[52,55,71,71a] Piles were circular with two flat faces reinforced with ⅝ in diameter (15.875 mm) longitudinal bars, and ³⁄₁₆ in (4.76 mm) rods for hooping, hooked around the longitudinal bars (Figure 7.7d).

BRC produced 12 in (304.8 mm) square piles for loads up to 40 t, and 14 in (355.6 mm) octagonal piles for loads up to 60 t, reinforced with longitudinal rods wrapped with sectional helical reinforcement, with reinforcing hoops at the top and intervals along the shaft (Figure 7.7e).

In-situ piles

Among the earliest *in-situ* piles were those of the Raymond Concrete Pile Company.[72–74] The earliest (US) patent was dated 1897, and A. Raymond followed this with a succession of improvements which made his one of the leading piling firms of the early 20th century (Figure 7.8). In its original form a tapered cast steel form or core was used 20 ft long, (6.1 m), 18 in (457 mm) diameter at the butt and 6 in (152.4 mm) at the point; it was in two sections fitted with a dove tail joint and secured by a key at the head of the pile. This core was fitted with a tapered shell of thin steel or iron plate. Once the pile had been driven the key was withdrawn and the core could then be collapsed and withdrawn and the shell filled with concrete. The shell was later reinforced with spiralled wire. Piles were also sunk using a water jet system,[74] a method used by Mouchel and other systems. The system was known in the UK at least from 1901 when its trial in Chicago was reported in *Engineering News*, and later described at a Municipal Engineers' meeting in 1905,[63] but its first British application is unknown. In the 1920s it was licensed to J.W. Stewart.

The Simplex pile (Figure 7.9) invented by Frank Shuman of Philadelphia in 1903, was well known in Britain before the First World War.[62,75–79] Early contracts included Tranmere Bay Development (1905) (Figure 7.7). With Simplex piles the (originally wrought iron) casing had a cast iron toe which was used to prevent the head opening when driving. To avoid the loss of these toes an 'alligator point' was developed. The jaws could be opened after the pile had been driven, and withdrawn once the concrete was poured into position. Problems arose when earth pressure caused the jaws to partially shut. Generally the pile casing would be withdrawn 2 ft (610 mm) and 3 ft (914 mm) of concrete inserted and rammed into position by a 600 lb (272 kg) drop hammer. Where the diameter was sufficient expanded metal 3 in (76 mm) reinforcing mesh at ⁵⁄₁₆ in (7.9 mm) could be used, and even precast piles were inserted through the case. By the use of a heavy mandrel the concrete could, where necessary, be compressed to form a bulb at the head. In some cases a precast concrete shoe was used on which the reinforcement rested. An outer casing was used when the tip was in water-bearing strata. Problems could arise when driving a group of piles, as driving could interfere with adjacent piles. Although J.W. Stewart had the original UK licence, Simplex soon set up their own offices.

Of the large number of *in-situ* systems developed on the continent at this time, the most important was perhaps the invention of the Belgian Edgard Frankignoul in 1909.[80,81] With Franki piles a dry concrete plug was placed in the casing, and driven by a drop hammer, the concrete pulling down the casing behind it. Once

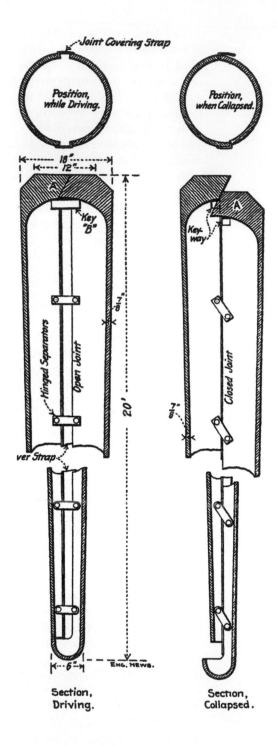

Figure 7.8   Steel core
Raymond pile, c. 1901.

the desired depth had been reached the pile was concreted, the shell being gradually withdrawn as successive layers of concrete were rammed in. This system gave
the pile a bulbous head and irregular profile which helped with bearing capacity
in granular soils for which it was ideally suited, although it could be a disadvantage where negative skin friction was a factor. It was soon discovered the system
could be driven at an angle of 25°, and a reinforcement cage could be added. It
was claimed that these piles could be used with advantage where there was a narrow band of relatively good ground beneath the surface on which the pile could
bear. The Franki pile was exhibited at the Brussels exhibition at the end of 1910,[82]
where a pile was test loaded with 70 t, and in 1911 Frankignoul established an

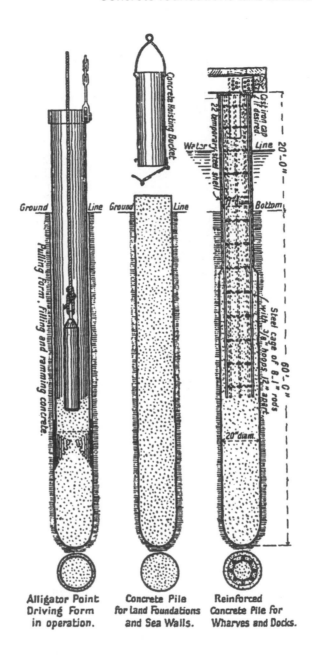

Figure 7.9   Simplex pile
system, c. 1906.

international firm, which became world famous. The UK subsidiary was incorporated in Liverpool in 1931. By 1957 it was the largest piling organisation in the Commonwealth. In 25 years it installed 400,000 Franki piles, 300,000 after the war. The first known contract was Codnor Reservoir, Derbyshire, followed by Gravesend sewage works. In 1935 Franki piles were used for the prestigious job at John Barnes store, Finchley Road. Their largest immediate post-war job was Port Talbot steelworks, claimed to be the largest piling contract in the world (initially 32,000 and eventually 90,000 piles) (Figure 7.10).

Case studies: prior to 1914

Second Tobacco Warehouse, Bristol (1906)[71a]

The early dominance of reinforced concrete piling L.G. Mouchel and the Hennebique system was broken in 1906, by the contractor William Colvin and Sons using Coignet piles for a large tobacco warehouse contract, involving over 600 piles,

Figure 7.10   Installing Franki piles at Port Talbot after the Second World War.

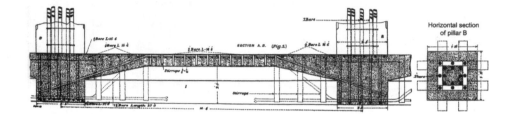

Figure 7.11   Raft foundations at CWS Warehouse, Newcastle-upon-Tyne, 1900.

in Bristol docks. Mouchel claimed that the Coignet system infringed Hennebique patents, and took the case to court. Initially it was determined there was a patent infringement but work was allowed to proceed on the basis that damages would be paid equivalent to a licence fee if the decision was upheld on appeal.[83,84] In the event the Court of Appeal overturned the verdict,[85] and their view was upheld by the House of Lords, the verdict being that the concept of a reinforced concrete pile had been anticipated by Brannon and others, and Hennebique/Mouchel could not claim exclusivity for their patents.[86] This may have helped other systems gain acceptance after this date.

CWS Warehouse, Newcastle-upon-Tyne (1900) (Figure 7.11)

Reinforced slabs and footings were also dominated by L.G. Mouchel and the Hennebique system prior to c. 1906, the first example of a foundation slab being at Newcastle-upon-Tyne c. 1900 (see Chapters 4 and 5 by Bussell). Plain concrete slabs, occasionally combined with timber piles, remained the dominant foundation form, with steel grillages being used after steel frames came in from 1904 onwards; to that extent reinforced concrete was an atypical material.

Extension work at Victoria Station[87] (Figure 7.12)

Works to the London–Brighton and South Coast Railway side of the Station involved extending the Grosvenor Hotel, and a considerable increase in the area of the station, obtained in part by covering in an area of the Grosvenor Canal. With

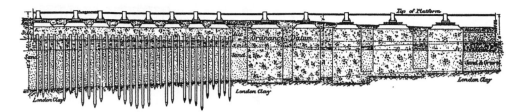

Figure 7.12 Foundations at Victoria Station, 1900.

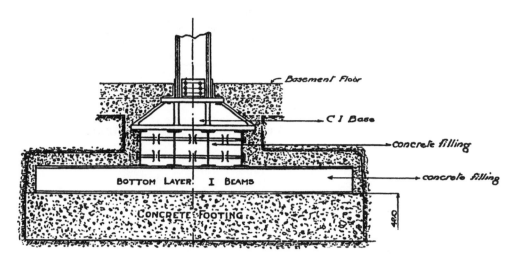

Figure 7.13 Ritz Hotel foundation grillage, 1904.

the proximity to the river ground was very varied, and the underlying London clay dipped from around 20 ft (6.1 m) below the surface on the Buckingham Palace Road side to 40 ft (12.2 m) near the South Eastern Railway station. Overlying strata included sand and gravel, peat and silt. Generally the foundations were carried down to the London clay. Where the clay depth did not exceed 35 ft (10.7 m) 7 ft 6 in^2 (2.285 m^2) concrete (1:6) column bases were founded on the clay, beneath that timber piles of 12 in × 12 in (304 mm × 304 mm) pitch pile were driven at 3 ft (914 mm) centres, supporting a 1:8 concrete slab 7 ft thick (2.134 m). The average load per pile was about 14½ t (44 kN/m^2). Beneath the Hotel annexe and Eccleston and Elizabeth bridges voids with brick arches and inverts were left, between 5 ft (1.52 m) and 10 ft (3.04 m) wide.

War Office Buildings, Whitehall[88]

The contract for the excavation and substructure comprising a 5 ft (1.52 m) thick plain concrete raft, and plain concrete retaining walls for a two storey basement, was completed in 1904. The wall footings' courses were above basement floor level.

Ritz Hotel[89] (Figure 7.13)

The Ritz Hotel was probably the first steel framed structure built in London. It comprised a seven storey steel frame above ground floor level with a lower ground and basement floor below. The structure was supported on 118 columns up to ground floor level, founded on cast iron base plates resting on a steel grillage, above 18 in (457 mm) thick concrete footings, about 13 ft (3.96 m) square, sunk in pits into clay below basement floor level. Part of the frame was supported on rocker bearings.

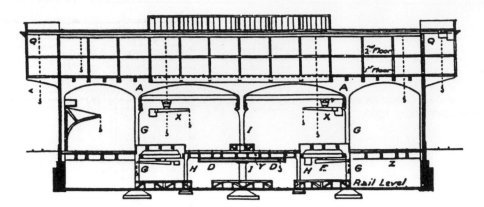

Figure 7.14   Section through North-Eastern Railway Goods Station, Newcastle-upon-Tyne, 1906.

The Institution of Civil Engineers[90]

The present ICE building erected 1910–13 is another early steel frame structure, with load-bearing external masonry walls on one elevation. The foundations comprise a 4 ft (1.22 m) thick mass concrete raft of 1 : 6 Portland cement : aggregate mix, containing an asphalt waterproofing layer, with some timber piling at the south west corner. The frame is supported on steel grillages contained within the concrete raft.

North-Eastern Railway Goods Station and Warehouse, Newcastle-upon-Tyne[91] (Figure 7.14)

This structure, completed in August 1906, was designed as a reinforced concrete frame, with overall dimensions of the building 430 ft (131 m) long, 178 ft 4 in (54.35 m) wide, and 83 ft 4 in (25.4 m) high, divided into four floor levels including a basement floor which served as the low level goods station. Foundations rested on boulder clay, the safe bearing capacity of which was estimated at $5 \, t/ft^2$ ($536 \, kN/m^2$). The structure was designed for warehouse loadings, ten rail tracks of goods traffic, and turntables, cranes, etc. To first floor level the framework comprised 70 wall and interior columns in five rows, 33 ft (10 m) apart longitudinally, and with 25 spans of 37 ft 2 in (11.33 m) and two of 52 ft (15.85 m) centre to centre between the rows, two of these rows were purely to support the ground floor goods station. At basement level there were 30 additional columns. The wall columns were supported on massive concrete retaining walls, which formed the walls of the lower level station. The footings for the central row of columns were 15 ft 6 in (4.72 m) square, those on each side 7 ft (2.13 m) square, and the outer bases 14 ft (4.27 m) square, loads on the columns varying between 224 (kN/m^2) and 1105 t. The latter columns were 1600 in in area (40 in × 40 in) ($1016 \, mm^2$). The footings for these were reinforced with horizontal bars laid in rows at right angles to each other, connected by vertical loops for resisting shear stresses.

Rowntrees Works, York: Melangeur Block (1907)[92] (Figure 7.15)

This block, 105 ft (32 m) × 76 ft 7 in, (23.34 m) was 93 ft 6 in (28.5 m) high, built on six floors including the basement level. The ground being compressible, it was decided to found the building on a reinforced concrete slab 12 in (304.8 mm) thick which projected 10 ft (3.048 m) in every direction beyond the outer walls, distributing the load over an area of $9700 \, ft^2$ ($900 \, m^2$). The slab was reinforced with round steel bars tied longitudinally and transversely near the top and bottom surfaces.

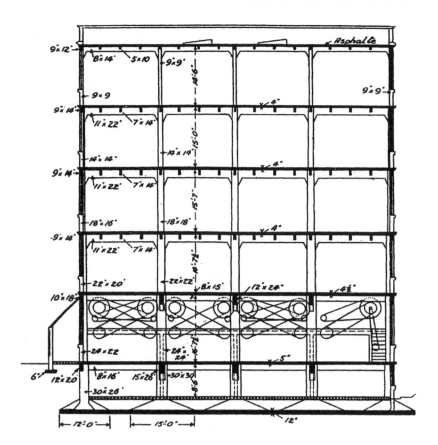

Figure 7.15 Section through Melangeur Block, Rowntrees, York, 1906–1907.

Columns were supported on bases resting upon the foundation slab. Bases supporting the outer wall columns were 12 ft (3.66 m) square and those supporting the internal columns were 15 ft 6 in (4.72 m) square, all rising to 2 ft 6 in (767 mm) above the slab. The reinforcement comprised round steel bars near the base to resist tension, with vertical stirrups of strip steel to resist shear. Bars forming the vertical reinforcement of the columns were carried into the base, and connections were made between the slab and column bases using reinforcement. Earth was spread over the foundation slab and rammed before a plain concrete slab 9 in (228.6 mm) deep was placed, to form the basement floor.

J.C. & J. Field's Factory, Rainham, Essex (1906)[93] (Figure 7.16)

These factory buildings had a brick superstructure but, presumably in view of ground conditions comprising 3 ft (914 mm) of hard silt resting on a peat bed 25 ft (7.62 m) thick, it was decided to use reinforced concrete for the foundations, and attempt to reduce the pressure on the ground by distributing the load. The Coignet system was employed, the second instance of its use in Britain. Foundations were required for a building 271 ft (8.23 m) × 47 ft 6 in (14.48 m), and a boiler house 92 ft 6 in × 72 ft 6 in (28.19 m × 22.1 m) containing three Galloway boilers.

The boiler house foundations comprised reinforced concrete footings 4 ft 8 in (1.422 m) wide and 3 in (76 mm) thick beneath two walls, and 4¾ in (121 mm) thick along the third wall, distributing the weight of the building at the rate of 3 cwt/ft^2 (16,088 N/m^2) over the ground. On these footings a beam 10 in (254 mm) wide and 12 in (304.8 mm) deep was formed to support the brick walls, and reinforced concrete columns supporting the roof. Reinforcing bars were placed in the lower half of the slab, connected to bars in the beams. The foundation beam was

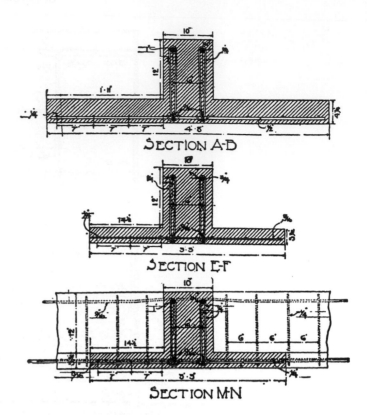

SECTION A-B

SECTION E-F

SECTION M-N

Figure 7.16   Section through foundation beam for boiler foundations, Rainham (Coignet system), 1906.

supplemented by two piles in one place and two transverse footings in another where the loads on the columns were excessive.

The boiler foundations comprised a slab c. 40 ft (12 m) square, to reduce the load to 3 cwt/ft² (16,088 N/m²). Care was taken not to disturb the silt strata, and a crushed brick layer was placed on this followed by 5 in (127 mm) of concrete with ¼ in (6.35 mm) reinforcing rods placed in the lower third near the external wall, but in the upper third beneath the boiler. On this slab were placed six parallel reinforced concrete beams 10 in (254 mm) thick, with their reinforcement looped into the foundation slab, and these beams supported the seating for the boilers. The footings for the shed comprised a reinforced concrete slab 4 ft 2 in (1.27 m) wide and 4 in (102 mm) deep supporting a reinforced concrete beam 10 in (254 mm) wide and 12 in (304.8 mm) deep. Reinforcing bars were between ¼ in (6.35 mm) and 1 in (25.4 mm) diameter, with the beam reinforcement looped into the foundation reinforcement every 6 in (152.4 mm).

Brooklands Motor Track: bridge over the River Wey[94]

Constructed on the Hennebique system the bridge was 200 ft (61 m) long and 100 ft (30.5 m) wide, designed for a uniform distributed load of 112 lb/ft² (5.4 kN/m²) and a point load of 2 t/ft² (214.6 kN/m²). The foundations included forty-two 14 in (355.6 mm) square piles, which were connected continuously with the beams and columns of the superstructure.

Tranmere Bay Development Works (1905–1906)[79] (Figure 7.17)

The engineering workshops of this scheme were largely founded on made ground. The main erecting and machine shop was 1035 ft (315.5 m) long, with overhead cranes 60 ft (18.29 m) above floor level, travelling along a 74 ft (22.56 m) span. The

Figure 7.17   Photographs of piling works at Tranmere Bay, 1906.

concrete blocks supporting the column-bases were carried on groups of Simplex piles — from 4 to 12 for each column. The tubes used for the casing were 40 ft (12.2 m) long, 16 in (406 mm) in outside diameter, of ½ in (12.7 mm) thick metal welded together from three lengths of tubing. This had to be imported from Germany as the English pattern with one joint and a riveted cover plate was unsatisfactory. The casing had an alligator point and was driven with a cast steel cap in the top of which was an elm or hardwood piece to bear the blows of a 30 cwt (152.4 kg) iron monkey, falling 6–8 ft (1.8–2.4 m). The casing was driven to a set of 1 in (25.4 mm) in 4 blows, penetrating 20 ft (6.1 m) of made ground and 10 ft (3.05 m) of the original strata. The casing was withdrawn using tackle with a capacity of 50 t (500 kN). Concrete (1 : 6 cement : aggregate mix) was tipped sufficient for 2 ft (609.6 mm) lengths of pile. The maximum load was 20 t/ft^2 (2146 kN/m^2) of pile head. Piles were driven at 3 ft (914.4 mm) centres, and in good ground there did not appear to be disturbance of newly driven piles by driving adjacent piles. Where a layer of spongy blue clay 6–10 ft (1.8–3 m) thick was present, however, piles were found to be deformed when exposed, and reduced to 11 in (279.4 mm) in diameter.[95] Here concrete footings were taken down below the layer of clay. Nine piles were driven a day per machine. Because of the large number of piles used and the cost of equipment, the foundation cost 15% more than a timber piled foundation, but it was felt the foundation would be more secure.

Reinforced concrete retaining walls[96]

Retaining walls provide one of the clearest examples of the liberating effect of reinforced concrete on civil engineering design. Mass concrete retaining walls of the type seen in Figures 7.3a–d were designed essentially as gravity structures, whereas reinforcement made possible the design of lighter structures of adequate strength of two main types — cantilever and counterfort, offering a considerable saving in cost. Methods of design were developed before the First World War.[97-99] Numerous examples of retaining walls of these types were erected in the United States from around 1904 onwards.[100-103] In Europe Hennebique again led the way. The retaining wall erected for the Paris exhibition of 1900 at the Quai Debilly, near the Trocadero in Paris, is perhaps the best known early example.[104,105] Here the wall was formed of slabs with ribs at the back, both being connected to the foundation slab, which in this case had a rib at the toes at the front (Figure 7.18a).

Other systems aside from Hennebique were used in Britain before the First World War. At Mappin and Webb's famous Queen Victoria site in the City (1911) a complicated construction was installed comprising a base, vertical slab, curved wall and top slab, reinforced by 'indented steel bars' (Figure 7.18b).[106,107] A more normal cantilever wall was built to the same system for the Royal Insurance Company offices, Piccadilly (Figure 7.18c).[106,108] Their office in Lombard Street was essentially a steel framed building (Figures 7.18d–f), but the Considère system of reinforced concrete was used in the foundations which included a cantilever retaining wall with an overall depth of 26 ft 6 in (8.08 m) to the base of its toes, with the wall varying in thickness between 21 and 10 in (533–254 mm).[109] The toe incorporated a reinforced concrete beam distributing the loads from the stanchions supporting the superstructure and was given increased depth to form a reinforced concrete foundation for stanchions at 14 ft (4.27 m) centres. Comparison of the thickness of the reinforced concrete with a brick retaining wall suggested the latter would need a maximum thickness of 13 ft (3.96 m) to give the same stability. Reinforcement comprised ¾ in (19 mm) vertical rods with horizontal stirrups ¼ in (6.35 mm) diameter to resist shear.

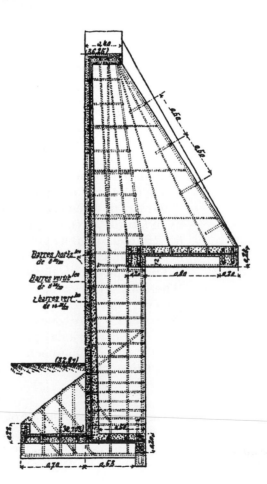

Figure 7.18a   Hennebique retaining wall at Quai Debilly, Paris, 1900.

The Expanded Metal system was used for retaining walls with buttresses or counterforts.[110],[111] The walls at Salford public baths were 8 ft 9 in (2.64 m) high, made of concrete 6 in (152 mm) thick reinforced with No. 10 expanded steel mesh, with a horizontal bar at the top of the wall; the buttresses were 9 in (229 mm) thick and 3 ft 6 in (1.069 m) deep at the base of the wall.[106] A more substantial wall at Guildford was 17 ft (5.18 m) high, 6 in (152 mm) thick at the top, 9 in (229 mm) at the base, with buttresses up to 9 ft (2.74 m) deep at the base, 9 in (229 mm) thick and at 5 ft 6 in (1.68 m) centres (Figure 7.18f). The wall reinforcement was No. 10 expanded steel mesh, and the base 15 in (381 mm) thick with No. 30 expanded

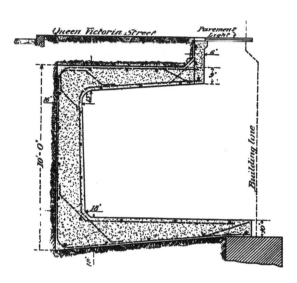

Figure 7.18b   Retaining wall at Mappin and Webb's, Queen Victoria Street (indented bar system), 1911.

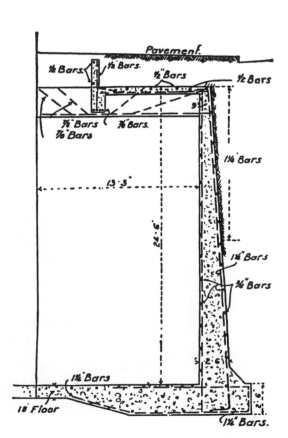

Figure 7.18c   Retaining wall, Royal Insurance offices, Piccadilly (indented bar system), 1907.

(d)

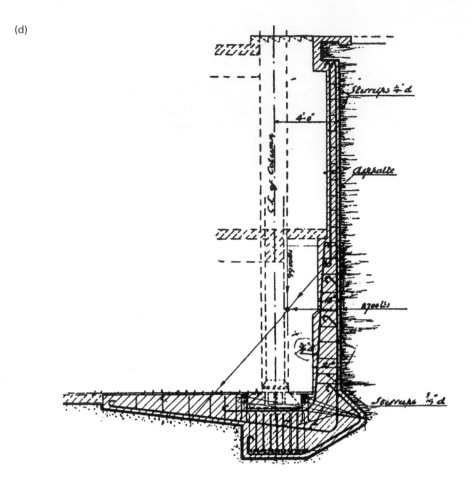

Figure 7.18d, e   Foundations, Royal Insurance offices, Lombard Street (Considère system), 1911.

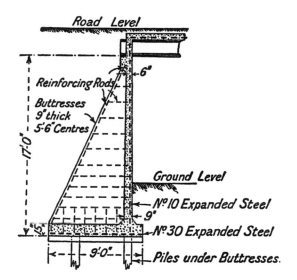

Figure 7.18f   Retaining wall, Guildford, c. 1911.

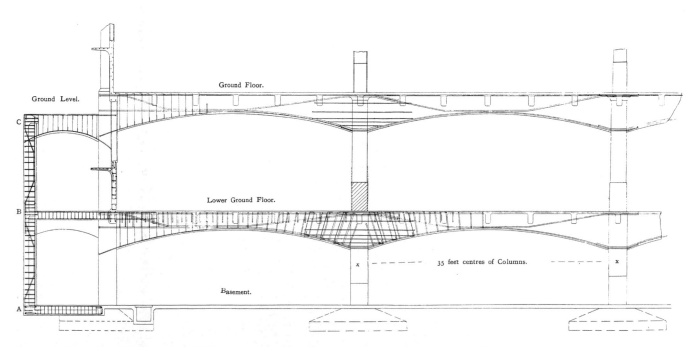

Figure 7.18g   Retaining wall at GPO extension, King Edward Street (Hennebique), 1907–10.

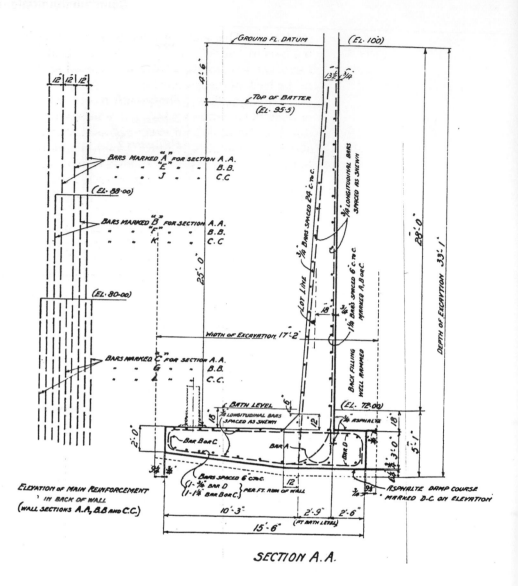

Figure 7.18h Retaining wall
at the RAC building, 1910.

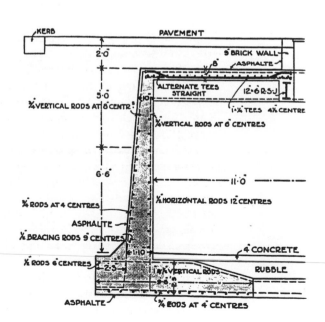

Figure 7.18i Retaining wall,
General Accident Assurance
building, Aldwych, 1909.

steel mesh. This wall was piled on timber piles beneath the buttresses. Reinforcing bars in walls of this type could vary in size from ½ in (12.7 mm) square at the top of the wall, to ¹¹⁄₁₆ in (17.5 mm) square at the base and 1¼ in (31.7 mm) square in the back of the counterfort.[106]

Retaining walls were frequently integrated with the main structure. In the case of the GPO extension (1907–10) retaining walls were 7 in (177.8 mm) thick at the top, and 8 in (203.2 mm) at the bottom, with an average height of 30 ft (9.144 m) above the footings (Figure 7.18g). Reinforced with ⁷⁄₁₆ in (11 mm) diameter horizontal rods at 4–10 in (101.6–254 mm) centres on the outside of the wall, and ⁵⁄₁₆ in (7.9 mm) diameter at 8–24 in (203–610 mm) centres on the outside, and vertical ⁵⁄₁₆ in (7.9 mm) diameter rods at 8 in (203.2 mm) centres with stirrups every foot in height, the counterforts, at 6 ft (1.83 m) intervals were 7 in by 14 in[112] (177.8 mm × 355.6 mm) with 1 in (25.4 mm) diameter reinforcement. The walls were stayed by struts and arches at the ground floor and lower ground floor levels. At Newcastle Goods stations the majority of the retaining walls were plain concrete, but on the eastern side the ground was unstable, and the retaining wall comprised stanchions built up from the basement supporting a thin reinforced concrete wall, and supported by beams at ground floor level and struts in the basement floor between the wall columns and those of the main building structure.

Indented steel bars were used in the reinforced concrete cantilever retaining walls for the 36 ft (10.97 m) high wall at the rear of the Royal Automobile Club, Pall Mall (Figure 7.18h).[113] The wall varied in thickness from 3 ft 9 in at its base to 14 in (355.6 mm) at ground floor level, with the toe from 15 to 22 ft (4.57–6.7 m) wide and 3 ft to 18 in (914.4–457.2 mm) thick. The reinforcement for the wall comprised 1¼ in (31.75 mm) and ¾ in (19 mm) indented steel bars, and the concrete was composed of 1:2:3 cement:sand:crushed gravel, the latter passing a ¾ in (19 mm) mesh. The earth was supported by timbering for six weeks before the formwork was removed.

The reinforced concrete for the General Accident Assurance Building, Aldwych was provided by the British Fireproof Construction Company, and included a conventional cantilever retaining wall (Figure 7.18i).[114] In the 1930s BRC recommended, on economic grounds, cantilever construction for walls supporting up to 15 ft (4.57 m) of earth, and the use of counterforts above this height.[115]

Piling between the wars: research

Despite early optimism concerning the use of reinforced concrete for piles, problems soon became evident. A particular problem contractors encountered arose when driving precast piles through a hard strata to a set in firm ground below.[116] Piles were damaged and even failed under these circumstances, and the Federation of Civil Engineering Contractors asked the Building Research Station to investigate. The results are reported in the reports of the Building Research Station in the 1930s, and summarized in papers of 1935 and 1938.[117,118]

Research revealed most failures were due to excessive compressive stress, with head failures most frequent, a major factor being compression of the head packing and unevenness in placing the packing in the head. Failures lower down the pile were normally the result of hard driving through dense stratum, although not necessarily at the toe, which might derive lateral support from the ground. Tensile cracks were found to develop when a hard stratum was encountered relatively close to the surface. Longitudinal reinforcement provided it was sufficient for the safe handling and transport of the pile, was found likely to be adequate for safe driving. On the other hand lateral reinforcement, particularly at the head and toe was important, and it was recommended that for a distance from the head and toe 2½–3 times the external diameter of the pile the volume of the lateral reinforcement should not be less than 1% of the gross volume of the corresponding length of pile. External bands at the head were found to be beneficial.

Wet curing of piles had a great influence on impact strength, and for Portland cement concrete it was recommended it should be of at least 14 days duration before driving. It was found that the impact strength of concrete was 50–80% of the cube compressive strength, and a maximum stress of 50% of cube compression strength was recommended for a driving factor of safety of unity. In view of the low factor of safety in pile driving it was stressed that care needed to be taken with driving, and best results were obtained with a heavy hammer and a head-cushion of lowest resistance, such as of rubber or asbestos fibre.

Considerable effort was also expended on an exposition of the mathematical theory of stress waves induced in piles during driving. Wave action had been identified in 1931,[119] but these were the first published solutions of the wave equation applied to pile driving. The full potential of this approach was not realized until the advent of computers in the late 1950s,[120] but is well-recognized today.[121]

From 1936 the Institution of Civil Engineers became involved in the research programme, which continued under the auspices of the Joint Committee on Piled Foundations after the war, when a considerable amount of data was gathered on pile loading tests, and work published on impact testing of concrete, and pile group behaviour.[122–126]

Piling between the wars: practice

In the 1920s British Steel Piling developed their own system of *in-situ* concrete piles, known as 'Vibro' piles.[127] A steel tube, usually 16 in (406.4 mm) in diameter, with the lower rim slightly thickened for tamping the concrete, was placed on a conical shoe, and driven to a firm base. A dry mix of concrete was placed in the tube which was then withdrawn by upward blows of the hammer in 1½ in (38 mm) stages. During withdrawal the tube was subjected to 80 blows a minute by the hammer, tamping the concrete and forcing it down and out against the ground. Reinforcement could be placed when necessary before commencing concreting. The completed piles were normally at least 17 in (431.8 mm) in diameter with a working load of 50 t (60 t with reinforcement). Over 700 of these piles were used at St James' Park Underground Station in the 1920s.

The system was extensively employed on the Royal Docks Approaches improvement scheme which began in 1929, and involved piled foundations for the Canning Town and Silvertown viaducts[128] (Figure 7.19). Canning Town viaduct was 4000 ft (1220 m) long, and supported on concrete columns resting on 3500 vibro piles. The piles were driven through a layer of ground into the underlying clay. The majority of the piles were 40 ft (12.129 m) in length, 17 in (431.8 mm) in diameter, with longitudinal reinforcement of 6¾ in (19 mm) diameter steel bars with spiral binding. They were designed for a working load of 40 t, with a safety factor of 3. Some of the piles were test loaded, a load of 110 t on a single pile producing a settlement of a small fraction of an inch. Some 14 in (355.6 mm) diameter piles were used, with ⅝ in (15.88 mm) diameter bar reinforcement, designed for a working load of 30 t. At Silvertown the average length of the piles was 35 ft (10.67 m) but some were driven 65 ft (19.8 m) using an 85 ft (26 m) high steel frame. The working load here was 50 t. Pile cap details are shown in Figure 7.19. BSP also marketed the Prestcore system, using precast sections grouted together under pressure in a bored excavation (Figure 7.20).

The (Francois) Cementation Company developed *in-situ* piling methods based on the technology of the 'cementation' process.[129,129a] For the 'Express' pile an 18 in (457.2 mm) diameter steel tube with a solid point was driven to a set, a reinforcement cage placed in position, and then concrete placed and tamped out

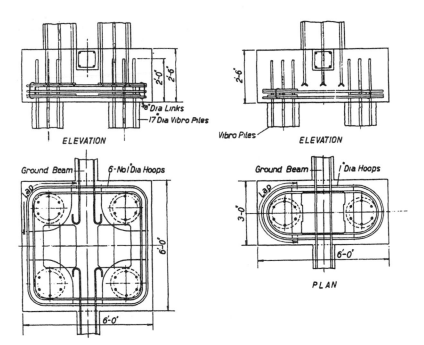

Figure 7.19 Pile cap details,
Royal Docks Approaches,
1929.

Figure 7.20 Prestcore Pile
Test.

below the end of the tube which was gradually removed. These were first used for a garage foundation in Durham in 1929. In the case of the 'Bored' pile ground was excavated by boring techniques as the tube was sunk, rather than being displaced. Reinforcement was placed as necessary, and a perforated injection tube. Ballast was inserted in small quantities and tamped by a drop hammer. Cement grout was then introduced through the tube under pressure. Bored 'Francois' piles were excavated in a similar manner, but consolidated concrete was placed rather than using aggregate and the cementation process. The system was first used in the UK in 1926 at Retford Gasworks. The firm also used 'Tapered shell' piles on a similar basis to the Raymond system. One thousand seven hundred piles of this type were used for Unilever House in 1930.

West's Piling Company marketed the Rotinoff system.[129,130] Precast concrete shells in 3 ft (914.4 mm) lengths, with steel bands at the joints, were threaded on to a steel mandrel, and a shoe fitted at the end. As the shoe was fitted to the mandrel it took most of the driving stresses. Holes in the shell units were available for placing reinforcement. When the desired set had been achieved the mandrel was withdrawn and the pile could be inspected before concreting. Pile diameters of 14⅛–24 in (356–609.6 mm) were used.

The 'Screwcrete' system was developed by Braithwaite as a concrete alternative to traditional cast iron screw piles.[130–132] A screw or helix made of reinforced concrete, steel or cast iron, was driven by a steel mandrel rotated by an electric capstan; a steel or concrete casing was attached to the helix, and could be rotated with it. When the required depth had been reached the mandrel was withdrawn and the pile constructed. Thrust bearings connected to the mandrel were required for the reinforced concrete casing. The screw could be 6 ft (1.83 m) or more in diameter. For 12–18 in (304.8–457.2 mm) diameter piles the pile was normally precast and driven by thrust bearings with just a central hole for the mandrel which was concreted after the mandrel was removed. Other inter-war developments included pressure piles, using augers and compressed air, and the 'Hawcube' piles made up of precast units.[129,130]

Post-war developments: bored piles

In the immediate post-war period research was carried out into the application of short-bored piles to house foundations in London Clay.[133] Further research followed, and as the 1950s progressed increasing use was made of bored concrete piles for more substantial structures in London Clay.[134–136]

By 1959 Skempton was able to report on results of experience with bored piles at ten sites in London Clay covering the years 1950–59, and produce conclusions regarding the end bearing and shaft bearing capabilities of such piles, and thus their ultimate bearing capacity, which could be used to establish the working load for a single pile.[137] Most of these examples were relatively small diameter (10–18 in) (254–457.2 mm), and to modest depths, but there was already increasing demand for suitable foundations for tall buildings in the London area, and large diameter bored piles excavated to a considerable depth in London Clay was one solution.

The use of large diameter 'pier' foundations can be traced back to the Chicago school of foundations developed before the First World War, and the Gow method of large diameter in-situ pile construction was well known.[138] In America these foundations were often taken down to bedrock, which was not an option in London, and the design of large diameter bored piles in London Clay attracted considerable interest in the engineering community.

By 1954 the Benoto system of excavation by hammer grabs had been used in the Midlands for shaft sinking,[139] and in 1956 large diameter piles were installed by Sir Robert McAlpine at Bradwell.[140] The best known early example of large diameter bored piles is that of the Shell Centre on the South Bank development. Here 4 ft 6 in (1.37 m) diameter bored piles, underreamed at the base to 9 ft (2.74 m) diameter, were used.[141,142] Soon after, in June 1959, work began on 3 ft (914.4 mm) diameter bored pile foundations, 65 ft (19.8 m) in depth for the 38 storey Millbank Tower.[143,144] These piles were not underreamed, and most subsequent bored pile foundations followed the same pattern.

Despite their increasing use engineers still felt concerned about the lack of knowledge about the design, installation, and behaviour of these foundations. Early experiences were shared at a 1961 conference, and an approach was made to the Institution of Civil Engineers Piling Committee about various industry concerns.[145] The BRE were commissioned to carry out research, and other investigations were carried out by the contractors.[146] In 1966 a further conference was held on large bored piles.[147] Much of the experience described was in London Clay, but one paper discussed the use of large bored piles in rock. These papers show practical problems involved, particularly when ground conditions were not ideal. Plant was not always reliable, and when there was a delay in placing concrete after the pile had been excavated there were concerns over what kind of pile would result in the presence of water bearing strata or other unfavourable ground conditions.

Prestressed concrete piles[148]

The potential of prestressed concrete piles was dramatically revealed when Freyssinet made use of them for his remedial works at Le Havre Harbour[149,150] in the 1930s (Figure 7.21). To have driven piles by ordinary methods would have been difficult without further risk to the foundations of the existing quay structure, already suffering alarming differential settlement, and Freyssinet decided to sink 100 ft long prestressed concrete piles using hydraulic jacks. Instead of 2000 conventional reinforced concrete piles, carrying 75 t each, only 475 prestressed piles were required, capable of carrying 300 t. The piles were hollow cylinders of 24 in (610 mm) external and 15 in (380 mm) in internal diameter, reinforced longitudinally by 8 wires of 8 mm diameter hard grade steel, with transverse loops of 6 mm diameter hard grade steel, the concrete mix being 1 : 1.33 : 2.66 using sulphate resisting Portland cement. The concrete was vibrated, hydraulic pressure of 285 lb/in^2 (1965 kN/m^2) applied, and the pile then steam cured and sunk. Stresses during sinking may have been as high as 7000 lb/in^2 (48,263 kN/m^2). The success was spectacular, and after the war a whole range of prestressed concrete piles became available from manufacturers.

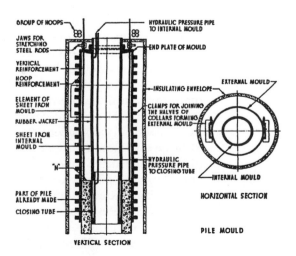

Figure 7.21   Making a Freyssinet Pile, 1935.

Two methods developed in Belgium were published in Britain in 1945.[151] By 1950 about 1200–1300 prestressed bearing piles had been driven in Britain, the largest, about 12 in × 12 in (304.8 mm) solid, and 18 in × 18 in (457.2 mm) hollow.[152] At that time Freyssinet, in his lecture at ICE, envisaged future piles being of tubular precast elements of high quality concrete, with holes in their walls for cables.[153]

Perhaps the earliest UK investigations were on a North Thames Gas Board site at Stanford-le-Hope, Essex, where driving tests for a 10 in × 10 in (254 mm × 254 mm) × 35 ft (10.67 m) prestressed concrete pile were compared with 14 in (355.6 mm) and 12 in (304.8 mm) square reinforced concrete piles on the same site.[154,155] Performance was satisfactory, and in early 1950 was followed up by their application for berthing dolphins at Brentford,[154] and then at Beckton[155] and Bromley-by-Bow. Solid prestressed piles 12 in (304.8 mm) square and 50 ft (15.24 m) long were test driven at Portishead B power station at this time,[156] and prestressed piles were also used at the Isle of Grain Anglo-Iranian Oil Refinery.[157] Hollow piles appear to have been first used at Falmouth in 1949 — of 18 in (457.2 mm) octagonal section and 60 ft (18.29 m) long, and again at Beckton for a coaling jetty in 1955.[154] It would appear that, particularly compared to the US, take-up of prestressed concrete piles was relatively low and only c. 20,000 had been driven by 1960.[154]

Typical details for Lee McCalls' solid and hollow piles c. 1952 are shown in Figures 7.22a–c, and relevant details are given in Table 7.1.[158]

Diaphragm walls[159–164]

Diaphragm walls were developed from the exploitation of bentonite slurries, of which the most famous early application was in drilling oil wells. The ability of bentonite, when placed in an excavation, to produce a thixotropic slurry, which resists overbreak into an excavated shaft, enabled deep wells to be drilled when other lining methods were impractical, and offered the potential of stabilising an excavation while material was placed to form a more permanent barrier. Research began in the United States before the Second World War to investigate the efficacy of bentonite to control seepage through levees on the Mississippi, suggesting it could be used for cut-offs.[165–167] In 1945 a cut-off of clay was installed at Trotters Levee by J.W. Black using a slurry to support the excavation.[159] A.D. Rhodes installed a cut-off wall several kilometres long for the Los Angeles Harbour levees in 1950.[159,168] This early example of the 'slurry trench' process involved the placing of selected soil material blended with bentonite slurry, and by 1968 more than 630,000 m^2 of material had been placed by this method by US and Canadian contractors, the largest single job being the Dead Sea Dikes.[163] In Europe as early as 1934 bentonite may have been used in the installation of a bored pile wall, and in 1948, 1 m diameter piles were installed after excavation using a bentonite slurry at Bone in Algeria.[159,168–170]

The technique was discussed by Professor Lorenz of Berlin in 1950,[171] and in 1951 he took out his first patent.[172] This was largely overshadowed by Veder's Italian patent of 1952.[173] Veder had carried out his first private tests of structural diaphragm wall techniques using trenches in 1948, and practical applications followed in 1950 at Fedala Dam.[173] This early Italian work gained the most publicity.[174] Following disastrous floods in the Po Valley in 1951, in 1952 the Italian Ministry of Public Works initiated a programme of flood defence works which involved placing of concrete 'diaphragms' within the clay dykes. Initially this was done by bored pile techniques,[174] but in 1954 a bentonite technique was tried out by the contractors ICOS, advised by Veder, excavating a 500 mm wide trench, which was filled with bentonite and then concrete placed. The success of the technique was immediate, and was successfully applied for dam cut-offs, foundations, basement walls and cofferdams.[175] Through the 1950s research continued led by

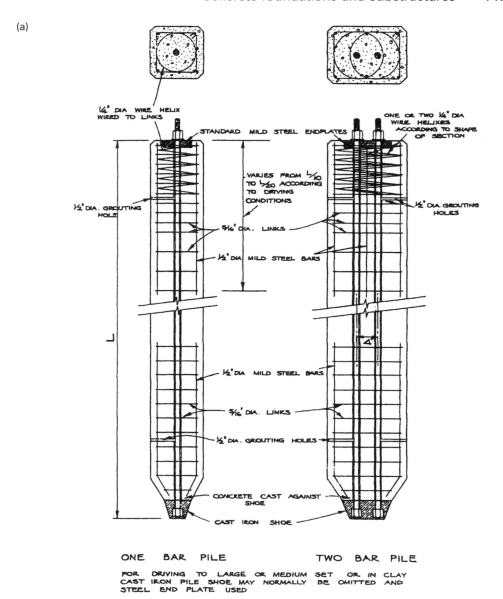

(a)

Figure 7.22 Macalloy pre-stressed piles, 1952: (a) solid, (b) hollow; and (c) open-ended tubular piles.

Lorenz and Veder, and specialist plant was developed for excavating the trenches and placing the bentonite and concrete. Soletanche developed the Radio Marconi system at this time, Soletanche being associated with Soil Mechanics Limited in the UK.[169]

The earliest British application of diaphragm walling was at Hyde Park Corner underpass in 1961.[176] ICOS were the contractors. The work was closely monitored, and much interest was attached to this success in London Clay. The same year a diaphragm wall was installed to take vertical and horizontal loading at an RAF base.[160] By the late 1960s the technique had become well-established in Britain. The success of the system in London clay saw its widespread use for the support of excavations and creation of deep basements as at King's Road (1964) for Boots' and Sainsburys' stores.[160] The largest application was at Seaforth Docks, Liverpool, where a major consideration was the extent to which frictional or adhesive forces between the ground and the wall could be mobilised. The wall here was keyed into the underlying sandstone. ICOS were again the contractors.[177,178]

Other early applications included a cantilevered ICOS diaphragm wall for the A412 widening scheme near Rickmansworth,[179] a curved retaining wall for the

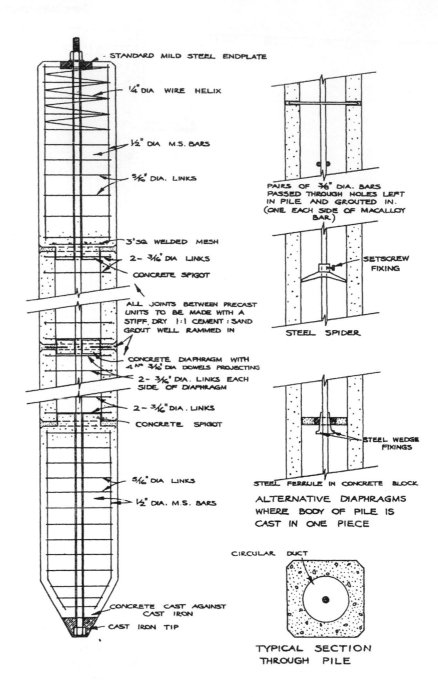

STANDARD MILD STEEL ENDPLATE

¼" DIA. WIRE HELIX

½" DIA. M.S. BARS

5/16" DIA. LINKS

3' SQ. WELDED MESH

2 - 3/16" DIA LINKS

CONCRETE SPIGOT

ALL JOINTS BETWEEN PRECAST UNITS TO BE MADE WITH A STIFF, DRY 1:1 CEMENT : SAND GROUT WELL RAMMED IN

CONCRETE DIAPHRAGM WITH 4 No 3/16" DIA DOWELS PROJECTING

2 - 3/16" DIA. LINKS EACH SIDE OF DIAPHRAGM

2 - 3/16" DIA. LINKS

CONCRETE SPIGOT

5/16" DIA LINKS

½" DIA. M.S. BARS

CONCRETE CAST AGAINST CAST IRON

CAST IRON TIP

PAIRS OF 3/8" DIA. BARS PASSED THROUGH HOLES LEFT IN PILE AND GROUTED IN. (ONE EACH SIDE OF MACALLOY BAR)

SETSCREW FIXING

STEEL SPIDER

STEEL WEDGE FIXINGS

STEEL FERRULE IN CONCRETE BLOCK

ALTERNATIVE DIAPHRAGMS WHERE BODY OF PILE IS CAST IN ONE PIECE

CIRCULAR DUCT

TYPICAL SECTION THROUGH PILE

Figure 7.22b

Victoria Circus development in Southend,[180] Bromley Library/Theatre complex,[181] London's Guildhall precincts redevelopment (1970),[181,182] London Central YMCA (1971),[182] Grangemouth pumping station perimeter wall (1970–71),[183] Peterhead quay wall (1974),[183] and Corsehouse dam cut-off (1973).[183] The first application of concrete diaphragm wall techniques for a British dam was for remedial works at Balderhead dam in the late 1960s.[184]

Bored piled retaining walls

Precast concrete sheet piles were used by Hennebique in the early days of reinforced concrete, and first used in Britain for a retaining wall on the Itchen at Southampton in 1897.[67] Between the wars concrete piles were driven either side by side, or at intervals supporting reinforced concrete slabs to form retaining walls.

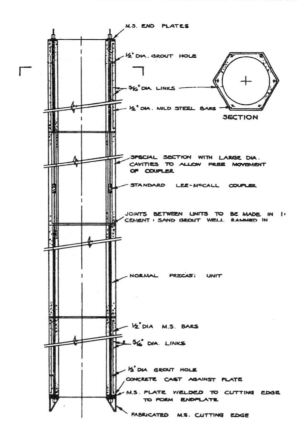

Figure 7.22c

Table 7.1 Macalloy bearing piles

Size (in × in)	Dia. of hole in	Wt/ft (lb)	No. of bars	Prestress lb in²		Maximum safe B.M. lb/in	Maximum safe load (t)	Maximum length for handling (ft)		
				Initial	Final			Maximum cantilever	·3ℓ	·2ℓ
Type 'A' solid*										
10 × 10		104	1–1″	830	695	107,000	56	12	40	57
10 × 10		104	1–1⅛″	1050	882	136,000	48	13	45	65
11 × 11		126	1–1⅛	860	723	150,000	67	13	43	61
12 × 12		149	1–1⅛	720	600	146,000	88	12	40	59
14 × 14		204	2–1″	830	700	317,000	112	14	48	70
14 × 14		204	2–1⅛″	1035	870	382,000	97	16	53	77
16 × 14		233	2–1⅛″	920	770	390,000	121	15	50	72
Type 'B' hollow†										
12 × 12	8	96	1–1⅛″	1100	935	220,000	43	18	60	86
14 × 14	9	131	1–1⅛″	806	675	244,000	73	16	54	77
16 × 16	11	161	1–1⅛″	655	550	297,000	99	16	54	77
16 × 16	11	161	2–1⅛″	1310	1100	594,000	61	23	76	109
18 × 18	12	214	1–1⅛″	493	413	330,000	144	14	49	70
18 × 18	12	214	2–1⅛″	985	825	660,000	106	20	69	99

* No. of bars at 42 t/in².
† No. of bars at 45 t/in².

Concrete sheet piles might have a joint such as a tongue and groove connection, or grouting grooves. The walls might be anchored using a tie rod connecting a waling to the anchor, or by the use of a slab connecting the retaining wall to raking piles driven behind the wall.[185]

As early as 1914 Dyckerhoff and Widmann constructed a retaining wall using the Strauss method of *in-situ* pile construction developed by the Russian mining engineer Anton Strauss from drilling techniques.[186] British interest in bored pile retaining walls grew in the late 1950s alongside the development of large diameter bored piles, Sir Robert McAlpine's used bored piles in the retaining wall at the TUC building c. 1955–58,[187] and Derrington advocated the use of such piles for retaining walls in 1961.[188] In that year Soil Mechanics Limited installed 24 in (609.6 mm) and 17 in (431.8 mm) diameter bored piles as a continuous diaphragm for the basement of the Audley Square multi-storey car park using the Soletanche system (Figure 7.23).[189,190] At the Stag Place redevelopment in the early 1960s a Hochstrasser bored pile wall was installed where the western boundary wall ran alongside the King's Scholars pond sewer. The piles were 76 cm in diameter and reinforced with secondhand bull-head rails, with a facing of 5 in (127 mm) of concrete over the inner face to complete the work.[191]

Bored pile walls were installed to two main patterns — contiguous and secant.[192] In the former case piles were installed more or less adjacently, in two alternative series, to give the concrete in the first series time to gain strength before boring the adjacent pile. With secant walls the piles were installed by boring and concreting at centres of less than two pile diameters, again in two series, creating an interlocking structure. Lilley Construction developed a method of construction along these lines in 1968.[193] In their case the first series of sites to be installed were known as 'female' piles, and the second series, installed midway between, and cutting a 'secant' section from the female pile, were known as 'male' piles. For lighter forms of construction the male piles were reinforced with mild steel bars, but for heavier work all piles were reinforced with UB sections. Relatively little British literature was produced on either technique before the mid-1970s. Examples of contiguous bored pile walls included the NatWest Tower,[194] and secant pile walls

Figure 7.23   Bored pile wall at Audley Square, London W1, 1961.

were used for the Heathrow extension of the Piccadilly line,[195] and for the British Library basement.[196]

Reinforced earth[196a1–8]

In the 1960s the French engineer Henri Vidal developed 'Terre armee' or 'reinforced earth' as an alternative to heavy retaining walls to stabilize earthworks built above existing ground level, using galvanized or stainless steel strips to strengthen or reinforce the soil by mobilizing the friction between the (in most early applications cohesionless) soil and the reinforcing elements. The concept of strengthening earth structures was not new, and traditional methods are described in C.W. Pasley's early 19th century work on fortification. There is little evidence however that these ideas had been taken forward in an analytical sense before Vidal began his investigations, apparently prompted by building sandcastles. British engineers use of the techniques was delayed by a patent dispute between Vidal and the British Government. Whilst the technique in its origin was not an application of concrete, as it developed the use of concrete facing units, whether structural or non structural, became an integral part of the system. The first application in the UK was in 1973, and research at TRRL and elsewhere publicized the system. Techniques of soil reinforcement developed rapidly, with the use of grids and geotextiles rather than steel strips for reinforcement.

Crib walls[196b1–20]

Crib walling is another technique whose origins can be traced back centuries, but which was reinvented in the last quarter of the 20th as another economic alternative to heavy retaining wall construction. At its simplest it takes the form of alternative layers of precast concrete elements acting as stretchers and headers. Such open bottomed timber boxes or cribs, filled with earth or stones, have a long history of use as a cheap means of slope stabilization where timber was in abundance, such as the Alpine areas of Europe. Cribwork was used extensively by North American engineers through the 19th century, the term being also applied to open cofferdams for piers and bridgework.[196b1–3] The rapid spread of the railway network and widespread availability of timber sleepers or ties provided a ready source for the stretchers and headers from which cribs could be constructed. In the early 20th century, as reinforced concrete beceame increasingly popular, precast elements replaced timber on American railways. Such elements were in use in North America by the time of the First World War, some taking the form of open boxes or bins.[196b4–6]

Cribwalls provided an economic alternative to heavier forms of retaining wall construction. They were easy and quick to assemble, although there were height limitations. A considerable research effort in Germany, New Zealand and particularly Austria where it began in the 1960s and continued into the 1980s, did much to improve the understanding of crib structures and develop their design.[196b8–12,15,16] One development, the 'New Wall', replaced the stretchers by anchoring straps, and was intended to address the height issue.[196b13,14]

The open nature of the cribs meant that such techniques could be combined with biotechical methods of slope stabilization, as seen in the 'Evergreen' wall. Some methods have been developed combining concrete elements with live vegetation.[196b8,10,19,20]

Crib walls have generally been developed outside the UK. An early British example, on the M5 through the Clevedon Hills, won a Concrete Society Award in

1973.[196b7] In the 1980s and 1990s research work was sponsored by the TR(R)L to encourage use.[196b18]

Design of foundations

In Britain geotechnical engineering, in the modern sense of the term, is largely a post-Second World War development. Some sense of the excitement felt by the early pioneers can be obtained from Sir Harold Harding's autobiography.[197] When one looks at the design of foundations by previous generations of engineers one must bear in mind, therefore, that they lacked many of the methods of site investigation, sampling, testing, analysis and design which are taken for granted today. Problems faced by the engineer before the war were highlighted by Terzaghi in his 1927 paper *The science of foundations — its present and future*.[198] This focused on specific shortcomings of foundation design at that time — selecting allowable soil pressure regardless of the area covered by individual foundations and the maximum permissible differential settlement of the superstructure, calculating the bearing capacity of piles by the *Engineering news* formula without regard to the properties of the soil, and using the bearing capacity of an individual pile as a guarantee of the bearing capacity of the whole foundation. The discussion on Terzaghi's paper provides a fascinating insight into state of soil mechanics of the time.

The question of an allowable soil pressure for the design of foundations appears to have developed on an empirical basis through the 19th century (Table 7.2).[199–201] One could regard foundation design of the time as a two stage process: having computed the superimposed load of the superstructure, design foundations of sufficient strength to sustain this load, while selecting the foundation type and dimensions to ensure that the load would not exceed the safe bearing capacity of the ground. It is apparent there was little consensus in the late 19th century as to what the safe bearing capacity might be. This dilemma was highlighted by E.L. Corthell in 1902 when involved in the design of deep foundations at Rosario harbour in Argentina.[202] The experienced contractors Schneider and Hersent proposed a foundation based on a load of $8 \, kg/cm^2$ ($7.3 \, t/ft^2$). This was rejected by the Board considering the design, and after considerable discussion an allowable bearing pressure of $3.5 kg/cm^2$ ($3.2 \, t/ft^2$) was determined upon, with consequent increase in the cost of the works. Corthell was dissatisfied with the lack of consensus among engineers as to safe bearing capacities of soils, and compiled a large amount of data to illustrate the situation (summarized in Table 7.3).

Table 7.2   Bearing capacities used in foundation design, 1830–90

Project	Ground conditions	Foundation type	Design loads (t/ft^2)
London Bridge (1831)	London Clay/Woolwich and Reading Beds	Timber piles	5.75*†
Nelson's Column	London Clay	Concrete block	1.3
Crystal Palace (1850)	Compact gravel	Concrete footings	2.5
Charing Cross Bridge (1860)	Gravel over London Clay	Concrete-filled cast iron cylinders	8*
Hownes Gill Viaduct (1861)	Debris over clay or gravel	Inverted arches	1
Westminster Bridge (1862)	London Clay	Timber piles and cast iron cofferdams	2
Tower Bridge (1886–90)	London Clay	Concrete and caissons	4

* Settlement known to have been a problem. At Charing Cross it was partially solved by pre-loading the foundations.
† This load was calculated after the bridge was completed. It is unclear what design load the Rennies used.

Table 7.3   Summary of Corthell's findings (1907) for deep foundations

Ground type	Loads (t/ft^2)	Average load (t/ft^2)	Number of examples
Fine sand	2–5.2	4	10
Coarse sand and gravel	2.1–6.9	4.6	33
Sand and clay	2.25–7.6	4.4	10
Soil and alluvium	1.3–5.5	2.6	2
Hard clay	1.8–72	4.6	16

Corthell was not the first to investigate the question.[199] In the late 1880s I.O. Baker had attempted, by examining a group of case studies, to compile some guidance on safe bearing capacities of various types of ground.[203] Even earlier British engineers in Bengal, confronted with numerous examples of settlement and cracking of buildings in Calcutta, carried out a series of experiments to establish the optimum load on the alluvial soil of the area, and the depth to which foundations should be dug, concluding that to avoid differential settlement the load should not exceed one ton per square foot ($107.3 \, kN/m^2$), and in undisturbed ground the foundation depth should be 4–6 ft[204] (1.2 m–1.8 m). In 1893 Sutcliffe[50] and Newman published some figures for various types of ground which bear many similarities to the recommendations of the 1950 Civil Engineering Code of Practice for Foundations.[209] The first statutory regulations appear to be those contained in the iron and steel frame regulations of the 1909 London County Council (General Powers) Act.[211] Over the next 30 years guidelines were published in various trade catalogues, some of which were more detailed than the LCC recommendations.[205–212] These various recommendations are summarized in Table 7.4.

Pile driving formulae

Another area discussed by Terzaghi was the value of dynamic pile driving formulae.[213] From the early 18th century various formulae were proposed by engineers and scientists to calculate the percussive effect of piling engines, and relating the force exercised by the ram to the set and the bearing capacity of the foundation. Much was written on the subject, and a large number of formulae are listed by Chellis.[214] Among the earliest formulae to come into widespread use were those of Woltmann[215,216] and Eytelwein.[217,218]

There is not much evidence to suggest these formulae were used by British engineers in the first half of the 19th century. It is possible that a crude formulae based on the velocity of the ram as described by Cresy in his *Encyclopaedia of Civil Engineering* (1847) was used.

In the second half of the 19th century A.M. Wellington developed the *Engineering news* formula.[219] This was apparently widely used, and continued to be in the early 20th century. All of these formulae were essentially developed before steam hammers were widely used, and were modified accordingly around the end of the century.

Of the formulae developed in the first half of the 20th century two attracted most comment. The Hiley formula was developed in the 1920s.[220–222] Dissatisfaction with this and other formulae led Oscar Faber to develop his own formulae, attempting to take account of the difference in behaviour between piles driven in clay and sand or ballast.[223]

His formulae attracted much interest at the time, but their value was immediately questioned particularly with reference to clay. As the science of soil mechanics has progressed and foundation technology changed such formulae have been replaced by more reliable methods of foundation design.

Table 7.4 Allowable bearing value in t/ft² (kN/m²) of soils, 1893–1990

Soil type	Baker (1889)[203]	Sutcliffe (1893)[50]	Newman (1893)[202a]	LCC (1909–30)[211]	Redpath Brown (1913)[205]	BRC (1918)[206]	BRC (1932)[207]	Redpath Brown (1938)[208]*	CECP4 (1950)[209]	BS 8004[210]
Rocks	10–20 (1000–20,000)	Beware of fissures	8–20 (860–2145)		<16 (1700)	<15 (1600)	5–30 (535–3200)	10–40 (1070–4300)		(2000–10,000)
Chalk			1–4 (107–430)				3–6 (320–640)	6 (640)	6 (640)	
Dense sand and gravel	8–10 (800–1000)	4–6 (430–640)	7–9 (750–965)	4 (430)	4–8 (430–860)	6–8 (640–860)	4–8 (430–860)	4 (430)	4–6 (430–640)	(>600)
Sand and gravel		2–3 (215–320)	6–7 (640–750)	4 (430)	3 (320)		4–6 (430–640)	4 (430)	2–4 (215–430)	(<200–600)
Compact sand	4–6 (400–600)	5–7.5 (535–800)	6–7 (640–750)	2 (215)		4–6 (430–640)		4 (430)		(>300)
Medium compact sand	2–4 (200–400)	2–3 (215–320)	3.5–5 (375–535)	2 (215)		3–4 (320–430)	3 (320)	2 (215)	2–4 (215–430)	(100–300)
Loose sand		1–1.15 (107–160)	2.5–3 (250–300)	1 (107)		1–2 (107–215)	1 (107)	1 (107)	1–2 (107–215)	(<100)
Dry compact clay	4–6 (400–600)	3–5 (320–535)	5–8 (535–860)	4 (430) (London)	4–5 (430–535)	4–6 (430–640)	2–4 (215–430)	3–4 (320–430)	4–6 (430–640)	(300–600)
Compact clay	2–4 (200–400)	2–3 (215–320)	3–6 (320–640)	2 (215)	2 (215)	2–4 (215–430)	2–3 (215–320)	2 (215)	2–4 (215–430)	(150–300)
Moist clay		1–1.5 (107–160)	1.5–2 (160–215)				1 (107)		1–2 (107–215)	(75–150)
Soft clay	1–2 (100–200)	0.5–0.75 (50–80)	0.25–1 (25–107)	1 (107)		1–2 (107–215)	0.5–1 (50–107)	1 (107)	0.5–1 (50–107)	(<75)
Alluvial soil/ quicksand	0.05–1 (5–100)	0.5–0.75 (50–80)	0.2–1.5 (20–160)				0.5–1 (50–107)	0.5 (50)		

Note: imperial units are precise.
*Similar to 1937 LCC Regulations.[212]

Conclusions

This review has covered a period of two centuries of the use of concrete underground, a period when engineers have become increasingly confident in its use and inventive in its application. For much of this century many of the new developments have been imported from the continent or the United States, by foreign born engineers, or enterprising contractors. One might almost say British engineers went to sleep between the wars, to be awakened under the impact of soil mechanics in the 1940s and 1950s. To an extent this is reflected in the literature, and there was no British textbook to compare with Patton[200] or Fowler[224] for 50 years until the works of Henry,[225] Tomlinson,[226] and Little[227] were published. Limitations of space have precluded a discussion of the impact of plant on design and construction, but many of the post-war developments are a direct result of rapidly changing plant technology. This is a story which needs to be written.

References

1. Holy Bible. Matthew 7: 24–27.
*2. Skempton, A.W., Landmarks in early soil mechanics. Proceedings of the VII European Conference on Soil Mechanics and Foundation Engineering: Design Parameters in Geotechnical Engineering, 1979, 5, 1–26.
*3. Skempton, A.W., A history of soil properties 1717–1927. Proceedings of the 11th International Conference on Soil Mechanics and Foundation Engineering, 1985, Golden Jubilee Volume, 95–121.
*4. Heyman, J., Coulomb's Memoir on Statics. Cambridge University Press, 1972.
5. Glossop, R., The invention and development of injection processes. Part I: 1802–1850. Geotechnique, 1960, 10, 91–100; Part II: 1850–1960. Geotechnique, 1961, 11, 255–79.
6. Glossop, R., The invention and early use of compressed air to exclude water from shafts and tunnels during construction. Geotechnique, 1976, 26, 253–80.
7. Peck, R.B., History of building foundations in Chicago. University of Illinois, Engineering Experiment Station, Bulletin Series 373, 1948.
8. Flodin, A., Broms, B., Historical development of civil engineering in soft clay. Developments in Geotechnical Engineering 20: Soft Clay Engineering. 1981: 28–156.
9. Kerisel, J., Down to Earth: Foundations Past and Present. Balkema: Rotterdam, 1987.
10. Skempton, A.W., Foundations for high buildings. ICE Proc., 4, Part III, 246–69.
*11. Dobson, E., A Rudimentary Treatise on Foundations and Concrete Works. Weale: London, 1850 (9th edn, Crosby Lockwood, 1903).
12. Law, H., The Rudiments of Civil Engineering, 4 parts in 1 Vol. Weale: London, 1848–52.
13. Hughes, T., A series of papers on the foundations of bridges in Weale, J. Comp. The Theory, Practice and Architecture of Bridges of Stone, Iron, Timber and Wire. Weale: London, 1839–43.
14. Skempton, A.W., The Albion Mill Foundations. Geotechnique, 1971, 21, 203–10. (This foundation can be described as buoyant, but it seems unlikely the concept was recognised before the latter part of the 19th century. The problem of foundations for large mill buildings was generally avoided by building on rock!)
15. Smeaton, J., A narrative of the building and a description of the Construction of the Eddystone lighthouse with stone, London, 1791.
*16. Huberti, G. et al., Vom Caementum zum Spannbeton. Band 1. Bauverlag: Wiesbaden, 1964: 5–36.
17. Belidor, B.F., de. Architecture hydraulique, II Partee, Tome 2. Jombert: Paris, 1753: 178–90.
18. Vicat, L.J., Recherches experimentales sur les chaux de construction, les betons et les mortiers ordinaires. Goujon Paris, 1818.
19. Vicat, L.J., Note sur un mouvement periodique observé aux voutes du Pont de Souillac. Annales de Chimie et de Physique, 1824, 27, 70–79 (abstracted in Mechanics Magazine, 1825, 3, 264–65).
20. Vicat, L.J., Notice sur le pont et Souillac construit sur la Dordogne. Bulletin des Sciences Technologiques, 1826, 6, 117–19.

*Particularly useful for general information.

21. Mary, –. De l'emploi du beton dans la foundation des ecluses. Annales des Ponts et Chaussees, 1832, 3, 66–105.
22. Semple, G., A Treatise on building in water, Dublin, 1776.
23. Godwin, G., Essay upon the nature and properties of concrete, and its application to construction up to the present period. Trans. Inst. Br. Architects, 1835–36, 1, 1–39.
*24. Pasley (Sir), Observations on Limes, Calcareous Cements Mortars, Stuccos and Concrete (etc.). Weale: London, 1838: 265–73, Appendix pp. 23–33. (This is the most authoritative source on the early 19th century use of cements and concrete.)
25. Crook, J.M., Sir Robert Smirke: a pioneer of concrete construction. Trans. Newcomen Soc., 1965–66, 38, 5–22.
26. On what materials and their properties are best to form concrete masses for the foundations of buildings. ICE Min. Conversations, 1830, 264–69 (unpublished).
27. Donaldson, T.L., Handbook of Specifications. Atchley: London, 1857: 416–27. (The concrete was to be thrown into the trench from a height of at least 10 feet, a practice condemned by Dobson.)
28. Colvin, H.M. et al., History of the Kings Works, Vol. 6. HMSO: London, 1982: 606.
28a. Clegg, S., Jr., A Treatise on the manufacture and production of coal-gas, 1844, etc.
29. Chrimes, M.M., Hugh McIntosh (1767–1840): national contractor, Trans. Newcomen Soc., 1994–95, 66, pp? This work draws heavily on the researches of Dr David Brooke and Professor A.W. Skempton.
30. Taylor, G.L., An Account of the methods used in underpinning the long storehouse at His Majesty's Dock Yard, Chatham, in the Year 1834. Trans. Inst. Br. Architects, 1835–36, 1, 40–43.
31. Taylor, G.L., The Auto-biography of an Octogenarian Architect, Vol. 1, 2 vols. Longmans: London, 1870–71: 167–72.
32. Metropolis Management and Buildings Act, 1878. HMSO: London, 1878.
33. Fletcher, B., The Metropolitan Building Acts: A Textbook for Architects, Surveyors, Builders, etc. Batsford: London, 1882.
34. On concrete. ICE Minutes of Conversation, No. 158, 4 March 1830, 531–32. The proportions at Archway were 1 cement: 2 sand: 7 gravel.
35. Paxton, R.A., The Influence of Thomas Telford (1757–1834) On the Use of Improved Constructional Materials in Civil Engineering Practice, MSc thesis, Heriot-Watt, 1975.
36. Skempton, A.W., Portland cements, 1843–1887. Trans. Newcomen Soc., 1962–63, xxxv, 117–52.
*37. Vernon-Harcourt, L.F., Harbours and Docks, 2 vols. Clarendon: Oxford, 1885.
*38. Bray, R.N., Tatham, P.F.B., Old Waterfront Walls. Spon: London, 1992.
39. Guillery, P., Building the Millwall Docks. Constr. History, 1990, 6, 3–22.
40. Rickman, J., (ed.), The Life of Thomas Telford. London, 1838: 156–57.
41. Smith, N.A.F., History of Dams. P Davies: London, 1971: 208–13.
42. Skempton, A.W., Historical development of British embankment dams to 1960. In: Clay Barriers for Embankment Dams. TTL: London, 1990: 15–52.
43. Binnie, G.M., Masonry and concrete dams 1880–1941. Industr. Archaeol. Rev., 1987, X, 41–58.
44. Bateman, J.F., La Trobe, The History and Description of the Manchester Waterworks. Spon: London, 1884.
45. Wegmann, E., The Design and Construction of Masonry Dams. Wiley: New York, 1888.
46. Gordon, G., Concrete dam for Geelong waterworks. Professional Papers on Indian Engineering. Second Series, 4(178), 402–404.
46a. James IGSE
47. Institution of Civil Engineers Engineering Conference, 1907. Trans. Sect. Railways, Docks Harb., ICE: London, 1907.
48. Shankland, E.C., Steel skeleton construction in Chicago. Min. Proc. Inst. Civil Engrs, 1896–97, 128, 1–27.
49. The High Court, Calcutta. The Builder, 30 October 1869, 27, 857.
50. Sutcliffe, G.L., Concrete: Its Nature and Uses. Crosby Lockwood: London, 1893.

51. Twelvetrees, W.N., Concrete-Steel Buildings. Whittaker: London, 1907, 337–54.

52. Jones, B.E., (ed.), Cassell's Reinforced Concrete. Cassell: London, 1913.

53. Twelvetrees, W.N., Concrete-Steel Buildings. 1907: 8–10.

54. Twelvetrees, W.N., Concrete-Steel Buildings. 1907: 105–07.

*55. Marsh, C.F., Reinforced Concrete. London, 1904. The third edition (1906) co-written with William Dunn contains minor amendments and updates.

56. Edgell, G.J., The Remarkable Structures of Paul Cottancin. Struct. Engr, 1985, 63A, 201–07.

57. Galbraith, A.R., The Cottancin System of Armoured Construction. Colwell: Ipswich, 1904.

58. Emploi du sable dans les foundations sur sol compressible. Annales des Ponts et Chaussees, 1835, 10, 171–214.

59. Mitchell, A., On submarine foundations; particularly the screw pile and moorings. Min. Proc. Inst. Civil Engrs, 1848, 7, 108–132.

60. Jopling suggested the use of concrete filled cast iron columns in 1834. Mechanics Magazine, 1834, 20, 424–28.

61. Concrete filled (square) cast iron piles are described. Architect Engr Surveyor, 1843, 4, 24, 359–62.

62. Walmisley, A.T., Reinforced concrete piles in tidal waters. Concr. Constr. Eng., 1906–1907, 1, 88–101.

63. Galbraith, A.R., Reinforced concrete piling. Proc. Assn Municipal County Engrs, 1904–1905, 31, 356–75.

64. Reinforced concrete work on the Coignet system in England. Concr. Constr. Eng., 1907, 2, 320.

65. Gow, C.R., History and present state of the concrete pile industry. Proc. Am. Concr. Inst., 1917, 13, 174–218.

66. Cummings, R.A., Reinforced concrete piles. Proc. Natl. Assn Cement Users, 1912, 8, 312–25.

67. L.G. Mouchel and Partners. Mouchel-Hennebique Ferro-Concrete, 4th edn. Mouchel: London, 1909–21.

68. Effect of steam curing on the crushing strength of concrete. Eng. News, 1907, 58, 249–50.

69. Wig, R.J., The Effect of high pressure steam on the crushing strengths of Portland cement, mortar, etc., concrete. United States Bureau of Standards, Technological Papers, 1912.

69a. Reinforced concrete systems: The Williams system. Builders J., 1908, 439–42. (This lists 13 contracts executed to the system.)

70. Spirally armoured concrete piles. The Engineer, 1908, 105, 476–77.

71. Berger, C., Guillerme, V., La construction en ciment armé. Dunod: Paris, 1902, 336–37.

71a. Edmond Coignet Limited. Reinforced Concrete Systems. Coignet: London, c.1911. (Coignet's London office was opened in 1905.)

72. A New system of concrete pile construction. Eng. News, 1901, 45(25), 450–51.

73. Concrete pile foundations at Aurora. Eng. News, 1902, 48(24), 495.

74. Concrete piles for sandy ground. Eng. News, 1903, 49, 275. (James Brunlees first developed a system of water jetting for the installations of cast iron piles across Morecambe Bay, with the contractor Philip Brogden in 185).

75. Shuman, C., The Simplex system of concrete piling. Proc. Engrs Club Philadelphia, 1905, 22, 347.

76. Mackellar, T., The Simplex system of concrete piling. Assn Eng. Soc., 1907, 39, 266–76.

77. Recent improvements in piles. The Engineer, 1906, 101, 79.

78. Recent improvements in piles. The Engineer, 1906, 101, 382.

79. Ellis, S.H., Tranmere Bay Development Works. Min. Proc. Inst. Civil Engrs, 1907, 171, 127–70.

80. The Franki Compressed Pile Company Limited. Franki Pile: London, 1957.

81. Braatvedt, I.H., Guide to Piling and Foundation Systems. Bramley, Transvaal, Frankipile, 1975.

82. Der Eisenbetonbau auf der Weltausstellung in Brussel. Beton und Eisen, 1911, 10, 1–4.

83. British patent rights in reinforced concrete piles. Concr. Constr. Eng., 1906–1907, 1, 282–85.

84. Tobacco Warehouse. General correspondence. Bristol Records Office.

85. British patent rights in reinforced concrete piles. Concr. Constr. Eng., 1907–1908, 2, 139–48; Ferro-concrete pile case. Engineering, 1907, 83, 354.

86. British patent rights and reinforced concrete piles. Concr. Constr. Eng., 1909, 4, 4.

87. The Enlargement of Victoria Station. Engineering, 1906, 82, 35–38.

88. Information provided by B.L. Hurst.

89. The Ritz Hotel. Concr. Constr. Eng., 1906, 1, 441.

90. Drawings and specification in ICE Archives.

91. Twelvetrees, W.N., Concrete-Steel Buildings. Whittaker: London, 1907: 72–93.

92. Twelvetrees, W.N., Concrete-Steel Buildings. 1907: 240–52.

93. Twelvetress, W.N., Concrete-Steel Buildings. 1907: 232–39.

94. Copies of drawings, etc., in Concrete Archive at ICE.

95. Problems of pile deformation were well-known before the First World War with *in-situ* techniques, see Gow's paper (ref. 65).

*96. Ketchum, M.S., The Design of Walls, Bins and Grain Elevators. Eng. News, New York, 1907. (This work was published in London by Constable and available at ICE from December 1907.)

97. Bone, E.P., Reinforced concrete retaining wall design. Eng. News, 1907, 57(17), 448–52.

98. Bone, F.A., Comparative costs of gravity and reinforced-concrete retaining walls. Eng. News, 1908, 59(13), 347.

99. Klein, A., Die Form der Winkelstutzmauem aus Eisenbeton mit Rucksicht auf Bodendruck und Reiburg in der Fundamentfuge. Beton und Eisen, 1909, 16, 384–87; Gilbrin, G., Die Wirtschaftlichste Form der Eisenbeton-Winkelstutsmauer. Beton und Eisen, 1912, 10, 233–35. (See for German work.)

100. The Subway of the Philadelphia Rapid Transit Co. Eng. Record, 1905, 51(8) 224–28.

101. Graff, C.F., High reinforced concrete retaining wall construction at Seattle, Washington. Eng. News, 1905, 53(10), 262–64.

102. Graff, C.F., Difficult reinforced concrete retaining wall construction on the Great Northern Railway. Eng. News, 1906, 55(18), 483–87.

103. American Railway Engineering and Maintenance of Way Association. Retaining walls and abutments. Proc. AREMWA, 10, Part 2, 1909, 1317–37.

*104. Christophe, P., Beton Armée. Beranger: Paris, 1902: 290–91.

105. Marsh, C.F., Reinforced Concrete. Constable: London, 1904: 66–67.

106. Cantell, M.T., Reinforced Concrete Construction: Advanced Course. Spon: London, 1912.

107. Reinforced concrete retaining wall in Queen Victoria Street, London EC. Concr. Constr. Eng., 1911, 6, 956–958.

108. A Retaining wall in Piccadilly. Concr. Const. Eng., 1908, 2, 509.

109. Reinforced concrete at the Royal Insurance Building. Concr. Constr. Eng., 1911, 6, 811–20.

110. Consolidated Expanded Metal Companies. A Handbook of Design. Braddock: PA, 1919: 189–203.

111. Expanded Metal Company. Expament Pamphlet, No. 5, 2nd edn. Expament: London, 1923, passim.

112. Watson, T.A., The Reinforced concrete construction of the new General Post Office Buildings, London. Trans. Jr. Inst. Engrs, 1909–1910, 20, 58–71.

113. Bylander, S., The Architectural and engineering features of the Royal Automobile Club Building. Trans. Jr. Inst. Engrs, 1910–1911, 21, 243–70.

114. Reinforced concrete in Aldwych: General Accident Assurance Building. The Builder, 1909, 97, 234.

115. British Reinforced Concrete Engineering Company. BRC Reinforcements, 9th edn. BRC: Stafford, 1932: 262–65.

116. Detailed records of hard pile driving were published in the US before World War I. Thompson, S.E., Fox, B., Cast reinforced concrete piles. J. Assn Eng. Soc., 1935, 1, 150–234.

117. Glanville, W.H. *et al.*, The Behaviour of reinforced concrete piling during driving. J. Inst. Civil Engrs, 1935, 1, 150–234.

118. Glanville, W.H. *et al.*, An Investigation of the stresses in reinforced concrete piles during driving. Building Research Station. Technical paper 20, 1938.

119. Isaacs, D.V., Reinforced concrete pile formulae. J. Inst. Engrs, Australia, 1931, 3(9), 305–23.

120. Smith, E.A.L., Pile-driving analysis by the wave equation. ASCE Proc. J. Soil Mech. Div., August 1960, 86, SM4, 35–61.

121. Goble and Associates. Wave Equation Analysis of Pile Driving: WEAP Program, 3 Vols. US Federal Highway Administration: Washington, DC, 1981.

122. Joint Sub-Committee on Pile Driving. Interim Report 1. ICE: London, 1938.

123. Joint Committee on Pile Driving. Chartered Civil Engineer, May 1950: 16–21.

124. Green, H., Impact testing of concrete. Proc. Conf. Mech. Prop. Non-metallic Brittle Mater., 1958.

125. Whitaker, T., Experiments with model piles in groups. Geotechnique, 1957, 7, 147–67.

126. Joint Committee on Piled Foundations. Progress report. Proc. Inst. Civil Engrs, December 1959, 14, N10–15. (In a sense the Committee's work was continued by CIRIA.)

127. McCarthy, M.J., Piling in general, with special reference to the vibro concrete piling system. Trans. Jr. Inst. Engrs, 1927–28, 38, 171–81.

128. British Steel Piling. Royal Docks approaches improvement scheme. BSP pamphlet, No.8, 1934. (The licensed contractors were John Gill Contractors.)

129. Faber, O. (ed.), Concrete Yearbook. Concrete Publications: London, 1936 edition.

129a. The Francois Cementation Company Limited. London, c.1934. (The company set up in England c.1919.)

*130. Dean, A.C., Piles and Pile Driving. Crosby Lockwood: London, 1935: (especially) 16–60.

131. Wilson, G., The Bearing capacity of screw piles and 'screwcrete' cylinders. Inst. Civil Engrs J., March 1950, 34, 4–93.

132. Braithwaite and Company Engineers Limited. Braithwaite: London, c.1950, Figures 56–73.

133. Ward, W.H., Green, H., House foundations: the short bored piles. Public Works and Municipal Services Congress, 1952, 373–97.

134. Glossop, R., Greeves, I.S., A New form of bored pile. Concr. Constr. Eng., 1946, 41, 344–51.

135. Meyerhof, G.G., Murdock, L.J., An Investigation of the bearing capacity of some bored and driven piles in London Clay. Geotechnique, September 1953, 3, 267–82.

136. Golder, H.Q., Leonard, M.W., Some tests on bored piles in London Clay. Geotechnique, March 1954, 4, 34–41.

*137. Skempton, A.W., Cast *in-situ* bored piles in London Clay. Geotechnique, 1959, 9, 153–173.

138. Gow, C.R., Concrete piles. J. Assn Eng. Soc., 1907, 39, 255–65.

139. Hunter, L.E., Large diameter piles. Concr. Constr. Eng., May 1954, 49, 165–68.

140. Derrington, J.A., (Discussion) Large Bored Piles Symposium. ICE, 1966, 139.

141. Williams, G.M.J., Design of the foundations of the Shell Building, London. Proceedings 4th International Conference on Soil Mechanics and Foundation Engineering, 1957, 1, 457–61.

142. Bored pile foundations. Roads Road Constr., July 1959, 37, 215. (This describes the crawler mounted excavator developed by Economic Foundations, a specially formed subsidiary of Sir Robert McAlpine's, to drive a heavy duty auger based on Texas drilling methods.)

143. Davis, C., Structural engineering aspects of the Millbank Tower Block, London. Struct. Engr, 40, January 1962, 3–20.

144. Kirkland, G.W., The Millbank Tower. Consul. Engr, March 1962, 21, 290–93.

*145. Symposium on large diameter bored piles, March 1961. Reinforced Concr. Rev., 1961, 6, 673–726.

146. Frischmann, W.W., Fleming, W.G.K., The use and behaviour of large diameter piles in London Clay. Struct. Engr, April 1962, 40, 123–31.

*147. Large bored piles: proceedings of the symposium organised by ICE and the Reinforced Concrete Association, ICE, London, 1966. (Skempton provided some first approximation design rules.)

*148. Andrew, A.E., Turner, F.H., Post-tensioning systems for concrete in the UK: 1940–1985. CIRIA Report 106, 1985.

149. Boase, A.J., Notes on inspection of structures in Europe. Am. Concr. Inst. Proc., 1937, 33, 521–26.

150. Freyssinet, E., Progrès pratiques des methodes de traitement mecanique des betons. La Reprise en sous-oevre des foundations de la Gare Transatlantique du Havre. Travaux, 1935, 199–227.

151. Magnel, G., Prestressed concrete. Some new developments. Concr. Constr. Eng., 1945, 40, 221–32, 249–54; 1946, 41, 10–20.

152. New, D.H., Discussion on pile driving in difficult conditions. ICE Eng. Div. Papers (Works Construction Division), 1950–51, 9, 26–27.

153. Freyssinet, E., Lecture on prestressed concrete 17 November 1949. J. Inst. Civil Engrs, February 1950, 33, 331–80.

154. Gardner, S.V., New, D.H., Some experiences with prestressed concrete piles. Proc. Inst. Civil Engrs, January 1961, 18, 43–66. (Discussion), April 1962, 21, 867–91.

155. Concrete piles at Beckton Gasworks. Concr. Constr. Eng., May 1950, 45, 169–70.

156. Morgan, H.D., Haswell, C.K., The Driving and testing of piles. Proc. Inst. Civil Engrs, Part 1, 1953, 2, 43–75.

157. Records in ICE Piling Committee archives.

158. McCalls, Macalloy Limited. PRC Note 7: Piles, 2nd edn, c.1952.

*159. Boyes, R.G.H., Structural and Cut-off Diaphragm Walls. Applied Science: London, 1976.

160. Consulting Engineer. Practical design for diaphragm walls. Consul. Engr, Suppl., September 1974.

*161. Institution of Civil Engineers. Proc. Conf. Diaphragm Walls Anchorages, 1974. ICE: London, 1975.

*162. International Society of Soil Mechanics and Foundation Engineering. British National Committee. Proc. Symp. Grout. Drill. Muds Eng. Practice. Butterworth: London, 1963.

*163. International Society of Soil Mechanics and Foundation Engineering, 7th International Conference, Mexico City, 1969. Proceedings of Speciality Session 14 and 15. Societe de Diffusion des Techniques du Batiment et des Travaux Publics, Paris, 1969.

*164. Xanthakos, P.P., Slurry Walls. McGraw-Hill: New York, 1979.

165. Davis, C.W. et al., Bentonite: its properties…utilization. US Bureau of Mines: Technical Paper, 609, 1940.

166. Efficacy of bentonite for control of seepage. US Waterways Experimental Station. Experiment Station Bulletin, 1938, 2(1), 2–6.

167. US Waterways Experiment Station. Technical Memorandum, 351–1, 1938.

168. Berthier, P., La paroi moulée dans le sol. Memoires ICF, September 1964, 9, 33–46.

169. Soil Mechanics — Soletanche Limited. Technical data sheets. Soil Mechanics Limited, London, c. 1960. (These cover works executed c. 1949–59.)

170. Florentin, J., Les parois moulées dans le sol. Proc. 7th Intl. Conf. Soil Mech. Foundation Eng., 3, 1969, 507–12.

171. Lorenz, H., Ueber die Verwendung thixotroper Flussigkeiten in Grundbau. Bautechnik, 1950, 27(10), 313–17.

172. Lorenz, H., Erfahrungen mit thixotropen Flussigkeiten im Grundbau. Bautechnik, 1953, 30(8), 232–36.

173. ICOS. Underground works. ICOS, Milan, 1968.

174. Veder, C., Method for the construction of impermeable diaphragms at great depths by means of thixotropic seals. Proc. 3rd Intl. Conf. Soil Mech. Foundation Eng., Zurich, 1953, 2, 91–94.

175. Sistonen, H., Montta and Seitakorva hydro power plants. 9th ICOLD Congr., 1967, 3, Q.34, R.34., 609–27.

176. Granter, E., Park Lane improvement scheme: design and construction. Proc. Inst. Civil Engrs, 1964, 27, 293–315. (Discussion), 1964, 33, 423–36.

177. Agar, M., Irwin-Childs, F., Seaforth Dock, Liverpool: planning and design. Proc. Inst. Civil Engrs, May 1973, 54, 255–74.

178. Cole, P.G. *et al.*, Seaforth Dock, Liverpool: construction. Proc. Inst. Civil Engrs, May 1973, 54, 275–90.

179. ICOS wall solves ground widening problem. Ground Eng., 1968, 1, 38–39.

180. Diaphragm wall schemes show progress in Britain and France. Ground Eng., 1970, 3(2), 30–34.

181. Recent diaphragm walls schemes in Britain. Ground Eng., 1970, 3(6), 22–24.

182. Little, M.E.R., An economic appraisal. Consul. Engr. Diaphragm Wall Suppl., September 1974, S.23–27.

183. Coats, D.J., Three examples of diaphragm walling. Consul. Engr. Diaphragm Wall Suppl., September 1974, S.28–31.

184. Vaughan, P.R. *et al.*, Cracking and erosion of the rolled clay core of Balderhead Dam, and the remedial works adopted for its repair. 10th ICOLD Congr., Montreal, 1970, 1, Q.36, R5, 73–95.

185. Wentworth-Shields, F.E., Gray, W.S., Reinforced Concrete Piling. Concrete Publications: London, 1938: 35–47.

*186. Emperger, F., (ed.), Handbuch fur Eisenbetonbau, 3 Auft. Band 3. Ernst, Berlin, 1922, 312–23 (ill p. 323).

187. Arup job record No. 633. Information supplied by M. Bussell.

188. Derrington, J.A., Concrete cylinder retaining walls. Reinforced Concr. Rev., September 1961, 5, 696–98.

189. Soil Mechanics Limited. Continuous bored pile diaphragms in civil engineering practice. Geotechnical Pamphlet No. 9, 1961.

*190. Bullen, F.R., Notes on the History of foundation engineering. Struct. Engr., December 1961, 39, 385–404; 400–401.

191. Mason, J., Frost, A.D., Stag Place development. Struct. Engr., November 1963, 41, 347–365 (Discussion) May 1964, 43, 169–172.

192. North-Lewis, J.P., Lyons, G.H.A., Contiguous bored piles. ICE Conf. Diaphragm Walls Anchorages, 1974–1975: 184–94.

193. Neal, D., The Effects of concrete mix design on secant piling. Concrete in the Ground: Proc. Concr. Soc. Sem., 123–32. (Veder had been using techniques like this in Finland in 1954.)

194. Frischmann, W.W. *et al.*, National Westminster Tower: design. Proc. Inst. Civil Engrs, August 1983, 74, 387–434.

195. Jobling, D.G., Lyons, A.C., Extension of the Piccadilly line from Hounslow West to Heathrow Central. Proc. Inst. Civil Engrs, May 1976, 60, 191–218; (Discussion) November 1976: 719–37.

196. Deep foundations for the British Library. Ground Eng., April 1984, 17, 20–26.

196a1. America Society of Civil Engineers. Earth Reinforcement. ASCE: New York, 1979.

196a2. Vidal, H., La Terre Armee Annales ITBTP, 223–24, July/August 1966, 887–938; 259–60, July/August 1969, 1099–1155.

196a3. International Conference on soil reinforcement: reinforced earth and other techniques. Paris: Association Amicale des Ingenieurs Anciens Eleves de l"ENPC, 1979.

196a4. Banerjee, P.K., Principles of analysis and design of reinforced earth retaining walls. Highway Engr., 1975, 22(1), 13–18.

196a5. Murray, R.T., Research at TRRL to develop design criteria for reinforced earth. TRRL SR 457, 1977.

196a6. Schlosser, F., Experience on reinforced earth in France, TRRL SR 457, 1977.

196a7. Ingold, T.S., Reinforced Earth. TTL: London, 1978.

196a8. Jones, C.F.P.J., Earth Reinforcement and Soil Structures, 2nd edn. TTL: London, 1996.

196b1. Bovey, H.T., Cribwork in Canada, Min. Proc. ICE, 1881, 63, 268–72.

196b2. Engineering News, Index 1890–1899, passim.

196b3. Fowler, C.E., Ordinary foundations, including the cofferdam process for piers, 4 eds., title varies, 1898–1920.

196b4. Anon., Precast concrete timber form crib to retain wall. ENR, 1918, 81, 763.

196b5. Anon., Precast concrete cribbing for retaining wall. ENR, 1923, 91, 718–19; 858.

196b6. Anon., Concrete cribbing for railway retaining wall. ENR, 1926, 96, 654–56.

196b7. Payne, D.F., Cribwalling in road construction (M5). DoE Construction, 1973, 7, 6–7.

196b8. Schiechtl, H., Bioengineering for Land Reclamation and Conservation, University of Alberta Press: Edmonton, 1980.

196b9. Brandl, H., Tragverhalten und Dimensionerung von Raumgitterstutzmauern, Austria Bundesministerium fur Bauten und Technik Strassenforschung, 1980, 141.

196b10. Anon., Changing techniques of cribwall planting, Landscape Australia, February 1981, 91–96.

196b11. Brandl, H., Raumgitter-Stutzmauern, Austria Bundesministerium fur Bauten und Technik Strassenforschung, 1982, 208.

196b12. Brandl, H., System von Raumgitter-Stutzmauern, 2 parts. Austria Bundesministerium fur Bauten und Technik Strassenforschung, 1985, 251.

196b13. Brandl, H., Slope stabilization and support by crib walls and prestressed anchors. 3rd Intl Geotech. Seminar, Singapore, 1985, 179–85.

196b14. Germany Forschungsgesellschaft fur Strassen und Verkehrwesen. Merkblatt fur den Entwurf und die Herstellung vom Raumgitterwanden und Wallen, 1985.

196b15. Brandl, H., Stutzmauersystem 'New' und andere Konstruktione…Austria Bundesministerium fur Bauten und Technik Strassenforschung, 1986, 280.

196b16. Hong Kong. Geotechnical Control Office. Geoguide 1: Guide to retaining wall design, 1993, 2nd. edn.

196b17. US Army Corps of Engineers. Retaining and Flood Works. ASCE: New York, 1994, Chapter 10.

196b18. Masterton, G.G.T. et al., A Literature and design review of crib wall systems. TRL Report 131, 1995.

196b19. Morgan, R.P.C., Rickson, R.J., Slope Stabilization and Erosion Control: A Bio-Engineering Approach. Spon: London, 1995.

196b20. Gray, D.H., Soter, R.B., Biotechnical and Soil Bioengineering Slope Stabilization. Wiley: New York, 1996.

   197. Harding, H.J.B., Tunnelling History and My Own Involvement. Golder Associates: Toronto, 1981. For an alternative view of the era see R.E. Goodman. Karl Terzaghi: the engineer as artist. Rston, ASCE, 1998.

  *198. Terzaghi, K., The Science of foundations — its present and future. Trans. Am. Soc. Civil Engrs, 1927, 93, 270–405.

   199. Hunt, R., The Supporting power of soils. Assn Eng. Soc. J., 1888, 7(6), 189–97. APC 1864.

  *200. Patton, W.M., A Practical Treatise on Foundations. Wiley: New York, 1893.

   201. Smith, J.A., Some foundations for buildings on Cleveland. Assn Eng. Soc. J., 1906, 36, 155–84.

  *202. Corthell, E.L., Allowable Pressures on Deep Foundations. Clowes: London, 1907; Wiley: New York, 1907 (abstract in ICE Min Proc., Vol. 165, 1905–1906).

  202a. Newman, J., Notes on Cylinder Bridge Piers. Spon: London, 1893: 24–33.

  *203. Baker, I.O., A Treatise on Masonry Construction. Wiley: New York, 1889. (I have been unable to trace an earlier UK copy than the 3rd edn. of 1890.)

   204. Leonard, H., Weight on foundations of buildings in Bengal. Profes. Paper. Indian Eng., 2nd series, 1875, vol.4, 319–31.

   205. Redpath Brown and Company. Handbook of Structural Steelwork. 1913: 289.

   206. British Reinforced Concrete Engineering Company. BRC Reinforcements. BRC: Stafford, 1918.

   207. British Reinforced Concrete Engineering Company. BRC Reinforcements, BRC: Stafford, 1932, 83.

   208. Redpath Brown and Company. Handbook of Structural Steelwork. 1938: 485.

   209. Institution of Civil Engineers and Others. Civil Engineering Code of Practice 4: Foundations, ICE: London, 1950.

   210. British Standards Institution. BS 8004.

   211. London County Council (General Powers) Act 1909, 9th edn, 7 Ch cxxx, 1909.

   212. London County Council. Bylaws for the Construction and Conversion of Buildings and Furnace Chimneys, No. 3319. LCC: London, 1937. (These bylaws also specified various grades of concrete.)

   213. Terzaghi, K., Peck, R.B., Casagrande, A., Discussion on pile driving formulas. Proc. Am. Soc. Civil Engrs, 68, 311–31.

*214. Chellis, R.L., Pile Foundations, 2nd edn. New York, 1961.

215. Woltmann, R., Beytrage zur hydraulischen Architektur, 4 vols. Gottingen, 1791–99 (Vol. 4, 1799, 371–89).

216. Woltmann, R., Recherches theoriques et experimentales sur d'effet des machines et outils ... principalment sur d'effet du monton. Dieterich, Gottingen, 1804.

217. Eytelwein, J.A., Handbuch der Mechanik fester Korper, 1801, 1842, etc. (Chapter v).

218. Eytelwein, J.A., Praktische Anweisung zur Wasserbankunst, 4 vols, 1802–08 (esp. Vols 2 and 3).

219. Wellington, A.M., Formula for safe loads of bearing piles. Eng. News, 1888, 20, 570–12.

220. Hiley, A., The Efficiency of the hammer blow and its effects with reference to piling. Engineering, 1922, 119, 657–58; 721–22.

221. Hiley, A., A Rational pile driving formula. Engineering, 1925, 119, 657, 721–22.

222. Hiley, A., Pile driving calculations. Struct. Engr, 1930, 8, 246–59; 278–88.

223. Faber, O., A New piling formula. J. Inst. Civil Engrs, 1946–47, 28, 5–86.

*224. Fowler, C.E., The Coffer-Dam Process for Piers. Wiley: New York, 1898 (2nd edn., 1907, etc.)

*225. Henry, F.D.C., Design and Construction of Engineering Foundations. Spon: London, 1956.

*226. Tomlinson, M.J., Foundation Design and Construction. Pitman: London, 1963.

*227. Little, A.L., Foundations. Arnold: London, 1961.

8 The early development of reinforced concrete shells

Peter Morice and Hugh Tottenham

Synopsis

This chapter discusses briefly the engineer's problem of creating structural systems to enclose spaces using non-tensile resistant materials and draws attention to Nature's solutions to similar problems. The development of reinforced concrete, a material which can be readily shaped and which has tensile as well as compressive strength, enabled a radical change in possible structural forms, with the thin shell being one major example. The early development of these forms of structure took place mainly outside Britain, probably due to our established traditions of metallic construction. However, the shortages of steel following the Second World War gave great impetus to their adoption and many interesting structures were designed and built by British engineers.

Before the industrial revolution, engineers had a limited range of structural materials of which the most permanent, stone, was available only in small pieces and had a relatively poor tensile strength. The solution to the problem of spanning gaps using masonry, one of the structural engineer's principal tasks, was to adopt the geometrical form of the arch, which has a primarily compressive internal stress system. This two-dimensional structure is readily extended to form the vault, or into its true three-dimensional form as the dome, with, again, a primarily compressive stress system. The dome, like the arch and vault, can therefore also be built from individual small elements with effectively zero tensile strength between the elements. We are well aware, nevertheless, of the feats of structural engineering achieved from Roman times to the present day using masonry in these structural forms, which have indeed had a significant influence on modern shell construction.

A way of visualizing how these structural systems behave, and the shapes which correspond to applied loads, is gained by reversing the structure to become a tensile form. Thus the hanging string or chain supporting vertical loads is the inverse of the arch and, as it has negligible bending strength, all the loads are necessarily carried by direct tensile stress. Similarly, a balloon can be considered to represent the inverse of a dome, although the loading by internal pressure is a little different from true vertical load and will lead to a slightly different geometry. Indeed, the fact that a true direct stress system requires a unique geometry is seen by the changing shape of a hanging string when a concentrated load is applied or when we apply a concentrated load to an inflated balloon. A flexurally stiff system will accommodate the tendency to distort under changing load by developing internal bending and shear stresses. The engineer Gaudi[1] used inverse models to develop the most appropriate geometries for his somewhat unusual compressive stress roof structures.

An analysis of the membrane stresses induced in shell structures was first given by Lame and Clapeyron in 1828.[2] They showed that loading on shells can produce a consistent direct and shear stress system; this may lead to tensile as well as compressive stress resultants in certain regions of the shell membrane according to the shape and loading conditions. It is to be noted, incidentally, that the membrane solution is, of course, a state in an elastic shell in which there are deflections resulting from the strains corresponding to the stress state.

The invention of reinforced concrete provided the structural engineer with a material which could be formed into a thin shell of any required geometric shape appropriate to its principal loading, thereby producing an efficient primary internal direct stress system. At the same time the reinforced shell membrane had the capacity to resist tensile as well as compressive stresses. It could also accommodate the reasonable bending stresses developed in a structure which had to cope with a range of loading and support conditions.

In many ways reinforced concrete mimics bone, the basic biological structural material. The evolution of animals required the development of highly strength to weight ratio efficient structural forms capable of coping with the range of loading conditions which arise from movement. Efficiency is achieved by using internal direct stresses as much as possible and reducing stress variations due to bending. It is an interesting structural phenomenon that two systems have evolved:

(a)   an articulated skeletal structure which can arrange its geometry to establish primarily direct stress systems under a given load arrangement;
(b)   a shell form in which external loading is resisted — primarily by internal direct stresses — due to its shape.

With the arrival of reinforced concrete, engineers could make use of this latter form of natural development and apply the structural shell in building construction with increasing success.

Theoretical work on the bending effects in shells was reported by Love[3] at the end of the last century, and it is from this work that most of the modern analysis has developed, but initially its practical application was restricted to the design of spherical domes where mathematical solutions were more easily obtained.

The early shapes of shell tended to follow the forms which had been used for masonry construction; however, the necessary support to resist spread of the boundary, previously provided by ring chains, flying buttresses and massive walls, were now able to be incorporated within tensile reinforced concrete boundary members. As a result such structures became very much lighter in construction than their masonry equivalents.

The need for rectangular plan forms with as large as possible column-free areas for the majority of buildings soon led to the end-only supported cylindrical barrel vault. For modest spans the section of a cylinder could be used with slight thickening at the edges or in the valleys to accommodate tensile stresses (Figure 8.1). For larger spans, however, the thin shell vault was made integral with a pair of edge beams spanning the full length (Figure 8.2). In simple terms these both form hollow beams, the compression 'flange' being the crown of the shell membrane itself and the tensile flange being the valley portion or the edge beams. However, in both cases there is an incompatibility between the 'membrane' stress and strain state of the shell at its boundaries (which is 'expecting' the remainder of the full cylinder to exist), and that actually provided by the valley connection or the edge beams. This incompatibility can only be accommodated by edge bending effects

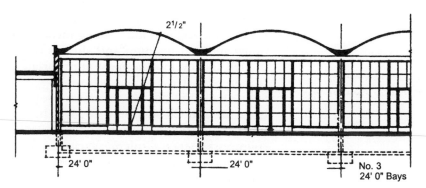

Figure 8.1   Cylindrical shells with thickened edges (May and Baker canteen, Dagenham), 1949.

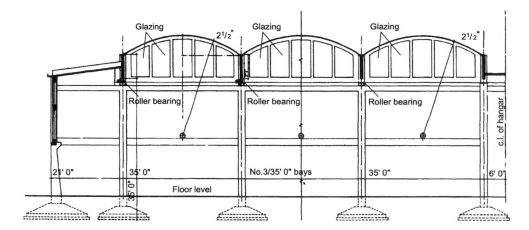

Figure 8.2   Cylindrical shells with edge beams (Karachi Hangar), 1947.

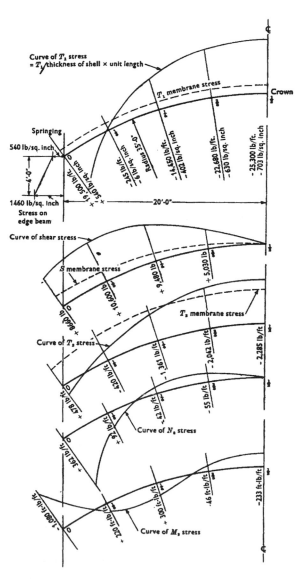

Figure 8.3   Cylindrical shell edge effect stress resultants.

in the shell membrane which will also produce shear and torsional influences on the edge beams themselves (Figure 8.3).

A whole panoply of theoretical studies was made of this problem, starting with the first practical approximate design analysis provided by Finsterwalder,[4] he assumed simple end support, using a Fourier series, and omitted end moments and twisting moments. Studies by Dischinger[5] extended the analysis to include previously omitted stress resultants, whilst Schorer[6] and Vlasov[7] introduced further simplifications to reduce the extent of design calculations. Jenkins[8] showed that a consistent general theory could be developed in a simple form including all the stress resultants. In all these cases the shell is described by eighth order partial differential equations, which accounts for a certain mathematical complexity. (Note that the standard analysis of an elastic beam requires a fourth order ordinary differential equation relating its deformation to its loading.)

Concurrently with the use of thin concrete shells for forming roofs, complete cylindrical shapes were used for liquid containment structures. The theoretical analysis of such structures was well established early on and is comprehensively set out in, for example, the work of Timoshenko.[9]

More exotic shell geometries were also being adopted with the realization that the primary stresses did not necessarily have to be compressive when using reinforced concrete. Doubly curved shells of negative Gaussian curvature started to be used; these are surfaces with the centres of curvature, in two directions at right angles, on different sides of the surface (Figure 8.4). They were adopted particularly in the form of the hyperbolic paraboloid, which could be cut into various ground plan shapes for roofs (Figure 8.5) or in total annular forms for cooling towers (Figure 8.6). These presented the theoreticians with a new range of analytical problems to solve. Other geometries such as elliptic paraboloids, cones, conoids (Figure 8.7) and translational surfaces have also been extensively used. Another popular arrangement has been the cylindrical shell formed as a north-light system (Figure 8.8), whilst various shapes of dome can be cut to give a rectangular planform (Figure 8.9).

The history of the thin concrete shell in the UK shows a slower initial progress than in some other countries. There are perhaps three reasons for this. Firstly, by the time that reinforced concrete had appeared as a practical competitive structural material, Britain was well into an age of metallic construction. Steel was abundantly available and had a firm grip on major structural work, and its ability to carry high stresses both in tension and compression in skeletal forms, which could be readily adapted to virtually all spanning requirements, gave it an undoubted lead. This is evidenced perhaps by the fact that the Concrete Institute, as a forum for developing reinforced concrete, was formed in 1908 outside the mainstream professional body of the Institution of Civil Engineers.

Secondly, the familiarity of steel implied lower costs in construction since reinforced concrete would require the retraining of the labour force, an increase in levels of specialist supervision of work, and longer construction periods due to the need to build complex formwork as well as the fixing of reinforcement and the casting and curing of the concrete itself. Indeed, despite early work on water to cement ratio theory and concrete creep, the design of really consistent high strength concrete only became well understood in the late 1930s.

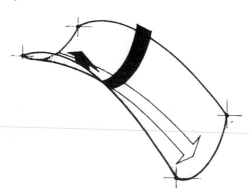

Figure 8.4   Negative Gaussian curvature surface.

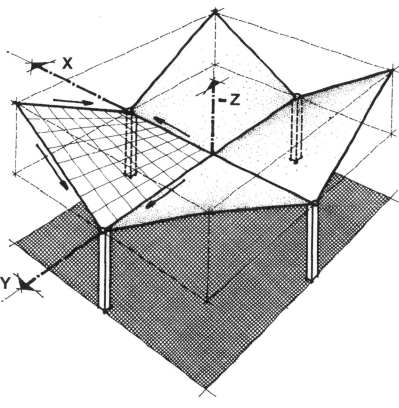

Figure 8.5   An arrangement of hyperbolic paraboloids on a rectangular planform.

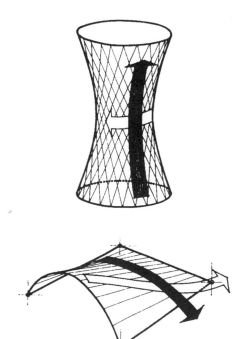

Figure 8.6   Annular hyperbolic paraboloid surface.

Figure 8.7   Conoid surface.

Thirdly, structural engineering in Britain during the early part of this century was not noted for its adoption of advanced theoretical procedures which new forms of structure required. It is true to say that, up to the Second World War, the teaching of civil engineering structural analysis, although it had encompassed reinforced concrete, was essentially restricted to two-dimensional analysis of skeletal frames, and even the theory of flat plates did not form a part of undergraduate studies.

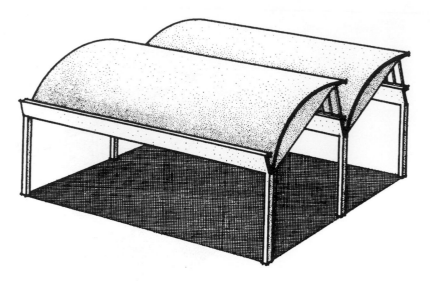

Figure 8.8   Northlight arrangement of cylindrical shells.

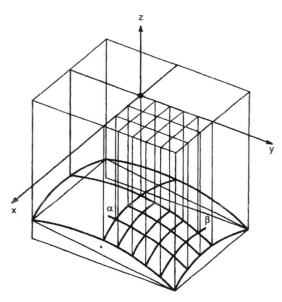

Figure 8.9   Doubly curved surface on a rectangular planform.

This is reflected in the contents of one of the most influential textbooks of the period by Pippard and Baker.[10] Important pioneering theoretical structural research was indeed taking place in Britain, but this was almost entirely restricted to aeronautical applications and the tools readily available to civil engineering designers and consultants were limited; in addition, strict building rules required full analysis of proposed structures, especially unusual forms. The result was that novel structures in new materials were few and far between. On the continent of Europe the steel industry had not acquired such a hold on construction, labour was generally cheaper and the results of structural research were being applied more rapidly in the construction industry. Shell development therefore went ahead much more quickly there.

One of the earliest descriptions of a reinforced concrete shell roof is that of the Armee-Museum in Munich, by Zollner in 1906,[11] which was soon followed by others at Düsseldorf and Leipzig in the following year. Many more domes were built in Germany during the next few years, with accompanying theoretical studies of stress analysis and, in 1915, a study by Zoelly[12] in Switzerland of shell buckling. Although few, if any, reinforced concrete domes were being built in Britain, a major structure consisting of three shells 15 cm thick, giving a free floor

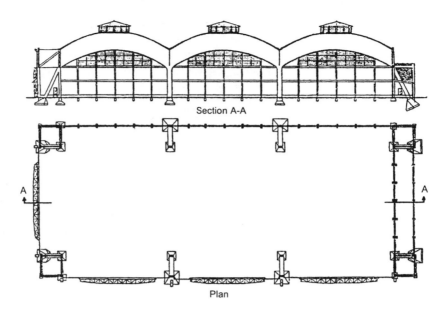

Section A-A

Plan

Figure 8.10   Multiple domes of the hangar at Reval, Russia, 1917.

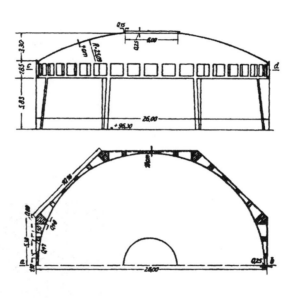

Figure 8.11   Dome of the Frankfurt electricity works, 1928.

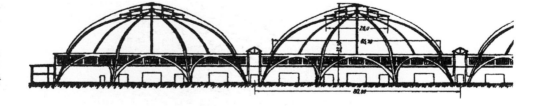

Figure 8.12   Octagonal domes of the Leipzig Market Hall, 1929.

area of 115 m by 50 m, was designed in Christiani and Neilson's London office and built in Russia in 1917 (Figure 8.10). An article describing it appeared in *The Builder* in 1920,[13] but it did not apparently excite sufficient interest to merit a report in either engineering institution journal. It is likely that the analysis was carried out on the basis of membrane stresses. A very large, very thin shell, 26 m in diameter and 4 cm thick, was built for an electricity works at Frankfurt (Figure 8.11) in 1928, whilst the octagonal domes of the Leipzig Market Hall (Figure 8.12) of 1929 gave a free floor area of 237 m by 75 m using shells 14 cm thick.

The Zeiss Dywidag system of shell construction was patented in Germany in 1920, and in 1926 Dischinger and Finsterwalder[14] described one of the first long cylindrical shells built for the Dusseldorf Exhibition. In 1927 the Frankfurt Market Hall (Figure 8.13) was constructed with long shells spanning 42 m. Thereafter long cylindrical reinforced concrete shells became a familiar constructional form on the Continent, with major shell structures being built for the Market Hall at Budapest and at Rheims railway station. In 1936 Torroja[15] built the large Fronton Recoletos Hall in Madrid (Figure 8.14). This consisted of two cylindrical vaults of different radii which shared one longitudinal edge but were continuously supported on their outer edges, providing a free area of 55 m by 35.5 m. The principal spanning was therefore in the radial direction. It seems it was designed essentially on the basis of the membrane theory plus the use of a one-tenth structural model which was tested to failure.

In the early 1930s some interest developed in France in the analysis and design of hyperbolic paraboloid shells and short cylindrical shells, reported mainly by Valette[16] and Laffaille.[17] However, this did not seem to lead to major developments. It is reported that the first contract in Britain for a thin concrete shell was let in 1936 for Doncaster Municipal Airport.

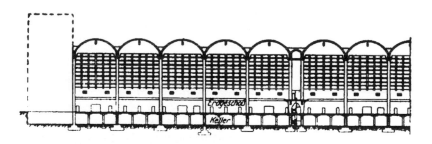

Figure 8.13 Cylindrical shells of the Frankfurt Market Hall, 1927.

Figure 8.14 Edge-supported cylindrical shells of the Fronton Recoletos, Madrid, 1936.

In the 1940s the advantages of reinforced concrete shells as a system economical in its use of materials were seen in South America, where the relative costs of labour and materials were more favourable. Candela[18] adopted the technique for a wide range of structures of many different geometries. He appears to have supplemented the membrane solutions with models and full scale tests.

The reconstruction after the devastations of the Second World War required forms of building which offered economy of material. This gave an enormous boost to the use of shell roofing in Britain as well as continental Europe, since materials, particularly steel, were in short supply everywhere. It stimulated further structural research in British universities, at the Building Research Station and the Cement and Concrete Association, and caused the rapid adoption of research results into design office practice. Many simplified design procedures were proposed and textbooks on shell theory and design became available in English.

The Zeiss Dywidag system was licensed to Chisarc and Shell 'D', who were responsible for a number of structures. Twisteel, under the guidance of Hajnal Konyi, were also major designers and reinforcement suppliers who produced many hundreds of shells. Blumfeld[19] described 'The development and use of barrel vault shell concrete' in one of the first important papers on the subject for the Institution of Civil Engineers. Ove Arup and Partners made a major contribution both in design, for example the factory at Brynmawr (Figure 8.15) which had nine rectangular planform domes each 25 m by 19 m, and in the theoretical analysis.[8]

Amongst the protagonists of shell construction in post-war Britain was the architect Mills,[20] who was responsible for a range of structures, and the structural engineers Cousins[21] and Snow.[22]

The mystique which had surrounded shell theory and its application to design was rapidly dispelled in the late 1940s and early 1950s by the reporting of detailed analyses of significant new structures. For example, Kirkland and Goldstein[23] described the substantial Bournemouth Bus Garage (Figure 8.16) with its clear span of 45.5 m, and Sexton[24] described the earlier Karachi Hangar (Figure 8.17), with a clear span of 39 m. In both these cases the edge beams were prestressed. To aid the spread of the technology a number of evening courses for practising engineers on shell theory and design appeared at this time.

The great post-war interest in shell roofing led the Cement and Concrete Association to hold the first symposium in 1952 devoted exclusively to the subject. This reports, in considerably more detail than is possible here, what were seen at that

Figure 8.15   Rectangular planform domes of Brynmawr rubber factory, 1952.

Figure 8.16   Cylindrical shells with prestressed edge beams used for the Bournemouth Bus Garage, 1951.

Figure 8.17   Cylindrical shells with prestressed edge beams used for the Karachi Hangar, 1947.

time as the major achievements to date, the constructional techniques currently adopted and the state of development and understanding of theory. A fairly comprehensive list of publications on shells, up to 1952, is included in the proceedings of this symposium.[25] This first specialist meeting was followed by a second in Norway in 1957,[26] which led to the formation of an international organization for the study and promotion of shell structures, the International Association for Shell Structures (IASS). This organization continued to hold regular meetings on aspects of shell studies.

References

1. Mainstone, R.J., Developments in Structural Form. Alan Lane: London, 1975, Chapter 5.
2. Lame, G., Clapeyron, E., Memoire sur l'équilibre intérieur des corps solides homogines. Mem. Pres. Par Div. Savants, (1828), 1833, 4, 465–562.
3. Love, A.E.H., A Treatise on the Mathematical Theory of Elasticity. Cambridge University Press: Cambridge, 1892.
4. Finsterwalder, U., Die Querversteiften zylindischen Schalangewolbe mit kreisseg-mentformigen Querschmitt. Ingenieur Archiv., 1933, 4, 43–65.
5. Dischinger, Fr., Die strenge Theorie der Kreiszylinderschale in ihrer Anwendung auf die Zeiss-Dywidag-Schalen. Beton Eisen, 1935, 34, 257–94.
6 Schorer, H., Line load action on cylindrical shells. Am. Soc. Civ. Engrs Trans., 1936, 62, 767–810.
7. Vlasov, V.Z., Thin walled elastic beams, 1st edn, 1940. General Theory of Shells and its Application in Engineering, 1949 (Moscow, Stroidzal).
8. Jenkins, R.S., Theory and Design of Cylindrical Shell Structures. O.N. Arup: London, 1947.
9. Timoshenko, S., Theory of Plates and Shells. McGraw-Hill: New York, 1940.
10. Pippard, A.J.S., Baker, J.F., Theory of Structures. Arnold: London, 1936.
11. Zollner, L., The reinforced concrete dome of the Armee-Museum at Munich. Bauzeitung, 1906, Nos 16–17.
12. Zoelly, R., The Buckling of Shells, Thesis. ETH: Zurich, 1915.
13. The Builder, 1920.
14. Dischinger, Fr., Finsterwalder, U., The Dywidag hall at the Hygene Exhibition, Dusseldorf. Bauingenieur, 1926, 1, 48, 929.
15. Torroja, E., Report on thin slabs in Spain. IABSE Final Report, Third Congress, 1948, 575–84.
16. Valette, R., Thin self-supporting roofs. Genie Civ., 1934, 104(4), 85–88.
17. Laffaille, B., Thin shells in the shape of hyperbolic paraboloids. Genie Civ., 1934, 104(18), 409–10.
18. Faber, C., Candela: the Shell Builder. Architectural Press: London, 1963.
19. Blumfield, C.V., The development and use of barrel vaults concrete. Institution of Civil Engineers Structural Division, 1948.
20. Mills, E.D., Reinforced concrete shell membrane structures. Archit. Bldg News, 1944, February, 94–98.
21. Cousins, H.G., Shell concrete construction. RCA Technical Paper No. 6, 1948, 32.
22. Snow, F.S., Shell concrete construction. Struct. Engr, 1947, 25(7), 265–86.
23. Kirkland, G.W., Goldstein, A., The design and construction of a large span prestressed concrete shell roof. Struct. Engr, 1951, 29, April, 107–27.
24. Sexton, C.G., Prestressed reinforced concrete hanger at the civil airport of Karachi. J. Instn Civ. Engrs, 1947, 29(7), 109–30.
25. Proceedings of a Symposium on Shell Structures. Cement and Concrete Association: London, 1952.
26. Proceedings of the Second Symposium on Shell Structures, Oslo. Cement and Concrete Association: London, 1957.

Concrete shell roofs, 1945–65

Robert Anchor

Synopsis

In the period 1945–65, considerable numbers of reinforced concrete shell roofs were designed and constructed for relatively routine jobs. This chapter considers the commercial background to this activity, and describes the layout of typical structures. The detailing of the concrete structure and the reinforcement is explained and examples of typical completed structures are given.

Introduction

Shell concrete was used before and during the Second World War for 'one-off' roof structures and also for power station cooling towers, but this chapter is concerned with the period between 1945 and 1965 during which shell roofs were used widely as a method of roofing over comparatively routine buildings. In the conditions prevailing in the UK after the Second World War, steel was only available through a rationing system devised by the government. This shortage, together with the need to replace war-damaged buildings, led to the use of concrete shell roofs. A roof designed as a shell would use less steel than the alternative, a steel truss roof. Shell construction provided a bonus, as the roof covering was also created. The marketing of concrete as a material reached unsurpassed heights in this period due to the efforts of the Cement and Concrete Association. Fashion also played a part in design, and no self-respecting architect at this time would be without a shell roof job. Typical uses were roofs for school halls, workshops, canteens, market buildings, garages, factories and swimming pools.

The description which follows is largely based on the author's own experiences as a senior engineer and later as Technical Director of GKN Reinforcements Ltd (formerly Twisteel Reinforcement Ltd) during the period under review. Other information has been obtained from Chronowicz[1] and British Reinforced Concrete Engineering Co. Ltd (BRC).[2] The diagrams are taken mostly from booklets which were printed at the time for sales purposes, both by Twisteel Reinforcement Ltd[3] and by BRC.[2]

The postwar construction industry

It is useful to review the commercial background to the shell era. In the late 1940s, the design and construction of reinforced concrete building structures was organized rather differently from the present. Traditional firms of structural consulting engineers were largely experienced in structural steelwork and masonry design (with notable exceptions in London). Reinforced concrete design for routine jobs was substantially in the hands of specialist commercial firms. Among these firms were the suppliers of steel reinforcement materials (bars and welded fabric) who operated structural design offices. For any particular project, they offered quotations to architects for the design of reinforced concrete structures, together with the supply of the necessary reinforcement. Such competitive quotations from nominated suppliers were incorporated into the bills of quantities as prime cost items. The reinforcement details for the concrete structures were provided by the supplier's office, and such firms became the source of highly experienced reinforced concrete designers and detailers who were in great demand in later years when consulting firms expanded into reinforced concrete design. Twisteel Reinforcement Ltd, which changed its name to GKN Reinforcements (GKNR) around 1955, and BRC were two such major firms, each with a similar organization. The job records of BRC have been destroyed, but a copy of the 13th edition of its sales booklet survives.[2] Examples of their work can be seen in the illustrations to the book by Chronowicz[1] and in the BRC book.[2] There were several other reinforcement firms operating at this time on a more regional basis, but these firms did not design shell roofs in any quantity. Twisteel established a structural design department in 1934 at New Malden, Surrey, and the department was finally closed by GKNR in 1979 when specialism in structural design faded away. In the 1960s, GKNR had works in London, Birmingham, Wigan, Belfast and Glasgow, but also supplied steel directly from the rolling mills in Cardiff. Design offices were located in London, Birmingham, Manchester, Southampton, Belfast, Glasgow, Cardiff and Middlesbrough. At the time of maximum output, the total technical drawing office staff exceeded 300. All the drawings prepared by GKNR for both shells and other jobs are still available on microfiche and have been donated to the ICE archive. There are records showing that in August 1944 (during the Second World War), a total of 25 shell designs had been prepared by C.V. Blumfield Consultants Ltd on behalf of Twisteel. Subsequently, Twisteel traded in the field of shell design as BVR (Barrel Vault Roof) Designs Ltd, and Dr Hajnal-Konyi was retained for some years from 1945 to advise on shell design. He later became a consultant in London and designed further shells on his own account, a particular example being a filling station canopy at Markham Moor on the A1 trunk road.

Twisteel and its successor (henceforth both referred to as GKNR) made available a shell design service throughout the UK but with special expertise in offices in London and Birmingham. In the period between 1948 and 1960, when shell construction was at its peak, it is believed that about 85% of all shell designs were prepared by GKNR and BRC. An estimate of the proportion of shell jobs undertaken by each of the two companies is not possible with any certainty, but, from discussions with those involved at the time, it is probable that GKNR designed about 50% of all shells and BRC about 35%.

From the records, it appears that a total of about 2500 shell schemes were prepared by GKNR. In one period of 6 months during 1949 a total of 212 shell designs were prepared in their offices, but by 1960 activity had reduced considerably. A study of the references to shell structures in the bibliography obtainable from the Institution of Structural Engineers Library shows that the period 1951–70 was the peak for articles, journals and books. There was a total of 179 references in this period. The theory followed the practice.

A second type of expert existed at this time. The general contractors of the period had little expertise in reinforced concrete frames, and the erection of the reinforced concrete structures for a building was often sublet to a firm with specialist experience. Well-known names at this time were Truscon, Laing, F.C. Construction Co., Peter Lind, Christiani & Nielsen, Tileman, Holst, Caxton and others. Again, the specialist contractors often had substantial design organizations, and would tender for the design and construction of reinforced concrete structures.

Shell layouts

The majority of shell roofs designed in the period 1945–65 were of single curvature in the form of part of a symmetrical cylinder or BVR (Figure 9.1). Many jobs featured multi-bay construction and sometimes continuous spans. The span of the shell (i.e. in the linear direction) could be up to 150 ft (45.7 m) and the width in the curved direction was typically half the span providing a column layout with a ratio of 2 : 1 (Figure 9.2). Typical sizes are shown in Table 9.1. These dimensions were chosen, together with a suitable radius of curvature, to avoid any tendency for the shell to buckle under compressive forces at the crown, and to limit the slope of

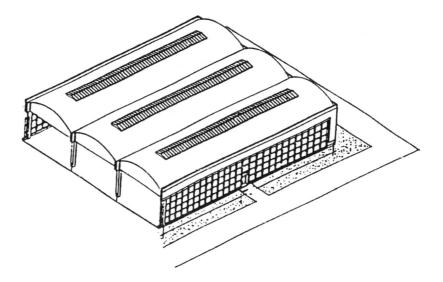

Figure 9.1 Barrel vault roof (BVR).

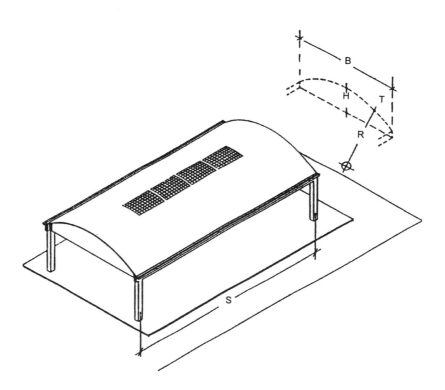

Figure 9.2 Leading dimensions of BVR.

Table 9.1 Typical layout dimensions of BVRs*

Span		Breadth		Rise		Radius		Thickness	
ft	m	ft	m	ft	m	ft	m	in	mm
180	54.9	50	15.2	18	5.5	40	12.2	3.0	76
140	42.7	40	12.2	14	4.3	35	10.7	3.0	76
100	30.5	50	15.2	10.5	3.2	35	10.7	3.0	76
80	24.4	40	12.2	8.25	2.5	35	10.7	3.0	76
60	18.3	30	9.1	6.0	1.8	30	9.1	2.5	63

*The values given in this table are taken from GKNR;[3] BRC recommended very similar values.[2]

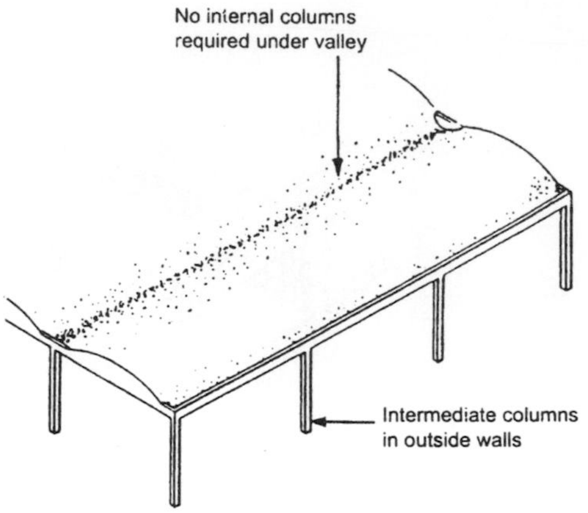

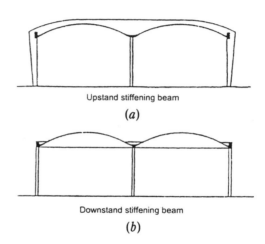

Figure 9.3   Outer bay of barrel roof.

Figure 9.4   (a) Upstand and (b) downstand stiffening beams.

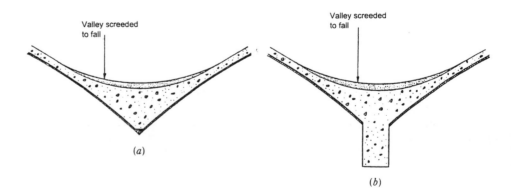

Figure 9.5   Details of: (a) feather edge valley; (b) downstand valley beam.

the upper surface of the shell springing to about 40°; otherwise freshly placed concrete would tend to move down the slope. Intermediate columns were provided under the beams at any external edge in a range of shells (Figure 9.3). This had the effect of preventing undue lateral movement. The chord beams forming the span supports were usually of solid concrete down to a level near to the column heads. Alternatively, a frame with a curved beam could be used (Figure 9.4).

The overall depth of construction from the soffit of the valley beam to the crown of the shell was usually one-tenth of the span. For spans over 100 ft (30.5 m), pre-stressing was sometimes used. The valley details between adjacent bays of shells could be 'feather-edged', flat, or could incorporate a downstand beam (Figure 9.5).

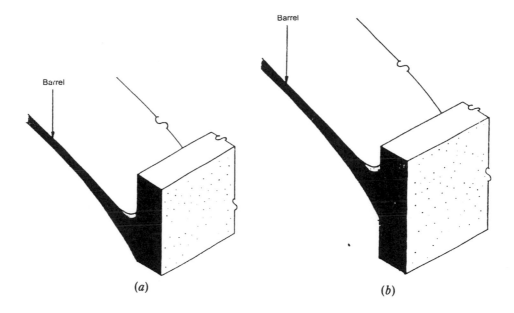

Figure 9.6   (a) Upstand edge beam; (b) dropped edge beam.

(a)   (b)

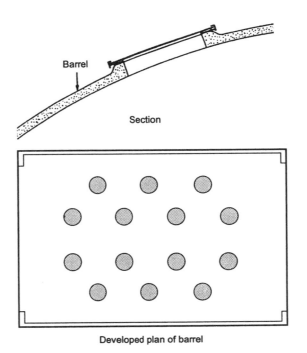

Figure 9.7   BVR showing typical arrangement of roof lights.

Developed plan of barrel

Similarly, the external edge beams could be upstand or downstand with respect to the shell (Figure 9.6). Roof lights were commonly provided in BVR jobs, usually of circular shape (Figure 9.7). It was also possible to provide a continuous rooflight along the crest of each shell (Figure 9.1). In this case, the two halves of the shell were strutted apart between edge thickenings.

A popular second type of singly curved shell was the northlight roof (Figure 9.8). Northlight shells had glazing incorporated on a straight sloping face and shells connecting the top of one slope to the bottom of the next (Figure 9.9). The upper and lower shell edges were propped by reinforced concrete posts at about 10 ft (3.05 m) centres. This arrangement allowed nearly uniform natural lighting in the work-space below. The column layout was usually in the ratio of between 2 : 1

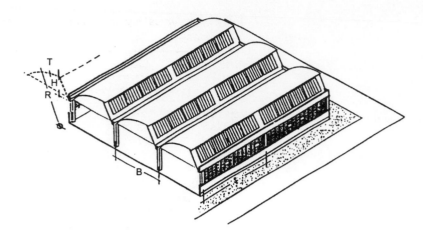

Figure 9.8   Northlight shells.

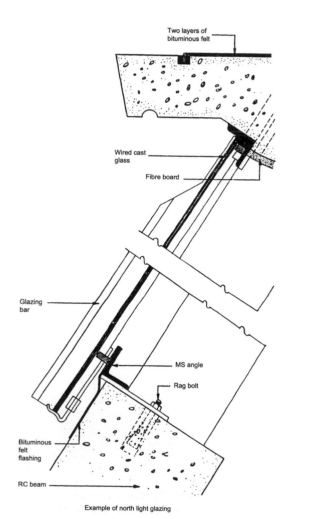

Example of north light glazing

*Figure 9.9   Example of
northlight glazing.*

and 3 : 2, and spans of 50 ft (15.2 m) to 90 ft (27.4 m) were often used. Typical sizes as used by GKNR are given in Table 9.2,[3] and sizes that were recommended by BRC are given in Table 9.3.[2]

A development towards the end of the main shell era was the use of doubly curved shells, and in particular hyperbolic paraboloidal shapes (see Figure 8.5).[4] An advantage of this shape (previously used vertically in power station cooling towers) was that the formwork could be formed with straight timbers laid at an angle

Table 9.2   Typical layout dimensions of northlight roofs, as recommended by Twisteel[3]

Span		Breadth		Rise		Radius		Thickness	
ft	m	ft	m	ft	m	ft	m	in	mm
60	18.3	40	12.2	16	4.9	40	12.2	3.0	76
50	15.2	33	10.1	14	4.3	40	12.2	3.0	76
40	12.2	27	8.2	12	3.7	35	10.7	2.5	63
30	9.1	20	6.1	10	3.0	30	9.1	2.5	63

Table 9.3   Typical layout dimensions of northlight roofs,* as recommended by BRC[2]

Span		Breadth		Radius		Thickness	
ft	m	ft	m	ft	m	in	mm
80	24.4	30	9.1	20	6.1	3.0	76
60	18.3	20	6.1	15	4.6	2.5	63
50	15.2	20	6.1	15	4.6	2.5	63
40	12.2	20	6.1	15	4.6	2.5	63

*The rise is not given as it may depend on the window opening.

to the edges. These shells were used singly, with two low corners and provision for resisting the outward thrust, and also to create a canopy by using three or four shells supported by a central column. Groups of these elements were used typically over petrol filling stations.

Building control

The concrete code of practice in these years was CP 114: 1948; it had no particular provisions for the design of shell roofs. Applications for bye-law approval were dealt with variously. In small towns, the authority would rather not know! In other areas, a certificate that the design was in accordance with CP 114 was required and willingly given. To the author's knowledge, no fire tests were ever made on shell roofs, and under the regulations no period of fire resistance was required; but a shell structure clearly has reasonable resistance to collapse in a fire situation.

Design

The evolution of mathematical shell design has been dealt with by Morice.[4] The bulk of the single curvature shells designed by GKNR used a system evolved by Tottenham[5] and Bennett.[6] Shell designs were originally prepared by a special group in London, but after about 1954, designs were also prepared in the Birmingham office. The design tools were then limited to slide-rules, eight-figure logarithm tables and mechanical calculating machines (which were made for accounting purposes). Shell structures tended to be designed and detailed by the more experienced members of the design team. At BRC, shell designs were largely prepared at the main design office at Stafford under the direction of A.P. (Pop) Mason and A. Chronowicz.[1] The shells designed by both firms were similar, but in detail family resemblances could be noticed.

The thickness of all types of shells was generally 2½ in (63 mm) in order to accommodate the layers of reinforcement with cover of ½ in (13 mm) on each face and to minimize the dead load. An allowance of $15 \, \text{lb/ft}^2$ ($0.75 \, \text{kN/m}^2$) on plan area was made for imposed loading together with a suitable allowance for any finishes. Occasionally the general shell thickness was increased slightly in order to reduce the compressive or shear stresses.

At the valleys along the sides of each bay the shells were thickened to 5 in (127 mm) to assist in resisting the lateral bending moments. The negative lateral bending moment in a shell is a maximum at the valleys and the maximum positive moment is at the crown of the shell. At the end stiffening beams, the shells were thickened to 3½ in (89 mm) on the upper surface for a distance of about one-tenth of the span from the end beams to enable the shear reinforcement and the L-bars connecting to the stiffening beams to be accommodated.

Detailing

Detailing followed the standard procedures of the time, although double elephant sized paper (1016 mm × 686 mm) would not always accommodate the developed plan of a shell. Copying drawings and pages of calculations using dye-line machines was relatively easy and usually part of the new graduate's induction course — after he (there were no girl graduates) had learned how to fold drawings correctly. The reinforcement in a shell generally consisted of two layers of welded mesh placed at the top and bottom faces and with suitable flying ends to enable overlaps to be made without increasing the number of layers to be accommodated. Special weights of welded fabric were available (Tables 9.4 and 9.5) to suit shell construction. Longitudinal ⅜ in (9.5 mm) high-tensile square twisted bars were placed at 12 in (305 mm) centres alternately in the top and bottom faces inside the layers of fabric. This provided steel to resist shrinkage cracking. Due to the shape of the shells, service cracking was not a serious problem. Diagonal bars were placed across each of the four corners of each bay of a shell to resist shear stresses. The main reinforcement resisting the tensile force due to the span bending moment was placed in the downstand beams at the valleys. Cold-worked square twisted bars were used in the earlier shells, with a transition to ribbed cold-worked bars later on. The working tensile stress was either 27,000 lb/in^2 (189 N/mm^2) or 30,000 lb/in^2 (210 N/mm^2). This reinforcement often occupied between three and five layers in depth, and as the beam length (which was the span of the shell) could be 100 ft

Table 9.4   Wireweld high-tensile barrel fabrics

Fabric no.	Size of mesh (in)	Gauge* of longitudinal wires	Gauge* of cross wires	Area of wires per foot width (in^2)		Laps (in)		Diameter of longitudinal wires (in)	Diameter of cross wires (in)
				Longitudinal wires	Cross wires	Longitudinal	Lateral		
216	12 × 12	5	5	0.0353	0.0353	12	12	0.212	0.212
217	6 × 12	6	10	0.0580	0.0129	11	6	0.192	0.128
309	6 × 12	4	9	0.0846	0.0163	14	9	0.232	0.144
428	6 × 12	2	8	0.1196	0.0201	16	9	0.276	0.160

*Birmingham wire gauge.

Table 9.5   Wireweld high-tensile barrel fabrics (metric version of Table 9.4)

Fabric no.	Size of mesh (mm)	Area of wires per metre width (mm^2)		Laps (mm)		Diameter of longitudinal wires (mm)	Diameter of cross wires (mm)
		Longitudinal wires	Cross wires	Longitudinal	Lateral		
216	305 × 305	75	75	305	305	5.4	5.4
217	152 × 305	123	27	280	152	4.9	3.3
309	152 × 305	179	35	356	229	5.9	3.7
428	152 × 305	253	43	406	229	7.0	4.1

(30 m) or more, normal spliced overlaps were not possible. The usual detail was of bars butted together along the span with the positions of the butt joints staggered by at least two lap lengths. Extra bars were added throughout the span to the theoretical number required so that the required full tensile strength could be maintained throughout the beam length.

Shell construction

In the early days, shell concrete construction required a certain amount of courage, and in order to give confidence to clients, architects and contractors, a loading test was carried out by the Building Research Station on a shell roof at Liverpool. The report said *inter alia* that: 'The maximum deflection of the roof at mid-span under an applied load of 150% of the design live load was ⅛ in (3 mm) only. This is about 1/6000 of the span'. The erection of formwork to a cylindrical shape and later to a doubly curved shape was not necessarily the most profitable type of work. However, there was always a feeling of pride (and relief) after satisfactory completion of a job. Removal of the formwork, starting at mid-span, could be a tense moment. Formwork was usually of plywood, but occasionally boards were used, laid on shaped timber bearers in turn supported by metal scaffolding. The vertical tubes were adjusted to height with a screw fixing, and the lateral pre-curved tubes were often referred to as 'banana bars'.

The normal concrete for shell roofs was a $1 : 1\frac{1}{2} : 3$ mix by volume with a required 28-day strength of 3750 lb/in^2 (26 N/mm^2). The actual shell section had ⅜in (9.5 mm) maximum-sized aggregate, and the concrete required a workability which was not too workable at the shell springings where the top surface was at an angle of about 40°. At the thin portion of the shell greater fluidity was necessary in order to work the concrete around the reinforcement in the depth available. Concrete was always mixed on-site, and it was the expertise of the mixer driver which ensured a satisfactory job. The water content of the mix was determined by eye, but the results were surprisingly accurate with an experienced and willing operative. Consolidation of the newly placed concrete was achieved by tamping with timber poles, which also required expertise and dedication. Vibrators were becoming available but could not be used easily on a shell only 2½ in (63 mm) thick.

Shells were usually finished on the interior face with a layer of insulating board which could be painted. For open structures (i.e. without walls) the shell soffit was either painted or left as struck. On the top surface of the roof, two layers of roofing felt were used (Figure 9.10). This provided waterproofing and also allowed some elasticity against shrinkage movement. Cracking in service was not generally a problem, and although carbonation of concrete was not seen as a problem in 1960, the finishes effectively prevented or delayed degradation.

Examples of jobs

Illustrations of actual jobs designed by BRC are given by Chronowicz[1] and in the BRC Handbook.[2] In particular, the Pannier Market at Plymouth, the Cattle Market at Gloucester, and the Terylene Works for ICI at Wilton, Teeside.

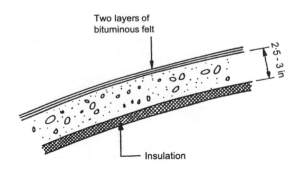

Two layers of bituminous felt

2·5 - 3 in

Insulation

Figure 9.10　Section of shell showing finishes.

The following descriptions are of jobs designed by GKNR in the years 1950–65. This is not to suggest that other firms did not design shell roofs of consequence, but the information available to the author is entirely from his work with this company.

Cadbury Bros., Moreton

In 1951, Cadbury Bros. decided to build a completely new factory at Moreton, Cheshire, to manufacture chocolate biscuits. It was built with concrete shell roof construction and covered an area of 30 acres (12 ha), and is shown in Figure 9.11. The factory buildings were of northlight construction in order to prevent sunlight entering the working area, and the boiler house and the canteen were covered with normal barrels. The shells were generally of about 50 ft (15 m) span and 25 ft (7.6 m) width. H block, which was constructed in 1958, was used for printing card-board packaging. The raw materials were stored at ground level, over which the printing floor was constructed of a 12 in (305 mm) thick flat slab floor spanning between columns at 25 ft (7.6 m) centres each way (Figure 9.12). The client requested a 50 ft (15 m) square layout of columns above first-floor level, to fit in with the machinery layout. This was achieved by providing a series of 50 ft (15 m) span frames, upstanding with respect to the shells. These frames were difficult to analyse with the equipment then available, their reinforcement was difficult to detail, and they were awkward to waterproof. However, the client's requirements were achieved.

In 1962, the last substantial block, for confectionery manufacture, was under construction. It covered an area of 510 ft (155 m) by 470 ft (143 m) and was a single-storey construction. The northlight shells again had a span of about 50 ft (15 m) and a width of 25 ft (7.6 m). Movement joints were introduced by providing double frames either side of the joint. The factory no longer belongs to Cadbury's, but is still in use and in good condition.

Figure 9.11   Aerial view of Cadbury's factory at Moreton, Cheshire, in 1958.

Figure 9.12   Northlight shell roof of H block, Cadbury's factory, Moreton, Cheshire, with column spacing 50 ft × 50 ft (15.2 m × 15.2 m).

Figure 9.13   Barrel vault roofs, Kidderminster Cattle Market.

Kidderminster cattle market

In 1961, a cattle market was constructed at Kidderminster, Worcestershire, and roofed with normal cylindrical barrel shells with a maximum span of 100 ft (30 m), as shown in Figure 9.13. There were three separate buildings. The pig building was roofed with five bays of shells 35 ft (10.7 m) wide and spanning 100 ft (30 m). The cattle building was similar. The market building has five bays 30 ft (9.1 m) wide and is continuous over spans of 65 ft (19.8 m) and 30 ft (9.1 m). The structures still stand in good condition.

Filling station, Harborne, Birmingham

In 1968, a petrol filling station canopy was constructed in Harborne, a suburb of Birmingham. It consisted of three identical hyperbolic paraboloidal structures, each supported by a central column and trimmed with upstanding beams. Each structure obtained support against wind forces by constraints provided by small props between the separated edges. The columns were mutually at 55 ft (16.8 m) centres. The canopy was still in good condition in 1997 when it was demolished in order to change the layout of the forecourt.

Swimming pool, Hatfield

One of the last major jobs, designed in 1966, was a roof to cover a swimming pool at Hatfield, Hertfordshire (Figure 9.14). It was constructed of four hyperbolic paraboloidal shells in a rectangular layout and contained from spreading with prestressing tendons at the level of the heads of four supporting columns. The columns were 2 ft 6 in (762 mm) diameter spaced at 69 ft × 61 ft (21 m × 18.6 m). Shells 1 and 3 were each 90 ft × 61 ft (27.4 m × 18.6 m) and shells 2 and 4 were each 80 ft × 69 ft (24.4 m × 21 m). The pairs of shells were placed so that glazed vertical triangular spaces occurred between them to provide natural lighting. The overall size of the roof was 180 ft × 160 ft (55 m × 49 m). Beams of width 28 in (711 mm) were provided along the edges of each shell, varying in depth from 24 to 12 in (610 to 305 mm). Secondary columns were provided along the external corners to limit vertical deflections. The structure was designed for the following loads:

Imposed load	15	0.75
Finishes	12	0.60
3-in Lightweight shell	25	1.25
	52 lb/ft^2	2.60 kN/m^2

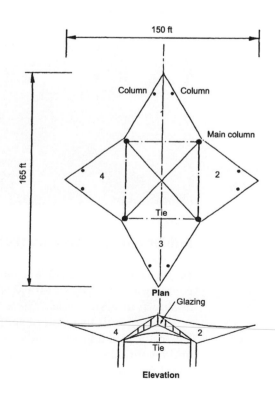

Figure 9.14   Details of shell roof, swimming pool, Hatfield.

Lightweight concrete was used for the shells in order to reduce the dead load and also to improve the insulation. A paper on this project was presented to a Symposium of the International Association for Shell Structures at Budapest in 1965.[7] The job is particularly interesting in that it has recently been refurbished (1996). During this work, it was found that the grouted prestressing cables had corroded and they had to be replaced. However, the shells and supporting structure were in a satisfactory condition and were retained.

Conclusion

Concrete shell construction in the UK ccased on any scale around 1965 when structural steel had become readily available, formwork costs had increased and architectural fashion had moved on. The shells described above have survived 35 years, and as far as the author is aware, have not required undue maintainance. Perhaps experience has a greater part to play in successful design than is presently acknowledged.

References

1. Chronowicz, A., The Design of Shells — A Practical Approach. Crosby Lockwood, 1959.
2. British Reinforced Concrete Engineering Co. Ltd., BRC Reinforcements, 13th edn. BRC: London, 1959.
3. Twisteel Reinforcement Ltd., Barrel Vault Roofing. Twisteel: London, 1953.
4. See Chapter 8.
5. Tottenham, H., A simplified method of design for cylindrical shell roofs. Struct. Engr, 1954, 32(6), 161–80.
6. Bennett, J.D., Some Recent Developments in the Design of Reinforced Concrete Shell Roofs. Reinforced Concrete Association: London, 1958.
7. Anchor, R.D., Shell Roof at Hatfield. Presented in Budapest at Int. Symp. Shell Structs, 1965. Struct. Concr., 1966, 3(2), 99–104.

10 Prestressing

Francis Walley

Synopsis   This chapter briefly reviews the early history of prestressed concrete. It discusses the materials used and the systems of pre- and post-tensioning in use, particularly in the early days. It then describes its uses in the building field and the type of structures in which it is likely to be found.

The early years   The subject of prestressing covers a relatively new process whose inventor and chief protagonist, Eugene Freyssinet, only died in 1962. Indeed, the major innovations in prestressed concrete and the notable structures that have resulted, have come about in the last 50 years.[1-4] To put the subject into perspective, it is necessary to start with the ideas and threads of knowledge that existed long before the successful applications of prestressed concrete.

In the first decades of this century, flexural cracks in reinforced concrete played a prominent part in the history of reinforced concrete in Germany.[5] In 1906 a treatise by Labes[6] laid down the following requirement for structures on the Prussian railways: flexural tensile stresses should offer a safety factor of 1.5–2.5 against cracking measured as direct tensile stress. This requirement effectively precluded the use of reinforced concrete for slabs and beams. At the same time, of course, it provided engineers the stimulus of an obstacle that had to be surmounted. The next year, in the same periodical in which Labes had laid down his requirement, Koenen[7] made, for the first time, the practical suggestion of giving the tension zone a preliminary compressive stress by putting steel rods into tension before concreting, using a stretching device that was removed after the concrete had hardened.

As a result of this suggestion, experiments were put in hand by a German Reinforced Concrete Research Committee and published before the First World War. In these experiments, two bars of 18 mm dia. were stressed and held stressed for 45 days up to the date of the test. These clearly showed that cracking was delayed as compared with similar beams with unstressed rods, that is, an artificial tensile strength had been given to the concrete. There was, of course, no increase in ultimate load and those beams that were prestressed but without mechanically anchored bars failed at about 70% of those with anchored bars. This result has often been observed subsequently when bond failure occurs.

This early work was not followed up for various reasons. First, no one could think of a sensible method of introducing the prestressing force and in particular the bent-up bars that Mörsch[8,9] had shown in 1907 to be necessary to resist shear. Second, Labes,[10] in the German Committee for Reinforced Concrete set up in 1907, modified his requirement by substituting the bending tensile stress for the direct tensile stress of concrete. Third, and more important as time went on, it was realized that calculations about factors of safety against cracking were largely illusory since the effect of shrinkage, particularly in the case of heavily reinforced beams, was of paramount importance. Indeed, to compensate for the shrinkage, high-tensile stresses, and therefore high-tensile steel, would have been necessary, and it was not available except at very high cost.

It was also found that, in some of the test beams kept for a considerable time, the prestressing no longer affected the shape of the load–deflection curve, which caused some puzzlement at the time before shrinkage and subsequently creep became better-known phenomena.

Economies were sought by using higher-strength steel and higher stresses. Unfortunately, although the elastic limit can be increased, Young's modulus cannot, so that stresses in the steel have to be limited to avoid large cracks in concrete. However, looking back historically, it was Freyssinet's realization that creep and shrinkage had to be overcome that enabled him to patent prestressing in 1928:

> At that time, so heavy was the intellectual oppression exercised by a handful of mathematicians, obsessed with their science and blind to reality, that an unquestioned belief in the constancy of Young's modulus for concrete was held, without a valid basis and in spite of many proofs to the contrary, by all our professors and indeed by all technicians.
>
> As a consequence, the properties of materials seemed to offer no difficulty whatever to the development of my idea — in fact, quite the contrary, since the fundamental contradiction between the elastic strains of steel and concrete in reinforced concrete appeared all the more startling, and the basic theory of this method of construction all the more absurd; from my very first acquaintance with this material it seemed a very temporary stage in the technique of associating steel with concrete.[11]

He had realized or suspected early on in his bridge over the Allier, in the first decade of this century, when he used prestressed ties in arch bridges, the importance of shrinkage and creep, but without any figures as to their ultimate values it was impossible to proceed to a definite conclusion.

By 1928 work had proceeded far at the Building Research Station on this subject, by Glanville and Faber in particular, which led to the publication of the famous Technical Paper Nos 11 and 12. This work confirmed Freyssinet's ideas. He realized that, if he used really high stresses in steel and compressed the concrete as high as possible, there would be sufficient prestress left in the concrete, after creep and shrinkage had taken place, to avoid any cracking under service loads; relaxation of steel was not yet thought of. The reinforcement was no longer passive, but active, so that Young's modulus was not a stumbling block. All earlier work tacitly assumed that the prestressing of beams would be done by rods and they would be straight, and hence some considerable attention was paid to vertical stressing of stirrups to reduce principle tensile stresses. Indeed, these were part of Freyssinet's patent. Although, in an addendum to the patent, anchoring by bond was included; this was not a process adopted until exploited by Hoyer, although Freyssinet had used it for powerline pylons in the 1930s.

The tacit assumption of using straight bars resulted in the need to use vertical prestressing in beams and also, where possible, the shaping of the longitudinal profile of a beam to a parabolic form to avoid high-tensile stresses in the upper fibres at the ends of the beams. There were devices described by Mörsch to stress additional short bars over the centre of spans, but they were clumsy.

The use of steel wire to form the tendons came in during the early years of the Second World War, when the Freyssinet jack was invented to anchor simultaneously 12 wires first of 5 mm and then 7 mm dia. Magnel developed a two-wire stressing system. These gave an unprecedented degree of design freedom.

One other major contribution was his research into the making of high-quality concrete — a subject that Freyssinet pursued up to the time of his death.

Wayss and Freytag were the licencees for Freyssinet's patents in Germany and, with the close tie-up between universities and industry, it was possible to integrate research and development work. More importantly, there was a programme of work for building autobahns so that prestressing could be applied to autobahn bridges as early as 1938, and some 10 km of bridge beams were made. During the war years, it became available for the construction of defence works and an enormous number of beams were used for the building of the U-boat pens.

The one breakthrough in 1934 that thrust prestressed concrete into the limelight was the maritime railway station at Le Havre. This was settling and in danger of being demolished when Freyssinet said he could save it using prestressed concrete. The choice was difficult — allow it to collapse or try this unproven method. The latter was chosen, and, as Freyssinet said, 'it had the good fortune to succeed. The risk had been great but I had broken the vicious circle which encloses all innovators.'

Early post-war use of prestressing

At the end of the war, German cities were virtually totally destroyed, and little was done for many years except clear up. In France the great need was for restoration of communications. In the UK the priority was the rebuilding of cities and providing accommodation. In France, therefore, the priority was bridges and five bridges over the Marne made up of precast units stressed together, were constructed in 1946. Figure 10.1 shows a prestressed water reservoir at Tours with offices underneath built in the late 1940s.

In Belgium, Professor Magnel was pioneering in both bridges and buildings. Figure 10.2 illustrates a large hangar at Moelsbruke with a 300 t beam. By 1949 he could boast of some 50 jobs constructed.

In the UK, of course, there were enthusiasts. Mautner, who had escaped from the Nazis in Germany, was a devotee of the Freyssinet system and did much for the subject. There was, however, a curious unease, a suspicion of other people, of infringement of patents (there were some 20 patents). It is difficult these days to describe the emotions, almost passions, that were raised. Although the Ministry of War Transport had had designed (by Mautner) some precast, prestressed beams for stockpiling to replace bridges during the war — these were never used until years later — no development on bridge works took place because there was no road programme for many years.

Figure 10.1   Water tower at Tours with offices below.

Figure 10.2   Hangar at Moelsbruke.

Figure 10.3   Three-bay single-storey workshop.

The major development work took place in the Chief Scientific Adviser's Division (later the Chief Development Engineering Division) of the Ministry of Works. It had its own Development Station, independent then of BRS where it pressed ahead with construction and testing of various units. It let development contracts with contractors (Costains) at Childerditch, Essex. It mounted exhibitions and provided lectures all over the UK. At the same time it selected certain buildings to demonstrate the use of prestressed concrete. Sighthill Stationery Office was a very early example (1948). This was followed by Kilburn Telephone Exchange (both being fully framed structures), numerous single-storey industrial buildings (Figure 10.3) and the providing of what became the temporary office building programme (Figure 10.4). Another notable example is the hangar at

Figure 10.4   Temporary office building.

London Airport designed by the Prestressed Concrete Co. (see Figure 2.31, p. 35).[16]

It is interesting to note that the annual UK production of prestressed flooring units rose from zero in 1950 to 1.3 million square yards in 1960.

In tackling any history, it is difficult at certain points to decide which way to proceed — whether to pick out salient events in succeeding years or to deal with various technical aspects historically. The latter has been chosen as being less confusing. The materials already mentioned hold the keys.

Steel

Although high-tensile steel wire is said to have been available as early as 1908, most of the early work used high-tensile steel of up to 14 mm dia., anchored using wedges in steel plates. The wire before 1940 had a high-tensile strength but a very variable elastic limit. This latter could be 'stabilized' if the wire was repeatedly stretched and relaxed, and indeed Freyssinet devised a method of doing this semi-continuously.

By 1940 high-tensile drawn wire was available. This was a high-carbon steel. In the UK it was 'patented' and drawn through dies giving a total reduction of 70% of area. This was wound off on small coils and gave no special difficulty provided that it was used for pre-tensioned work. Although Freyssinet had patented prestressing by bond in his addendum to his patent in November 1928, Hoyer in Germany succeeded in obtaining a patent for a long-line system of prestressing provided that his wires did not exceed 2 mm dia. and were not stressed more than 50%. For this type of work and this wire diameter, small coils were not a problem. Site work was a different matter. In the early days one of the first 'tools' or pieces of machinery to come on site was a wire straightening machine, without which it was impossible to make up tendons, as the wires always wanted to return to their small coils. Even as late as 1955, BS 2691 had the phrase 'when so required by the purchaser wire shall be supplied in coils of sufficiently large diameter'. It was with the next revision that large-diameter coils that paid out straight became the norm.

The production of steel for prestressing was a highly cooperative affair, and one that can be looked back on with some pleasure. After some initial hesitation as to whether prestressing was a 'flash in the pan', there was an enthusiastic response from wire drawers, and a creative dialogue took place between the designers and manufacturers, leading through heat-treated wire on large-diameter coils, to low-relaxation steels, and then to strands.

The relaxation or creep of wire came into prominence as a result of work by Professor Magnel at Ghent University in 1946–47.[12,13] It was an aspect of prestressing that Freyssinet, I feel, missed. Indeed, Freyssinet, as late as 1949 in his lecture to the ICE already quoted from,[11] played it down, saying that it was not important. One of the points to which Magnel drew attention was that, by tensioning a wire, releasing and tensioning again, the relaxation was reduced, and on early jobs this was written into the specification. However, since Continental European steel was made in a different way from British steel, some experimental work on relaxation was put in hand at the Ministry of Works experimental station at Thatched Barn in 1950. These gave values for British steel that could be used in design. They also showed that wire that had been heat-treated and supplied in large coils had a lower relaxation and that stressing and destressing did not really affect the relaxation. Consequently, this part of the specification was omitted. Work was continued by BRE, and since that time the effect of both low and high temperatures and radiation has been investigated in many places.

Although wire has been referred to at length, this history would be incomplete if it did not refer to Lee–McCall bars which, once rolled threads had been introduced, gave little trouble. In the early days of cut threads the lengths of bars had to be calculated precisely and the position of the nut was critical if one did not wish to risk a snapped bar. As the load required for tendons increased, attention turned to strand, and one of the first major uses was the bridge in Perth in 1959. To give some idea of the way in which prestressing increased in use, the weight of prestressing wire used increased from 4000 t/year in 1950 to 40,000 t/year in 1960.

Concrete

The advent of prestressed concrete made it advantageous to use high-strength concrete. Whereas in reinforced concrete it is not easy to take full advantage of high-strength concrete, in prestressed concrete there is almost no limit to its use. In the early days virtually the sole criterion was to obtain an adequate factor of safety.

It was not long, however, before its other properties of low creep and shrinkage were also appreciated. There is no doubt that prestressing provided a stimulus to the production of better and stronger concrete.

Pre-tensioned concrete

In the case of pre-tensioned work the stress in the steel has to be transferred directly to the concrete. In the early years (the late 1940s and early 1950s), following German practice, 2 mm dia. wire was considered to be the maximum that should be used. (The early German rule that pre-tensioned units should not be less than 3 m long or used for rolling loads was not adopted.) For this reason the *Stahlsaitenbeton* in Germany, with its multitude of small wires, became the *strangbetong* in Sweden. Eventually similar pre-tensioning beds were installed in Britain at Iver by Holland, Hannen and Cubitt. It was considered that the bond would not be good enough if a larger-diameter wire was used. However, people began to use 5 mm dia. wire, first with crimps and indentations and then, when these proved satisfactory, with plain wire. In the late 1950s, the Cement and Concrete Association carried out research work to determine transmission lengths, which were available to the Code Committee. By 1962, 0.7 in (17.8 mm) dia. strand was being used with a transmission length of 500 mm.

Stressing systems (post-tensioning)

The earliest stressing system to be used in Britain was the Freyssinet system which used a double-acting jack. The first action of the jack was to tension eight or 12 wires. The second action was to drive home a cone which consisted of a steel tube surrounded by high-alumina cement which locked the wires into the anchorage. The tube was used for injecting grout after stressing had been completed. Although appearing crude, it was an effective anchoring device. The smaller anchorages could receive eight, 10 or 12 wires of 0.2 in dia., but there were larger ones which could anchor twelve 0.276 in dia. wires and even twelve 0.315 in wires. These are shown in Figure 10.5, and a typical longitudinal section at an end in Figure 10.6. Figures 10.7–10.9 show a typical anchor block and cone, the spring spacer and a Freyssinet jack.

These were imported into the UK by the Prestressed Concrete Co., which set up PSC Equipment Ltd which itself developed systems largely using single wire stressing but had available a two-wire system which was used for the Intergrid school system. Table 10.1 shows the systems available in the 1960s.

The Magnel–Blaton system (invented by Professor Magnel of Ghent) came shortly afterwards and depended on stressing two wires at a time but building them up into quite large anchorages. Each anchorage was capable of receiving eight wires but could be built up to eight in number, giving 64 wires. Originally only 0.20 in dia. wires were used but the anchorages were later modified to receive 0.276 in dia. wires. The distribution plate to transfer the load to the concrete was of cast steel and the largest (64 × 0.276) weighed 95 lb. Details of this anchorage are shown in Figures 10.10 and 10.11. This system was eventually marketed in Britain by Stressed Concrete Design, which produced its own anchorage systems (Table 10.2).

The BBRV system was also introduced from Switzerland. In this system for post-tensioning, the wires had 'buttons' formed at their ends to anchor them to a steel anchorage device. This was not extensively used in the UK.

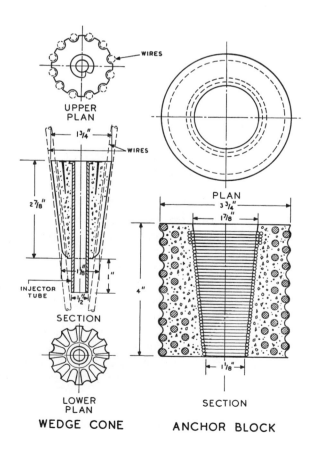

Figure 10.5 Details of Freyssinet anchorages.

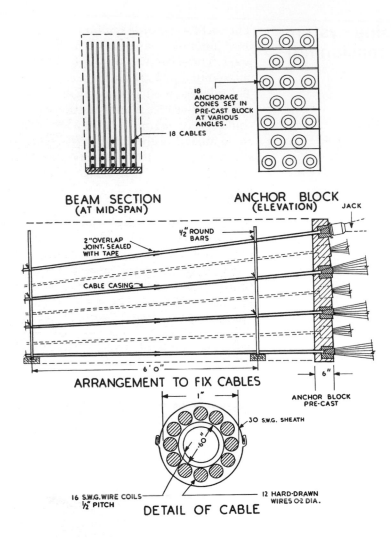

*Figure 10.6 Typical
arrangement of Freyssinet
cables and anchorages.*

*Figure 10.7 Freyssinet
anchorage block and cone.*

Figure 10.8 Freyssinet cables, showing spring spacer.

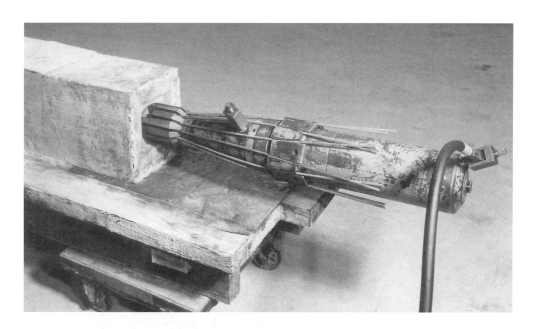

Figure 10.9 Freyssinet jack.

The Lee–McCall (Macalloy) system was developed by Donovan Lee in association with McCall & Co. Ltd of Sheffield in the early 1950s, and comprised an alloy steel bar with threaded ends, with a nut which transferred the force in the rod to a distribution plate at the end of the beam. In the early years the thread was cut, but later, after isolated failures of bars, was rolled on. Details of earlier anchorages are given in Table 10.3.

The CCL system developed from a small grip they produced for the end anchorage in a pre-tensioning system used by the Ministry of Works was used initially in the Gifford–Udall–CCL systems for post-tensioning. These broke down into the CCL system (Table 10.4) and the Gifford–Udall and Gifford–Burrow systems (Table 10.5).

Table 10.1 The Freyssinet/PSC system

Type	Initial force (lb × 10³)	Number and size (in) of wires or strand	Duct size (in)	Anchorage size (in)	Trade description	Jack type	Jack extension (in)	Retracted jack length (in)	Maximum diameter (in)	Jack weight (lb)
Single wire	5.0	1 × 0.200		1 7/32 × 1 7/32	Mono-wire	PSC Mono-wire	4 or 6	13½–17¼	3½	7–9
	9.4	1 × 0.276		1 7/32 × 1 7/32						
	9.9	2 × 0.200	⅝ dia.	1 3/16 × 2⅜						
	18.8	2 × 0.276	⅝ dia.	1 3/16 × 2⅜						
	19.8	4 × 0.200	¾ × ¾	2 13/32 × 2 13/32						
	37.4	4 × 0.276	¾ × ¾	2 13/32 × 2 13/32						
	75.2	8 × 0.276	1 9/16 dia.	4¼ × 4½						
	112.8	12 × 0.276	2 dia.	5 × 5						
12 wire	59.2	12 × 0.200	1 3/16 dia.	3¾ dia.	Freyssinet Multi-wire	Freyssinet Multi-wire (1953)	10–12	30	9	109
	112.8	12 × 0.276	2 dia.	4¾ dia.	Freyssinet Multi-wire	Freyssinet Multi-wire (1953)	10–12	30	9	109
	146.6	12 × 0.315	2 dia.	6 dia.	Freyssinet	Multi-wire	8	30	9	
Single strand	10.85	1 × 5/16	⅝ dia.		Mono-strand	Mono-strand				
	14.7	1 × ⅜	⅝ dia.							
	19.6	1 × 7/16	⅝ dia.							
	25.8	1 × ½	⅝ dia.							
	58.0	1 × 0.7	1 3/16 dia.	4¾ (int. anch)	Freyssinet strand		8	28	8	200
	76.0	1 × ⅞	1 3/16 dia.	5⅝ (ext. anch)						
		1 × 1	1⅝ dia.							
	129.0	1 × 1⅛	1⅝ dia.							
	103.0	7 × ⅜	2 dia.		Mono-strand	Mono-strand				
	137.0	7 × 7/16	2 dia.							
	181.0	7 × ½	2 dia.							
Multi-strand	310.0	12 × ½	2¾ dia.	8¼ dia.	Freyssinet multi-strand		11¾	38	12	270
	428.0	12 × 0.6	2¾ dia.							

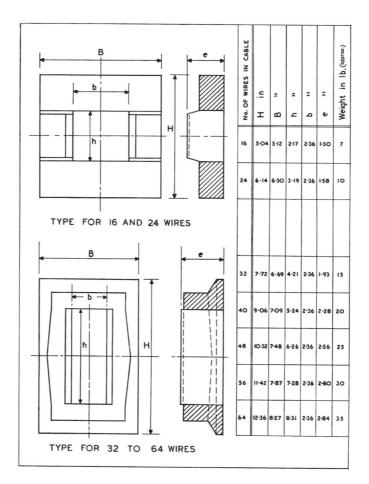

No. OF WIRES IN CABLE	H in	B "	h "	b "	e "	Weight in lb. (approx.)
16	5·04	5·12	2·17	2·36	1·50	7
24	6·14	6·30	3·19	2·36	1·58	10
32	7·72	6·69	4·21	2·36	1·93	15
40	9·06	7·09	5·24	2·36	2·28	20
48	10·32	7·48	6·26	2·36	2·56	25
56	11·42	7·87	7·28	2·36	2·80	30
64	12·36	8·27	8·31	2·36	2·84	35

TYPE FOR 16 AND 24 WIRES

TYPE FOR 32 TO 64 WIRES

Figure 10.10 Distribution plate details for 0.2 in dia. wires (Magnel–Blaton system).

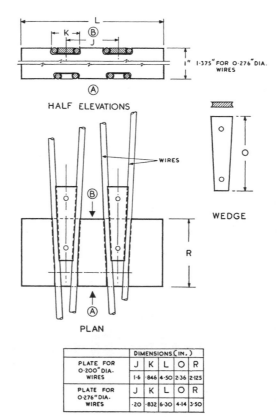

HALF ELEVATIONS

WEDGE

PLAN

	DIMENSIONS (IN.)				
PLATE FOR 0·200" DIA. WIRES	J	K	L	O	R
	1·6	·846	4·50	2·36	2·125
PLATE FOR 0·276" DIA. WIRES	J	K	L	O	R
	·20	·832	6·30	4·14	3·50

Figure 10.11 Magnel–Blaton sandwich plate and wedges.

Table 10.2 The Magnel–Blaton system

Type	Initial force (lb × 10^3)	Number and size (in) of wires or strand	Duct size (in)	Anchorage size (in)	Trade description	Jack type	Jack extension (in)	Retracted jack length (in)	Maximum diameter (in)	Jack weight (lb)
Double wire	19.7	4 × 0.200	2½ × 1	2¼ × 3⅛	Magnel–Balton system	Mark I	10	44	6⅞	70
	29.6	6 × 0.200	1½ × 1½	3 × 3½		Mark II 'S'	10	53	8⅞	112
	39.4	8 × 0.200	2 × 2⅛	3¾ × 3⅝		Mark II 'L'	18	69	8⅞	168
	78.8	16 × 0.200	2.16 × 1.97	5 × 5.1						
	118.2	24 × 0.200	2.16 × 2.95	6.15 × 6.3						
	157.6	32 × 0.200	2.16 × 3.94	7.73 × 6.7						
	197.0	40 × 0.200	2.16 × 4.92	9 × 7.1						
	236.0	48 × 0.200	2.16 × 5.90	10.3 × 7.5						
	277.8	56 × 0.200	2.16 × 6.89	11.4 × 7.9						
	317.2	64 × 0.200	2.16 × 7.87	12.4 × 8.3						
	37.6	4 × 0.276	1½ × 1½	3 × 4						
	56.4	6 × 0.276	2½ × 1¼	4 × 4¾		Mark II 'S'	10	53	8⅞	112
	75.2	8 × 0.276	2½ × 1¼	5 × 4¾		Mark II 'L'	18	69	8⅞	168
	150.4	16 × 0.276	2½ × 2½	5¾ × 8						
	225.6	24 × 0.276	3½ × 2½	7¼ × 10¼						
	300.8	32 × 0.276	5 × 2½	9½ × 10¼						
	376.0	40 × 0.276	6 × 2½	11½ × 10⅝						
	451.2	48 × 0.276	7 × 2½	13½ × 8⅞						
	526.4	56 × 0.276	7½ × 2½	14½ × 11½						
	601.6	64 × 0.276	8½ × 2½	15½ × 11¾						
	676.8	72 × 0.276	9½ × 2½	17½ × 12½						
	902.4	96 × 0.276	12½ × 2½	23½ × 12½						
Single ½ in strand	77.4	3 × 0.5	1¼ × 2½	7 × 3½	Magnel–Blaton stress-block system		12	28		37
	154.8	6 × 0.5	2½ × 2½	8 × 6			20	36½		47
	232.2	9 × 0.5	2½ × 2½	8½ × 8½						
	309.6	12 × 0.5	3½ × 2½	9 × 11						
	387.0	15 × 0.5	5 × 2½	9½ × 13¼						
	464.4	18 × 0.5	5 × 2½	9½ × 15½						
	541.8	21 × 0.5	6 × 2½	10 × 17½						
	619.2	24 × 0.5	6 × 2½	10¼ × 19¾						
	696.6	27 × 0.5	7 × 2½	10½ × 21¾						

Table 10.3   The Lee–McCall (Macalloy) system

Type	Initial force (lb $\times 10^3$)	Number and size (in) of bar	Duct size (in)	Anchorage size (in)	Trade description	Jack type	Jack extension (in)	Retracted jack length (in)	Maximum diameter (in)
Bar threaded	60.5	1 × ⅞	1½	5 × 5	Macalloy	Mark VIII	6	27	7½
	78.5	1 × 1	1½	5 × 5					
	101.0	1 × 1⅛	1½	5½ × 5½ (6 × 6)*		Mark VII	12	36	8¾
	123.0	1 × 1¼	1⅝	7 × 6 (7 × 7)*					

*Using wedges.

Ducts and grouting

Post-tensioning implies that the tendons are stressed after the concrete has hardened and therefore have to be free to move. Spacers were inserted if the tendon was not stressed as a whole since the stressing of, say, two wires out of 12 could lock the whole tendon. Sufficient clearances also had to be left to allow grout to penetrate the whole length of the duct. Grouting was deemed necessary to protect the tendon from corrosion and also to enhance the ultimate failing load of the beam. The ultimate strength of an ungrouted beam was usually of the order of 70% of a fully grouted one. The complete grouting of tendons has not always been successful, even if apparently every care has been taken. This has not led to any great troubles except where units have been exposed to weather, and in particular to aggressive conditions such as deicing salts and often in association with bad filling in joints in segmental construction. Where units have been 'protected' and subsequently demolished, the prestressing steel has been found in a pristine state.

Pre-tensioning as a method of stressing is not discussed at length in this chapter (units stressed by this method will be distinguished by having no end anchorages), it is interesting to record that the CCL system, developed by modifying one of their grips for joining cables to produce a small grip for the end anchorage in a pre-tensioning system used by the Ministry of Works.

Stressing of circular tanks

Nearly all of the above systems have been modified by using opposed anchorages to stress circular tanks. The only other system used was the American pre-load system which depended on stressing a continuous wire by passing it through a die which stressed the wire and reduced its diameter. In these cases no anchorages will generally be found except at the top and bottom of the tank.[16]

Prestressed concrete units

This chapter deals only with prestressed units likely to be found in building and general civil engineering.

Prestressing, by post-tensioning, was used for many shell structures, particularly in edge beams but occasionally in the shell itself, and these are discussed in Chapters 8 and 9. Prestressing both by pre- and post-tensioning has made a major contribution to bridge building, and details of these are to be found in Chapter 12. Prestressing as applied to maritime structures is to be found in Chapter 13.

By far the major contribution to the building programme has been in its use for flooring units, and these almost without exception have been produced by pre-tensioning. In the early days (post-war) the incentive was undoubtedly the fact that prestressing wire, unlike structural steel and reinforcement, was not rationed, but the units proved to be very reliable and trouble-free. The fact that in general no tension was allowed in the concrete meant that they were crack-free and corrosion of the wires was not a problem.

Table 10.4 The CCL system

Type	Initial force (lb × 10^3)	Number and size (in) of wires or strand	Duct size (in)	Anchorage size (in)	Trade description	Jack type	Jack extension (in)	Retracted jack length (in)	Maximum diameter (in)	Jack weight (lb)
Single wire	9.4	1 × 0.276		2 × 2	CCL compact	Mark I wire	5	16⅛	2⅝	14
	37.4	4 × 0.276	¾ or 1½	3 × 4	plate system		10	21⅛	2⅝	18
	75.2	8 × 0.276	1½	4 × 5			15	26⅛	2⅝	22
	112.8	12 × 0.276	1⅝	5 × 5½			5	20¾	2⅜	18
	37.4	4 × 0.276	¾ or 1½	2½ × 2½ × 3¼	CCL spiral	Mark II wire	10	26¾	2⅜	24
	75.2	8 × 0.276	1½	3 × 5½ × 7	wire system		15	32¾	2⅜	30
	112.8	12 × 0.276	1⅝ or 2	4¾ × 4¾ × 7			20	38¾	2⅜	36
	14.7	1 × ⅜	¾	2¼ × 2¼	CCL standard	Mark II small	10	30⅜	3¾	72
	25.8	1 × ½	¾	3 × 3	strand system	strand	20	40⅜	3¾	97
	58.8	1 × 0.7	1½	3½ × 3½						
	76.0	1 × ⅞	1½	4 × 4		Mark II Large	15	3½	10½	380
	102.0	1 × 1	1½	4½ × 4½		strand				
	129.0	1 × 1⅛	1½	5½ × 5½						
Single strand	58.0	1 × 0.7	1½	4½ × 4½ × 7	CCL spiral	Mark II large	8	19 3/16	7⅝	95
	76.0	1 × ⅞	1½	4½ × 4½ × 7	strand system	strand				
	102.0	1 × 1	1½	4½ × 4½ × 7						
	129.0	1 × 1⅛	1½	4½ × 4½ × 7						
	67.4	3 × ½	2	4½ × 4½ × 7	CCL 3-strand	Mark II small	10	30⅜	3¾	72
					spiral	strand	20	40⅜	3¾	97
	103.2	4 × ½	2	4½ × 4½ × 7	CCL 4-strand	Mark II large	8	19 3/16	7⅝	95
					spiral	strand				
	137.00	7 × 7/16	2	6 × 8	CCL 7-strand					
	181.0	7 × ½	2	6 × 8	system					
	250.0	7 × 0.6	3¼	9 × 9						
	406.0	7 × 0.7	3¼	10 × 10						

Table 10.5   The Gifford–Udall and Gifford–Burrow systems

Type	Initial force (lb × 10³)	Number and size (in) of wires or strand	Duct size (in)	Anchorage size (in²)	Trade description	Jack type	Jack extension (in)	Retracted jack length (in)	Maximum diameter (in)	Jack weight (lb)
Single wire	75.0	8 × 0.276	1½	3½	Gifford–Udall	Tube	5	14½	2¼	10
	112.8	12 × 0.276	2	4			10	19½	2¼	13
	124.0	8 × 0.354	2	4						
	146.6	12 × 0.315	2	6						
	186.0	12 × 0.354	2	6						
	197.0	21 × 0.276	2	6						
	269.0	22 × 0.315	2	8½						
	341.0	22 × 0.354	2½	8½						
	357.0	38 × 0.276	2¾	8½						
Multi-strand	181.0	7 × ½	2	6	Gifford–Burrow	0.5 Small strand	20	34	3	
	224.0	4 × 0.7	2	6			12	22	3	
	249.0	7 × 0.6	2	6						
	310.0	12 × 0.5	2¾	8½		0.7 Small strand	10	18½	4¼	
	392.0	7 × 0.7	2¾	8½						
	429.0	12 × 0.6	2¾	8½						
Large strand	58.0	1 × 0.7	1½	4	Gifford–Burrow	0.7 Small strand				
	58.0	1 × 0.7	1½	5						
	102.0	1 × 1.0	1½	4						
	102.0	1 × 1.0	1½	5		Large strand	10	24½	6½	96
	129.0	1 × 1⅛	1½	4						
	129.0	1 × 1⅛	1½	5						

Figure 10.12   Combined vertical and horizontal spacer (Magnel–Blaton system).

The use of prestressed concrete for railway sleepers started during the war and has continued ever since. It is estimated that over 35 million have been made in the UK alone. In general these have been made in dedicated factories but often, alongside, additional beds were laid down to produce beams and flooring units; flooring units tended after a while to be cast in dedicated factories. Great emphasis was placed in the early days on producing the lightest units possible, and Bison produced a prestressed equivalent of its standard hollow-core unit; outwardly it could be extremely difficult to say whether a particular unit was made of prestressed or normal reinforced concrete, unless one had access to its ends.

Another unit was an inverted trough which was more often used as a roofing unit. This was originally developed by the Ministry of Works but subsequently became the Milbank floor. It would have been difficult to make such a unit in reinforced concrete (Figure 10.13).

A popular unit was the X-joist which was originally created as a purlin for farm buildings, again by the Ministry of Works. This was taken up by the French particularly, in the first instance, for the rebuilding of Caen. The shape lent itself to the use of steel pipes for the side moulds which could be used for steam curing, thus getting a quick turn-around on the stressing beds. This was reintroduced into the UK to become the Pierhead floor. Unfortunately, back in the UK the pipes were not used for steam curing and, after a time of using normal cement, high-alumina cement was used to achieve a quicker turnaround in the moulds until the high-alumina scare of the 1970s. But undoubtedly many floors constructed with these joists will still be found. These were used with hollow pots, usually of concrete (Figure 10.14). One of the difficulties with all these units was the differential camber which occurred between them. Although strict quality control was in place, some units hogged more than others under prestress, partly because the 'small' section modulus proved sensitive to this effect and partly because early release from the pre-tensioning beds was an economic necessity.

This problem of differential camber was, in one sense, overcome by what was known as the 'Stahlton floor' which originated in Switzerland in 1945. Originally it was made up of discrete clay tiles grooved on the top. Pre-tensioned wires ran

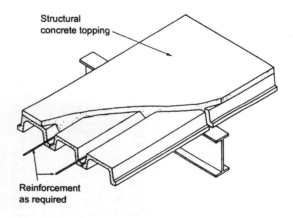

Figure 10.13   The Milbank inverted trough prestressed unit (shown in a composite floor). Reproduced by permission of the British Constructional Steelwork Association Limited.

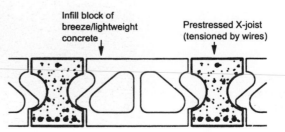

Figure 10.14   Typical prestressed X-joists.

in the grooves and mortar was placed between the ends of the tiles and in the grooves to achieve bond; being uniformly prestressed and flexible, they did not suffer from the problem of differential camber. These planks were supported at 5 ft centres and hollow pots were inserted between them. *In-situ* concrete and any additional reinforcement was then placed (Figure 10.15). These were introduced into the UK by the Costain Concrete Co. in 1952. In Scotland, and later in England, the planks were made in concrete so it would not be easy to recognize them as a prestressed concrete floor.

This form of construction was extended to form complete flat-slab floors, by using wide planks up to 4 ft in width for the tension zone of main beams, with the Stahlton-type planks butting up against them transversely, *in-situ* concrete with additional steel for shear and continuity being placed on top.

Samuely used and extended the principle for a large number of schools, using much larger planks and trough units between them, to achieve spans up to 40 ft, although later he tended to favour prestressed concrete tees rather than planks and achieved 1/36 depth to span ratios for floors and 1/45 for roofs (Figure 10.16).

More recently there has been a tendency to use wide hollow-cored prestressed concrete units which give an immediate working surface if the job justifies cranage being available. Extruded prestressed concrete slabs were being made in

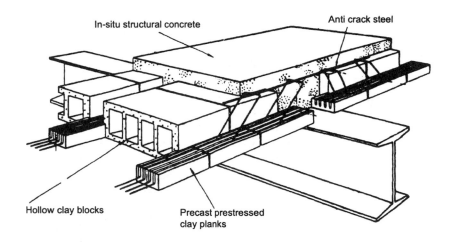

Figure 10.15 Stahlton floor units. Reproduced by permission of the British Constructional Steelwork Association Limited.

Figure 10.16 Prestressed concrete tee units designed by Felix J. Samuely & Partners.

Germany during the Second World War; 30 years later they were introduced into Britain. Annual production is running at about 2.75 million square metres at present.

Prestressed concrete trusses

It will be appreciated that a number of these systems are in fact simply providing the tensile resistance of a beam or a slab using prestressed concrete or tile instead of steel reinforcement. This principle has been logically applied to trusses, where the tension members were made of prestressed concrete and the compression members of reinforced concrete. Some of the lighter stressed tensile members were also made of reinforced concrete.

This method was usually employed for one-off jobs, and a truss with smaller than usual concrete tension members should be considered as a possible prestressed concrete truss. This method was applied in a more general form in the Intergrid and Laingspan systems. They were specifically developed for the post-war school building programme but were applied with modifications to other buildings such as in the rebuilding of Aldershot. The beams in the system basically consist of short precast concrete units either with grooves on the upper surface of the lower flange or with holes in them in which prestressing wires were inserted on site and pre-tensioned (Figures 10.17 and 10.18).

Prestressed concrete frames

In general these have been confined to simple portals, most often using precast columns and beams stressed together to give continuity. The likely distinguishing feature in these cases is the slenderness of the top member — the verticals often carried crane rails so that their slenderness would be masked. These were used in power stations as well as workshops. Figure 10.4 illustrates a workshop at Pyestock which also used Milbank units for the roof. As far as the author is aware, only two framed office-type buildings were erected, both in the London area.

Figure 10.17 Laingspan system of construction.

Figure 10.18   Intergrid system of construction.

Piles

Prestressed concrete has been used for precast piles on a number of jobs, but usually on sites such as oil refineries where a large number are required, and where it is economic to set up pre-tensioning beds on site to make them. Most piles in this situation really only require 'reinforcing' for handling and driving, and for this prestressing is ideal. Occasionally piles were made of precast units stressed together with Macalloy bars for use in confined situations.

Special applications

Containment vessels and high-pressure pipes have often been made of prestressed concrete. In the former category nuclear reactor pressure vessels are a good example. In these cases the prestressing tendons are often capable of being withdrawn and inspected for safety reasons. In the latter cases the tangential prestressing is often induced by winding on prestressing wire under tension — longitudinal stressing, often by post-tensioning, being provided largely for handling stresses.

Turbine blocks have also been prestressed largely to ensure their integrity under complex loading, as have large industrial presses.[14]

Ground anchors which are used extensively are essentially holes drilled into the ground into which prestressing tendons are inserted — the hole is grouted and, when sufficient strength has been reached, the tendon is stressed and locked off.[15]

Test floors have often been made using prestressed concrete, particularly in laboratories where the floors act as the lower half of test frames if access is possible or required below them. If steel members are not embedded in such floors it is likely that prestressed concrete has been used. Examples of these are to be found in university and government laboratories.

A few examples exist of prestressed concrete 'lift' slabs where the roof and floors were cast on top of each other at ground level, with holes left for the columns. The slabs were then jacked up to their appropriate level and attached to the columns using steel connections.

Following the successful use of prestressed concrete beams by the Germans in the construction of their U-boat pens, they have been used on several occasions in the UK in explosive situations, particularly in government research laboratories.

References

1. Thomas, F.G. (ed.), Prestressed Concrete: Proceedings of the Conference held at the Institution of Civil Engineers. ICE: London, 1949 (lists 50 patent and 289 references).
2. Walley, F., Prestressed Concrete: Design and Construction. HMSO: London, 1953 (50 references).
3. Walley, F., The progress of prestressed concrete in the United Kingdom. Prestressed Concrete Development Group (Cement and Concrete Association): Slough, 1962 (41 references).
4. Andrew, A.E., Turner, F.H., Post-tensioning systems for concrete in the UK: 1940–1985. CIRIA: London, Report 106, 1985 (illustrates and describes numerous systems used in this period).
5. Emperger, F., Handbuch für Eisenbetonbau, Bd 1, Entwicklungsheschichte, 3rd edn. Ernst: Berlin, 1921.
6. Labes, Die Anwendung des Eisenbetonbaues für Eisenbahnzwecke. Glasers Annln, 1906, 1 December.
7. Koenen, M., Wie kann die Anwendung des Eisenbetons in der Eisenbahnverwaltung gefoerdert werden? Centralbltt Bauverwaltung, 1907, 520–22.
8. Moersch, E., Versuche uber die Schubwirkungen bei Eisenbetontraegern. Dtsche Bauzeitung, 1907, 13 April, 207; 20 April, 223; 1 May, 241.
9. Moersche, E., Spannbetontraeger. K. Wittwer: Stuttgart, 1943.
10. Labes, Bestimmungen für die Ausfuhrung von Konstruktionen aus Eisen-beton bei Hochbauen von 24 Mai 1907, Berlin, 1908. Germany Concrete Regulations.
11. Freyssinet, E., Prestressed concrete: Principles and applications. J. Instn Civ. Engrs, 1950, 33(4), 221–380.
12. Magnel, G., Creep of steel and concrete in relation to pre-stressed concrete. Am. Conc. Inst. J., 1948, 19, 485–500.
13. Magnel, G., Le flauge des aciers et son importance en beton precontraint. Sci. Tech., 1945, 2.
14. Prestressed Concrete Development Group. Symposium on the Application of Prestressed Concrete Machinery Structures. PCDG: London.
15. Hannah, T.H., Design and construction of ground anchors. Report 65. CIRIA: London, 1980.
16. Preload Ltd. Prestressed concrete tanks. 1965, 1969 edns. London and Northampton.

11

The development of concrete bridges in the British Isles prior to 1940

Mike Chrimes

Synopsis

This chapter traces the use of concrete for bridges in the British Isles from its 19th century origins to the outbreak of the Second World War. It describes the introduction of reinforced concrete bridges by specialist firms in the early 20th century, and some of the more notable bridges designed in the interwar period when British engineers established a reputation independent of imported systems. The most important bridge types are identified, and provision for movement and considerations of appearance are discussed. The development of standard bridge loadings is traced, and the performance of some of the early bridges is outlined. The chapter also comments on the economic background, and contrasts progress in the British Isles with that overseas.

Background

Before the Second World War comparatively few bridges in the British Isles were built of reinforced concrete; at least 75% of the Department of Transport's concrete bridge stock has been built since 1960.[1] Reinforced concrete bridges came to Britain 20 years later than the continent, the first British textbook not appearing until 1913,[2] and the majority of the early bridges were built to methods and systems first developed abroad.[3] This backwardness is surprising, as British engineers had been using concrete regularly since the early 19th century.[4]

Early British development

In the early 19th century, British engineers began to use concrete for bridge foundations and substructures.[5] This development was largely indigenous and Lamande's use of hydraulic lime concrete for bridge foundations (1802–15) and Vicat's important research and concrete foundation work at Souillac bridge (1818–23) apparently had no immediate British emulator.[6] The earliest application of Roman cement to a bridge superstructure traced hitherto is for a road bridge over the Birmingham–Derby Railway line (c. 1837) on the Cliffe Whateley Road. Described as an 'arch in cement', the bridge compridsed an arch with a Roman or proto-Portland cement mortar in combinaton with engineering bricks as seen in sunivals. Bridges on this line were described as more economical than those on the earlier London–Birmingham Railway suggesting the motive for the innovation. The type was common on the Trent Valley line built in the 1840s.[6a] The earliest mass concrete bridge was designed by Thomas Marr Johnson, for Sir John Fowler, on the District Line near Cromwell Road.[7] A shortlived affair, the original lime concrete structure (1867) failed when the centering was struck and was replaced by a Portland cement concrete bridge, demolished by 1873. Although Fowler and Baker were later responsible for experiments on expanded metal reinforced arches (1895–96) no other concrete bridge designs by them are known.[8] However other British engineers began to use plain concrete for bridge superstructures in the last quarter of the century.

One patentee, Philip Brannon, erected a three span concrete arch at Seaton in Devon (1877, 50 ft middle span).[9,10] Railway engineers were also active.[11] Blockwork was used on the Callender line (1878) and 'rubble' concrete (layers of rubble set in concrete, with alternating layers of rubble concrete and plain concrete) used for the fill on the Dochart viaduct (1886), which had plain concrete arches.[11] The best known extensive use of mass concrete for bridges, on the West Highland Railway, is surprisingly late (1897–98).[12] By this time the London and South Western Railway had started using mass concrete, at Holsworthy and elsewhere.[11]

Overseas developments

On the continent Coignet was building substantial concrete arched structures in the late 1860s.[13–15] Monier began experimenting with reinforced concrete arches from 1873[16–18] and from the 1880s there were rapid developments in the theory and application of reinforced concrete on the continent and in the USA.[19–22] By the turn of the century their understanding of reinforced concrete was well advanced. These developments, although known in Britain through international exhibitions and engineering literature, were initially overlooked. British engineers' use of reinforced concrete for bridges was particularly backward. There was only one isolated instance of the use of iron embedded in concrete for bridges, Homersfield bridge.

Early reinforced concrete bridges in Britain

The bridge over the River Waveney at Homersfield on the Suffolk/Norfolk border comprises a wrought iron cage infilled with concrete, and is of uncertain structural action (Figure 11.1). The 50 ft span bridge was built in 1870 by T&W Philips using their patented fireproof system.[23–25] Hardly an example of reinforced concrete, it was more than 30 years before the introduction of foreign systems led to a rapid expansion in the use of reinforced concrete for bridges in the decade before the First World War.[3]

These early bridges were built without design standards, and specifications were provided by the designers without reference to national guidelines.[26] The structural forms of these bridges had all been developed abroad.[27–28] A resume of the principal types in use before 1940 is to be found in Table 11.1. This reveals strik-

Figure 11.1 Homersfield Bridge, before restoration work (photograph © E.A. Labnum).

Table 11.1 Typical bridge types 1900–1940:[29,30,99,178,224] although terminology varies, reinforced concrete bridges built before the Second World War can be classified broadly as given below

Bridge type	Section	Earliest UK examples
Beam and slab: parallel longitudinal beams with flat (a) or haunched (b) soffits below the roadway	*(diagram — (a) Ties, Top bars in beam, Main bars in slab, Top bars in slab, Distribution bars, Bottom bars; (bi), (bii))*	(a) 1902 (b) 1909
Beam and slab: parallel longitudinal beams with soffits curved to resemble arches	*(diagram)*	1904
Beam and slab: parapet beams supporting a transverse spanning roadway structure	*(diagram — Parapet Girder, Granite Kerb, Road Metalling, Deck slab, Transverse Beam; Top bars, Ties, Distrib. bars, Bottom bars)*	1906
Variable depth trusses (a) *or bowstrings* (b) with a transverse road structure (note: the terminology is used loosely in many reports)	*(diagram — (a); (b) Arch ribs, Hangers, Secondary beams, Ties, Roller bearing)*	(a) 1903–1904 (b) c. 1925
Slab vaults with earth or other fill retained by outer spandrel walls	*(diagram — Filling; Half section at crown, Half section at abutments)*	1904–1905

Table 11.1   (Continued)

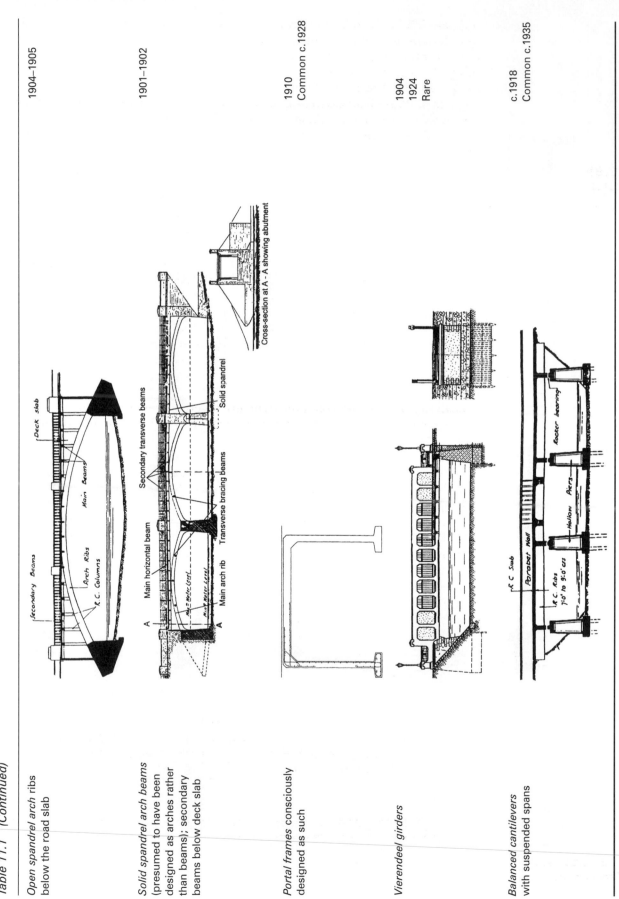

Open spandrel arch ribs below the road slab	1904–1905
Solid spandrel arch beams (presumed to have been designed as arches rather than beams); secondary beams below deck slab	1901–1902
Portal frames consciously designed as such	1910 Common c.1928
Vierendeel girders	1904 1924 Rare
Balanced cantilevers with suspended spans	c.1918 Common c.1935

ing parallels between bridges in more traditional engineering materials and the new designs in reinforced concrete. Engineers had been using arch ribs for more than a century in cast iron. Even more conservative were the earth-filled vault slab design for arch bridges. With masonry spandrel walls there is little to distinguish them from the thousands of stone and brick structures, familiar to generations of engineers. Without access to original drawings it is not easy to establish the structural principles behind the design of some early bridges, particularly arched bridges with solid spandrels, which may have been designed as beams rather than arches in some cases.

Mouchel-Hennebique Bridges[29–32]

The earliest examples of reinforced concrete bridges in Britain were all the work of L.G. Mouchel and Partners, the UK agents for the Hennebique system. By the end of the First World War they had been involved in over 33 bridges, viaducts and similar structures, the majority in the period 1907–15 as bridge construction virtually halted during the war. This was something like 80% of the reinforced concrete bridges erected in the British Isles at the time.

Some contracts were confined to foundations. Others involved strengthening or widening existing structures, such as Telford's cast iron bridges at Stokesay and Cound (1918), where reinforced concrete arch ribs were added.[33] The majority were road bridges although they designed nearly 30 footbridges. Most bridges were of modest span (6–18 m). They were characterized by monolithic construction methods which ensured that all the elements of a structure were combined with one another, with reinforcement of main girders anchored in adjoining members, and stirrups used to help bind the structure together. Very thin structural sections were frequently used for parapet beams, perhaps as little as 6 in with nominal 1 in of cover over the reinforcement.

Mouchel designed the first 'modern' reinforced concrete bridge erected in the British Isles, Chewton Glenn (1901) in Hampshire, a modest (18 ft) span skew arch with the concrete ribs concealed behind brickwork.[34] It was soon followed by a beam and slab bridge, on a slight skew, over Sutton Drain, Hull (1902) (Figure 11.2).[35] This had a 40 ft span but was 60 ft wide. The first reinforced concrete bridge erected in Scotland was a 28 ft span road and rail bridge in Dundee (1903), the first British reinforced concrete railway bridge. It was 40 ft wide, supported on 4 solid spandrel arched beams, and secondary transverse beams at 4 ft 4 in centres. The first so called bowstring girder was erected at Purfleet pier in 1903–1904, a 60 ft span (Figure 11.3).[36,37] The Mellor Street bridges in Rochdale, completed in 1905, were reinforced concrete slab vaults and were Mouchel's first such structures not to include ribs or beams in their construction. Bridges of this type had no deck slab. The Rochdale bridges were the first with abutments and foundations also all in reinforced concrete.[37a]

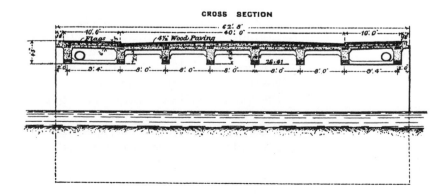

Figure 11.2   Cross-section of Sutton Drain Bridge, Hull (1902).

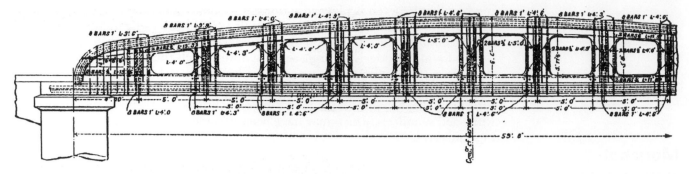

Figure 11.3   Truss girder, Purfleet Pier (1903–1904).

Figure 11.4   Berw Bridge, Pontypridd (1907).

For longer spans Mouchel used open spandrel rib arches. Berw Bridge at Pontypridd (1907–1909) comprised three spans of 25 ft (7.62 m), 116 ft (35.36 m) and 25 ft (7.62 m) (Figure 11.4).[38]

Of similar form was the recently demolished three span (36 ft, 89 ft, 36 ft) Jackfield Bridge (1909).[38a] The largest such arch bridge of this period was the Floriston bridge over the Esk in Cumbria, (1914) with a central arch rib span of 175 ft and two side spans of 147 ft 6 in.[38b] The arrangement of reinforcement here (Figure 11.5a) can be compared with that in other span spandrel rib designs of the period — the Kahn Bridge at Farnworth (1911) (Figure 11.5b) and the Considere Bridge at Bridgend (1910) (Figure 11.5c) Crewe Park Bridge (1907) is an example of Mouchel's solid spandrel arch design. The reinforcement details (Figure 11.5d) show it comprised an arch rib, with transverse deck beams and a continuous deck slab which also provided lateral bracing. The reinforcement in the spandrel walls comprised stirrups connecting the rib reinforcement with the decking, and dealt with shear, temperature and other stresses.[38c] Floriston replaced a cast iron structure of similar spans, a reminder British reinforced concrete bridges still had not exceeded spans achieved in cast iron a century earlier; they were also relatively modest compared with contemporary foreign bridges.[39]

The overall length of some was, however, impressive. A highway viaduct, erected in 1905 along the River Suir near Waterford, was 720 ft long, supported on 106

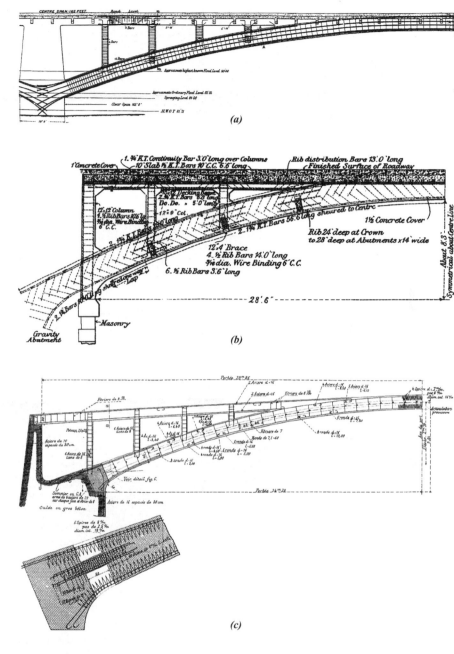

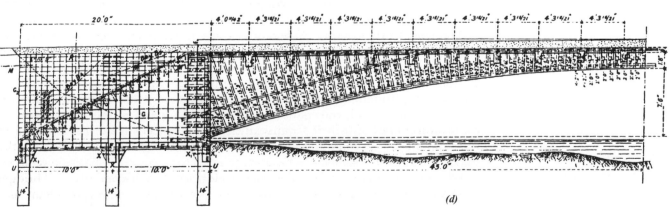

Figure 11.5   Reinforcement details: (a) Floriston Bridge (1914); (b) Farnworth Bridge (1911); (c) Ogmore Bridge, Bridgend (1910) and (d) Crewe Park Bridge (1907).

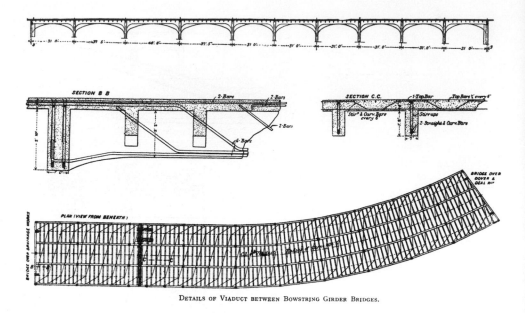

DETAILS OF VIADUCT BETWEEN BOWSTRING GIRDER BRIDGES.

Figure 11.6   Elevation and plan, from below, of Street Viaduct, Dover (1913–22).

piles.[40] Longer still was the 1000 ft Street Viaduct in Dover (Figure 11.6), with associated bowstring girder bridges; work began on this in 1913, but, due to the war was not completed until 1922.[41] Both were partly curved in plan.

Other systems

Other reinforced concrete systems were in use by 1914.[42] The Monier system had been marketed since 1902 by the Armoured Concrete Construction Company, but no bridges are known by them. The (American) Kahn System, originally intended for buildings, soon branched into bridge construction with the UK agents Truscon.[43,44] The earliest contracts were at Lilburn (1907) and Lucker (1906) in Northumberland, replacing two masonry arches washed away in a flood of 19 May 1906. Both were open spandrel rib arches. Of rather more interest is the 74 ft 9 in span structure at York Race Course, supported on parapet girders, with four curved crossbeams over, and cross beams beneath the deck (Figure 11.7).[45]

Hartlake Bridge (Tonbridge, 1910)[46] comprised a 36 ft sloping approach span, and a slightly arched (12 in rise) river span, with transverse beams to edge beams utilizing the parapets. Their first multispan bridge was possibly Kings Bridge Belfast (1909–1910), with four horizontal spans (40/50/50/40 ft).[47] The arrangement of the reinforcement for the beams (Figure 11.8a) can be compared with Mouchel's near contemporary structure over the Wansbeck at Stakeford (1909)[47a] (Figure 11.8b).

The work of the English branch of Considere and Partners reflects the thinking of Considere and his partner Caquot in France, and represents perhaps the first clear break with Mouchel Hennebique methods. The earliest Considère design was a 17 span beam and slab structure over the Great Eastern Railway at Angel Road, Tottenham (1908) (Figure 11.9), subsequently incorporated in the North Circular Road; the spans were 42 ft 9 in long and movement joints were employed every 200 ft.[48–50] A more attractive structure was the 134 ft span arch over the Mersey at Warrington (1909–1915). This bridge, 80 ft wide, was built in two sections.[51,52]

Edmund Coignet & Co. also designed a few bridges in this period.[53,54] The most important early structures were those at Kings Cross over the Metropolitan Railway,[55] and two bowstring girder railway bridges at Bangoed.[56] The latter were on a colliery line, the larger being a 56 ft 9 in skew span (Figure 11.10). The Bridge

Figure 11.7 York Racecourse Bridge (1908).

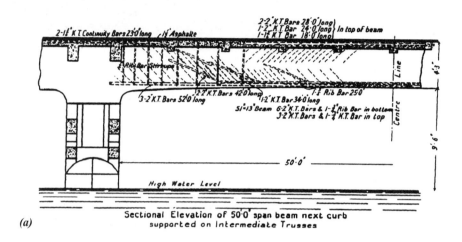

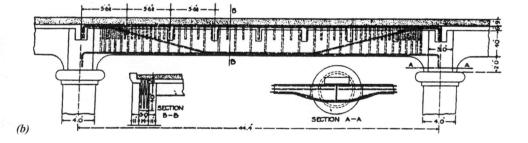

Figure 11.8 (a) King's Bridge, Belfast (1909–10) and (b) Stakeford Bridge (1909).

at Kings Cross was a beam and slab design, 130 ft long in two principal spans of 53 ft and 39 ft, and 60 ft wide, supporting a roadway and two tramway lines. The bridge was supported on columns at the junction between the principal spans.

The Indented Bar Company were beginning to emerge as an important design firm just before 1914, and would have been more so if their proposed bridges on the Metropolitan and Great Central Railways north of Watford had been executed.[56a]

Whereas Mouchel made use of specialist approved contractors, it is interesting to note that D.G. Somerville & Co. were agents for the Kahn, Considere and Coignet systems, advertising standard arch and beam/slab bridge types which could be built to any system.[57,58]

Figure 11.9   Angel Road Viaduct, Tottenham (1908).

Figure 11.10   Bargoed Bridge (1911).

As articles appeared describing the early system bridges, and textbooks became available, it became easier for local authority and other engineers to draw up their own designs.[59,59a] One of the earliest was Thorverton Bridge, Devon (1907).[60,61] This was designed by S. Ingram, surveyor for the north of the county. This was an 85 ft span structure supported on four arch ribs with secondary transverse beams supporting the decking. Reinforcement included trussed and plain bars and expanded metal mesh. Expanded metal[27,59a] was also used for a simple reinforced slab 12 in deep of 18 ft span, 40 ft wide at Withycombe Rd, Exmouth (1910); the reinforcement diagram suggests this was at least unconsciously an early portal frame (Figure 11.11).[62,63] This and a neighbouring bridge were designed by the local engineer, S. Hutton.

In Somerset a group of 24 reinforced concrete bridges (1909–1914) were designed by E.J. Stead, Assistant County Engineer.[64–67] The majority of these were slab vaults

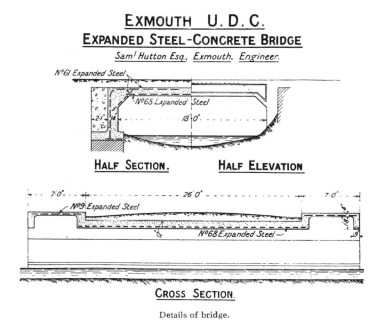

EXMOUTH U.D.C.
EXPANDED STEEL-CONCRETE BRIDGE
Sam.ᵗ Hutton Esq., Exmouth. Engineer.

HALF SECTION.   HALF ELEVATION

CROSS SECTION.

Figure 11.11 Withycombe Road Bridge, Exmouth (1910).

Details of bridge.

with masonry spandrels although Donyatt North Bridge was a beam and slab design with haunched external beams (1909).[68] Interestingly Stead had worked on a reinforced concrete arch bridge in Natal previous to coming to Somerset in 1908.[65] The arches were designed using conventional graphical methods of the type described by Cain[69,70] and Marsh,[59] and incorporated a bitumen water-proofing membrane above the arch ring.

Road bridges between the wars

Reinforced concrete established itself as a cheap alternative to steel or masonry for bridge building in Britain before the First World War. It was not used as widely as elsewhere partly as the result of a well-developed rail-based infrastructure, and also because initial concerns over durability restricted the funding arrangements available from the Local Government Board.[71]

In the inter-war period the spread of the motor car and the popularity of cycling encouraged a road building programme which was frustrated by the depression.[72] The Road Fund was established in 1920 and the Unemployment (Relief Works) Act (1920) provided an additional incentive to road building. Concrete bridges were found to be relatively easy to construct with an unskilled direct labour force and where a structure was not immediately required this gave the material an advantage over steel,[73] although concrete often lost out for major crossings. By 1930 there were something like 2000 reinforced concrete bridges in the country.

Nevertheless Britain lagged far behind leading continental countries and the United States in developing its road network, and one scheme approved before 1914, the Great West Road, was still incomplete in 1939. This was symptomatic of the period and British engineers' awareness of overseas developments in bridge building techniques[74] and sophisticated road junctions,[75] counted for nothing in the absence of equivalent investment. The road programme and associated bridge works virtually stopped during 1931–34. There were, however, some outstanding bridges built between the wars and notable designers like Sir Owen Williams emerged.[76]

The Royal Commission on Transport (1930) mentioned a target of 1000 bridge strengthening schemes a year. Although only 200–300 schemes had been approved by 1939, this was one area where concrete was regularly used. An interesting

Figure 11.12   White's Bridge, Leek (1930–31).

Figure 11.13   Royal Tweed Bridge, Berwick (1928).

example is White's Bridge in Leek (1930–31). Here, because of the skew angle and restricted channel width, and to avoid the stress concentration at the acute corner of the abutment, an arch rib and slab solution was preferred to a slab vault. Special consideration was given to the connection between the outside rib and the abutment where 'the steel was carried well back into the abutment' (Figure 11.12).[71]

Many early 1920s bridges resembled prewar designs. These include several major arch rib structures built in the 1920s by Mouchel such as Atcham Bridge carrying the A5 over the River Severn (1924–27),[78] and the Royal Tweed Bridge, Berwick,[79] completed in 1928. Berwick's scale was remarkable for British bridges of the time, comprising spans of 167, 248, 285 ft and 361 ft 5 in plus approach viaducts of 199 ft and 144 ft 6 in. Distinctively the spans increase as the north shore is approached and the bridge incorporated expansion joints over the piers and abutments. The arch ribs of the three longer spans were hollow. The bridge encap-

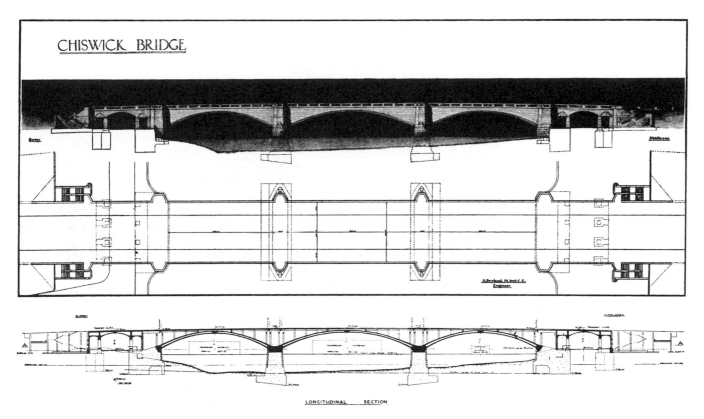

Figure 11.14   Chiswick Bridge (1933).

Figure 11.15   Twickenham Bridge (1933).

sulates all the best characteristics of Mouchel's work with open spandrel arch ribs (Figure 11.13).

The two arch bridges designed by Considere and Partners at Chiswick and Twickenham completed in 1933 are of interest.[80] Chiswick comprises a three span solid spandrel barrel arch, with the 9 in roadway slab resting directly on the vault-

Figure 11.16   King George V Bridge, Glasgow, closing of central span.

ing at the crown, and on reinforced concrete columns elsewhere (Figure 11.14). The whole was of cellular construction. At the crown vertical steel plates separated the half vaults so hydraulic jacks could be inserted to reduce the effects of creep and shrinkage.[81–83]

At Twickenham the three hinged barrel arch form of the three main spans was made explicit (Figure 11.15). The deck was supported on columns over the piers and abutments, but otherwise supported on longitudinal diaphragms.[84]

Some features of these bridges had been anticipated by the King George V Bridge in Glasgow (1924–28), although this was a continuous girder of arched appearance, 423 ft 6 in in overall length, resting on cast steel roller bearings.[85] The structure comprised 12 parallel girders joined by a vault slab below and the roadway slab above, with transverse stiffening walls. A central 15 ft strip was omitted during construction to avoid stresses in the deck slab and concreted monolithically with the rest of the structure as the final phase (Figure 11.16). The Ellon Bridge Aberdeenshire (1939–40) also had continuous girders simulating arch ribs, but in this case the vault slab was omitted, and the deck slab formed a top flange.[86]

New bridge forms

Cantilever bridges

Sir Owen Williams' replacement of the wrought iron suspension bridge at Montrose (Figure 11.17) (1928/30) is a twin 'balanced' cantilever truss with a central drop in section, the spans comprising 2 at 108 ft and a centre opening of 216 ft. Its appearance was in conscious imitation of its predecessor. Williams justified the use of reinforced concrete on cost grounds, although his design was criticized for seeking to imitate a structural form which would have been better engineered as a steel suspension bridge.[76,87,88]

Cantilever construction was hardly a novelty with such an example as the Forth Railway Bridge, and in the late 1930s several cantilever bridges were built. The balanced cantilever of the type we would recognize today was developed around 1930 on the continent and in Brazil. At Alveley Bridge (1937) cantilevered centering was used for balanced construction by BRC.[89,90] Greenfield Lock Bridge, Chester (1939–40), a 49 ft 4 in skew span, had a suspended span with 26 ft long longitudinal beams carried on cantilevers, with balance arms of cellular construction loaded with selected fill.[91] The idea of anchoring a shore span, using

Figure 11.17   Montrose Bridge (1930).

counterweights to balance a cantilever arm, and finally inserting a drop-in simply supported central span, may well have appeared much earlier.

Chettoe and Adams (1933) suggested the use of cantilever and suspended spans, particularly in multispan girder bridges where there was the possibility of settlement of piers, or for curved viaducts.[92] The Lindfield Bridge, Sussex, widening (1938), was a three span slab with the two side spans cantilevered out over the pile cap beam supports to support the central slab.[93] Inverscaddle Bridge, Argyll (1939–40) is a three span structure with end slabs counterbalanced at the abutments, cantilevering over the piers, and supporting a suspended central span.[94] The Bridge of Orchy, Argyll, (1937–39) was also a three span structure.[95] The elevation of its beams was that of a flat arch. There were four main beams to each span, with the side spans fixed at the abutments, and cantilevered out to support a 40 ft suspended span.

Rigid and portal frames

Rigid or portal frame type bridges were probably first used for a railway over bridge at Markersbach in Saxony by the Hennebique agent Max Pommer (Leipzig) in 1902.[96] It is unclear when Mouchel first used this form, which had obvious attractions where beam depth was limited by waterway or traffic clearance.[97,98] There were apparently early British examples at Exmouth (Figure 11.10) and Dugdale in 1910.[62] The latter was designed by the Indented Bar Company.[56a] Owen Williams realized their value and designed a bridge of the type at Shepherdslea Wood on the A2 over the Southern Railway (1927)[99] and from the late 1920s such bridges became increasingly common.[100,101]

Mouchel designed a bridge at Wisbech with a 92 ft 6 in span (1929–30), the clearance being required for navigation (Figure 11.18).[102] Two hinged Portal frame designs were also used in connection with railway works on the Larne line between Carrickfergus and Whitehead (1929–34).[103] These works also included the first and the Greenisland loop line flat slab highway bridge in the British Isles,[104] designed by Truscon.[104a,104b]

A two span design was used at Ystradgynlais (Figure 11.19) (c. 1932).[105] Steel portal frames were encased in concrete at Water Eaton, Oxfordshire (Figure 11.20) (1936).[106] An outstanding group of cellular portal frame bridges were designed by F.A. Macdonald in the late 1930s in Scotland.[107–109] That at Aberuthven on the A9 was a 40 ft wide three hinged structure with 73 ft main span (Figure 11.21). Either side of the crown hinge the slab was solid for 15 ft, increasing in thickness from a minimum of 14 in. It then divided into a deck and vault slab connected by

Figure 11.18   Wisbech Bridge (1929–30).

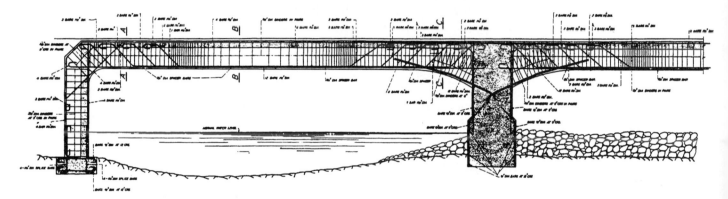

Figure 11.19   Ystradgynlais Bridge, showing reinforcement (1932).

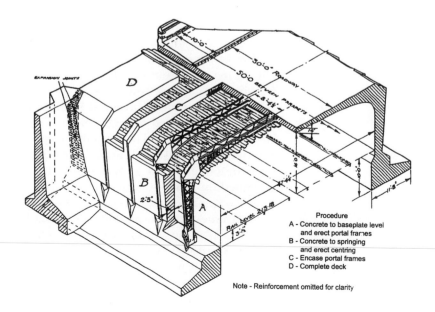

Procedure

A - Concrete to baseplate level and erect portal frames
B - Concrete to springing and erect centring
C - Encase portal frames
D - Complete deck

Note - Reinforcement omitted for clarity

Figure 11.20   Water Eaton Bridge (1936), (courstesy of the Structural Engineer).

Aberuthven Bridge, Perth and Kinross.
CELLULAR PORTAL FRAME.

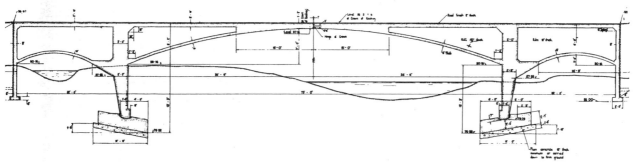

Figure 11.21   Longitudinal section of Aberuthven Bridge (1939).

Figure 11.22   Stow Bridge (1925–26).

ribs 9 in thick at 9 ft centres. Side flood arches were constructed monolithically to provide stabilizing moments. The bridge was faced in natural stone.

Other forms

Bowstring girders became more common from the mid 1920s.[110] Several early examples such as Stow Bridge (1925–26) (Figure 11.22) were built by Peter Lind., or K. Holst, and may reflect these firms' overseas origins.

Owen Williams appears to have been the first British engineer to use a Vierendeel truss, for his collaboration with Ayrton at Findhorn (Figure 11.23) (1924–26).[76,111,112] He used two pairs of beams, 98 ft span, separated over the central pier and 36 ft apart. This type had been anticipated by Wayss and Freytag at Krapiza in Austria–Hungary in 1900 and the idea had attracted some attention in Germany in the early part of the century.[113–116] Mouchel's early truss at Purfleet (Figure 11.3) bears a striking resemblance to this type.[36]

Williams designed a mass concrete arch, for aesthetic reasons, at Wansford (1930). The bridge was fitted with three temporary steel hinges which were grouted up when the construction was complete.[76,111] Earth filled solid spandrel slab vaults continued to be used throughout the period, but there is little evidence of the type

Figure 11.23   Findhorn Bridge (1924–26).

of vault arch used on the continent with open spandrels and the deck supported on columns or transverse walls.[117]

Waterloo Bridge[118]

The present Waterloo Bridge (1938–42) in London in many ways exemplifies the best of British interwar bridge design, although its planning was dogged by controversy. It was designed by Buckton and Cuerel of Rendel, Palmer & Tritton, with the architect Sir Giles Gilbert Scott. The bridge is a five span box girder structure, with four main beams, two on each side of the bridge continuous over two spans and a centre span comprising two cantilever arms and a suspended girder.

The maximum central spans are 250 ft. For architectural reasons the spandrels were faced with Portland stone, and Scott's desire to bring lightness to the underside of the bridge meant that the main box beams had to be located under the footways rather than the road pavement, which had to be supported on secondary T beams and slabs (Figure 11.24). These architectural requirements, including restraints on the width of each beam, demanded ingenious engineering solutions. In the end all the main reinforcement was welded, as lapping was impractical. One notable innovation was the use of prestressing in the shore cantilevers, at the top of the bearing walls, and around the centre span expansion joints, where high shear stresses appertained. The bars, with ends upset and screwed, were placed in steel tubes with projecting end connections. Once the concrete had gained strength the bars were stressed by passing steam through the tubes, taking up the thermal extension by turning the end nuts, so that the required stress (30,000 lbs psi) was induced on cooling. The steam connectors were used to grout up the bars in the tubes. To deal with differential settlement jacks were built into the piers.

When one considers the tribulations of many concrete bridges in the post-war period, Waterloo is an outstanding testimony to all involved.

Footbridges

Several footbridges incorporating reinforced concrete arches were built over the London and South Western Railway c. 1904.[119] If one excludes links between factory buildings, Mouchel's first footbridges were erected in 1908.[29] With little road traffic there are few early examples of pedestrian over road bridges, a notable exception being the 96 ft arch bridge over the Brighton Road at Reigate Hill

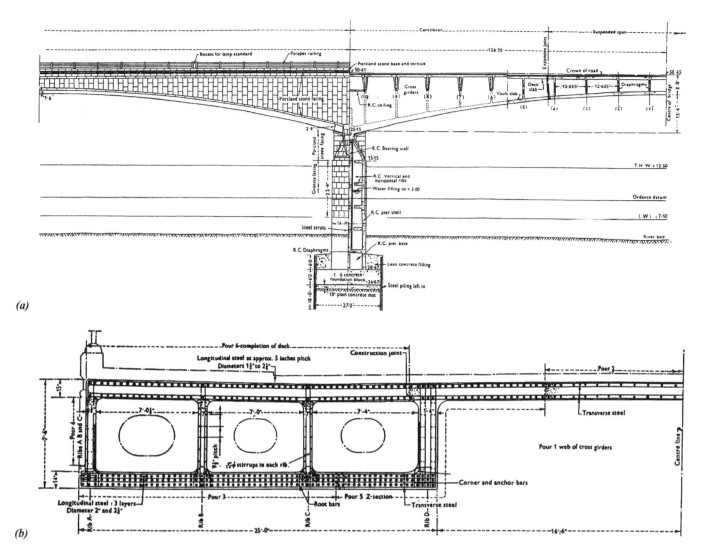

Figure 11.24   Waterloo Bridge (1938–42): (a) part-longitudinal view showing part of continuous girder and part of suspended central span and (b) half cross-section at crown.

(1908–1910).[120] The 67 ft arch span bridge at Alum Chine in Bournemouth spanned a steep-sided valley (1908) (Figure 11.25).

A large proportion of footbridges built in this period were over railways, generally at stations. These structures, because of the load gauge and lateral spacing could well be considered early 'standard bridges', and early precasting techniques were used.

The concrete girder usually reflected plate girder appearance. Kew Garden Station footbridge (1911–12) was an exception which used a concrete bow string for the main truss with diagonal stiffness provided by thin concrete infill panels (Figure 11.26).[121]

Footbridges were also required across large railway complexes like marshalling yards. Their parapets were often 6 ft high and formed the structural member. Such bridges could have spans of up to 130 ft. That over the Great Eastern Railway at Enfield Lock had an overall length, including two 63 ft spans and approaches, of 388 ft (1909).[121a]

Figure 11.25   Alum Chine
Footbridge (1908).

Figure 11.26   Kew Gardens
Footbridge (1912).

Railway bridges

Some railway engineers were notable exponents of reinforced concrete,[122-124] but its use for underline bridges was restricted. Although few new lines were built, bridges were being replaced, and there was some concern as late as 1929 that the potential of concrete was being ignored.[125,126]

Plain concrete continued to be used. Carrington viaduct (1903),[11] attracted comment for its expansion joints (Figure 11.27). In the 1930s precast vibrated concrete voussoir blocks were used in the replacement of one of Brunel's timber viaducts at Trenance; elsewhere the concrete superstructure was faced in natural stone.[127] The first reinforced concrete railway bridges to be built for main line traffic were those designed by Mouchel in the Bristol area (1907–1908).[29,128] Coignet's work at Bangoed was more or less contemporary.[56] These were exceptions, however, and reinforced concrete was normally restricted to overbridges,

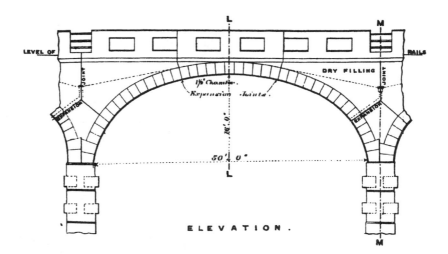

Figure 11.27 Elevation of Carrington Viaduct, showing expansion joints (1903).

Figure 11.28 Construction of the Viaduct, Valentine's Glen, showing a main arch after steel centring has been struck (c. 1932).

pedestrian and highway, rather than underline bridges. In the 1930s there was more widespread use of reinforced concrete, seen in arched viaducts on the LMS in Ulster (Figure 11.28), one built on a curve,[129,130] and increasing use of precasting techniques.[131,132]

Precast concrete

The first use of precast concrete in bridges is uncertain. Its advantages were obvious, particularly when access to a site was limited and pressure of time imperative. One early application was for railway footbridges; allegedly the Southern Region modelled theirs on that designed by Mouchel's at Oxshott (1908–1909).[133] An example is the replacement bridge precast at the Southern Railway's concrete depot at Exeter in 1923, and erected *in situ* by the use of a locomotive and steam crane (Figure 11.29).[133a]

The outstanding early use of precast concrete was at Mizen Head, Cork. The most spectacular of the early non-system bridges (1908–1910), it was a through arch of 172 ft span. The ribs were precast close by, and cantilevered out from the

Figure 11.29   Installation of precast concrete footbridge near Exeter (1923).

Figure 11.30   Mizen Head Bridge (c. 1909).

abutments (Figure 11.30).[134,135] Mouchel used precast beams for St John's Hill Bridge, over the London and South Western Railway in Clapham in 1915 (Figure 11.31), and at Wellesley Park Bridge, Gunnersbury, again over the L & SW Railway (1922).[136,137] 'L' beams were used for widening Limeworks railway overbridge near Doncaster (1938),[130] and more unusually precast arch ribs were used on the North Circular to cross LNER sidings at Neasden (1938).[139] The Midland Railway (Northern Counties Committee, subsequently the LMS (NCC), used precast beams before the First World War in Ulster, and with the availability of heavier duty cranes in the late 1930s were able to make use of longer span T beams. The expertize of the design team, led by W.K. Wallace, was transferred to mainland Britain with him after 1933. It was reflected in the use of large precast units on a rail underline bridge at Northampton (1938) and the development of prestressed beam units in the war (see Smyth, M.H. Gould, The concrete work of the LMS NCC, (1997), Innovation in civil and structural engineering, 41–46).

Kirkcudbright Bridge (1927) had raft foundations due to poor ground and the piers were made of precast reinforced concrete shells with mass concrete hearting.[140]

Figure 11.31   Placing precast concrete beams at St John's Hill Bridge, Clapham (1915).

Appearance matters

The appearance of reinforced concrete bridges was a subject for discussion almost before they were used in UK.[141] Cement manufacturers promoted concrete's adaptability from early in the century.[142] Its mouldability meant that it could be used to enhance unattractive structural forms. A feature of many early bridges was the use of moulded architectural features in contrast to today where the lines of the structure would be used to achieve an aesthetic solution. Decoration could enable a bridge to blend in with its surroundings. Decorative treatment could either be carried out *in situ*[143] or with precast elements such as parapets[144] supplied by firms like Empire Stone.[145] Precast blocks were also used to simulate natural stone. Coloured concrete, to blend in with local stonework was achieved by the use of appropriately coloured aggregate and dyed cement.[144,146,147] Natural stone finishes were simulated mechanically with pneumatic punches and bush hammering.[146–150] Shuttering marks were removed by carborundum bricks.[151] White finishes were created by a final cement wash,[152] and from the 1920s by the use of Snowcrete.[142]

With all these expedients available it is surprising that natural stone was employed at all, but on several prestigious bridges, and in specific locations, it continued to be employed to conceal the reinforced concrete structure.[153,154] Well-known examples of masonry cladding are King George V Bridge, Glasgow, faced with Dalbeattie granite, and Waterloo Bridge. False arches were a regular feature of designers of the period, King George V bridge being an obvious example.

The use of selected aggregates, natural stone and other features was encouraged by official policy from the mid 1920s. Possibly prompted by the quality of design which had hitherto prevailed, it was decided to take the appearance of the bridge into account in all applications for grant aid, a decision which apparently had the desired effect on subsequent applications.[155,156] The Royal Fine Arts Commission advised on major crossings.

One can possibly see the effects of these changing policies in London's arterial roads. The pre-war Angel Road viaduct (Figure 11.8) had no aesthetic pretensions,[48] but the neighbouring Lea Valley Viaduct (1926)[157,158] the associated Lea Navigation Bridge,[159] and the Barking Viaduct (1924–27),[160] with functionless approach towers (Figure 11.32) reminiscent of Francis Thompson's redundant pylons at the Britannia Bridge, attracted much favourable comment at the time. The Lea Valley Viaduct was the work of Owen Williams and Max Ayrton, and their willingness to experiment with reinforced concrete structural forms attracted considerable attention. The raked supports for the Wadham road viaduct, are a welcome alternative to vertical column supports of the beam and trestle type.[161]

Movement matters

Permanent hinges had been used in metal bridges through the 19th century, and were introduced for masonry arches in Saxony in 1880.[162] The idea was developed rapidly on the continent in the next 20 years for both stone and concrete[163] but hinged arches were rarely used before the 1920s in Britain. The oldest type used in concrete was a lead strip,[164] coated in copper to avoid contact with the concrete, placed to provide contact in the middle of the section and taking the line of thrust, with a filler of bituminous felt or cork at each side. Dowel bars were necessary to anchor the arch rib to the abutment. More efficient hinges were developed of curved contact type, made of cast iron or steel, with dowel bars permitting some rotation. Additional reinforcement was necessary transverse to the line of the hinges to spread the load concentration in both the arch and abutments. Where individual ribs were involved a pin bearing in a saddle was used.

A two hinged arch was designed over the Thames near Datchet in 1924.[165] This was a two span Truscon structure of unequal span. There were concerns about stresses from uneven loadings and temperature changes particularly on the centre pier. A number of three hinged arches had been designed in reinforced concrete

Figure 11.32 Barking Viaduct by L.G. Mouchel and Partners (1927).

by this time, and hinges were also employed in a long viaduct where settlement was anticipated, dividing the viaduct into rigid frames and short cantilever spans, thus obviating the need for expansion joints.[166]

The Considère type of hinge was in reinforced concrete with a view to permitting the arch to act as a three hinged determinate structure during construction, and finally be closed up to become a fixed arch for live loads.[50] It was intended to deal with much of the rib shortening due to dead load before closure, much of the shrinkage, and some plastic yield. By placing the hinge eccentrically in larger structures an initial bending moment could be created in the arch rib to counteract some of the severest moments likely to be experienced under service conditions. Considère had advocated their use for the relief of secondary stresses in arch bridges before the First World War[167,168] and his English subsidiary used them at Warrington (1909–1915) and Ogmore, Bridgend (Figures 11.4, 11.33) (1910).[49] Scott, one of Considère's engineers, publicized their use further in his standard text book,[169] and in an article published in 1924.[170] Hydraulic jacks were later used to compensate for shrinkage and creep, as at Chiswick, to avoid the necessity for springing hinges which would have been difficult to leave open with the masonry facing.[83,171]

Aside from the specific circumstances of arch bridges, there was need to provide for movement in bridges and viaducts generally due to shrinkage, temperature effects, and possible foundation movements.[92,169,172] Various types of bearings were used such as sliding plates, not particularly suitable for heavy concrete structures, segmental cast iron rockers well suited for deck girder and through girder bridges, heavier iron/cast steel designs, and cast steel roller bearings for heavier structures (Figure 11.34). Considère made use of reinforced concrete rocker bearings (Figure 11.35).[173]

Figure 11.33   Temporary reinforced concrete hinge at Ogmore, Bridgend (1910).

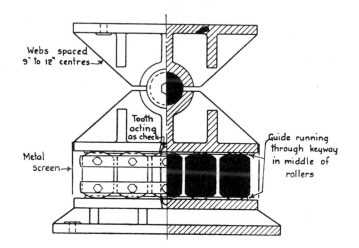

Figure 11.34   Cast steel roller bearings.

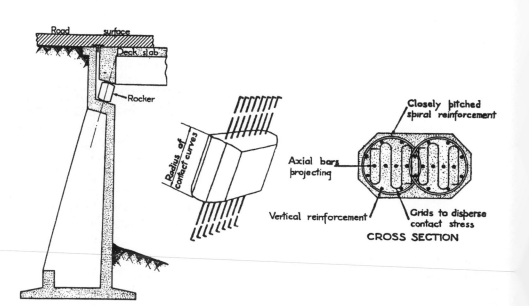

Figure 11.35   Reinforced concrete rocker bearings.

The suitable application of each type is described by Chettoe and Adams.[174] With girder bridges space was left between the girder and abutment to permit movement (Figure 11.36), with copper and steel sliding plates on the bearing surfaces, and some method of covering the space beneath the pavement. With a concrete pavement steel protection angles were used. On longer bridges expansion joints had to be provided. In multiple spans this would normally be done over piers. Some filler could be placed in the wearing surface, but the slabs themselves would have a plate sliding over steel angles (Figure 11.37). The five span bowstring girder bridge at Kirkcudbright (1927) had bitumen filled expansion joints, with the girders resting on cast steel rockers to accommodate settlement.[175] In some cases interlocking joints were used, as on Barking viaduct.[160]

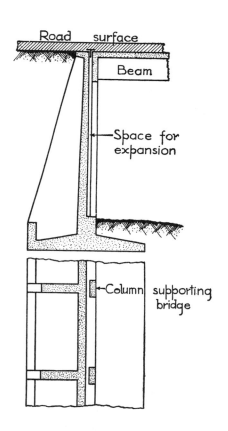

Figure 11.36   Movement joint for a concrete girder bridge.

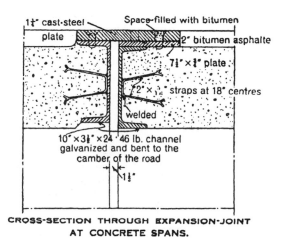

CROSS-SECTION THROUGH EXPANSION-JOINT AT CONCRETE SPANS.

Figure 11.37   Cross-section of expansion joint in the concrete spans, Kincardine-on-Forth Bridge.

Loading

Prior to 1910 there were no nationally agreed design standards and highway bridge loadings were specified by local authorities and usually related to 'normal traffic of the district'.[176,177] The structure's acceptability was verified by test load, generally steam rollers. The variety of early design loads is striking. The beam and slab bridge at the Sutton Drain was designed for four wagons carrying 25 t on two arches 8 ft apart while that at Kings Cross was designed for a uniformly distributed load of $0.2 \, t/ft^2$, and two moving loads of 8 t each at 6 ft centres, or 16 t on each axle. Footway loads were calculated for a uniformly distributed load of $0.0625 \, t/ft^2$. Kings Bridge, Belfast was designed for a uniform distributed load of $0.05 \, t/ft^2$, and a moving load of a traction engine and three tractors, whereas Wansbech Bridge, Stakeford was designed for a uniform distributed load of $0.05 \, t/ft^2$ but a moving load of a 30 t steam roller or two 15 t steam rollers. To address this situation the Concrete Institute and local government engineers set up a joint committee on highway loads. Their recommendations were published in 1918.[178] In 1922 the Ministry of Transport defined a standard train of loads which comprised a 20 t tractor with three 13 t trailers plus a 50% impact factor (Figure 11.38).[179,180,181]

These minimum standard loadings included a recommended standard width of 10 ft for each lane of traffic. Their application to bridge design was by no means straightforward. The justification for such heavy loading criteria, unlikely to be met in practice, was to produce bridges which would last 100 years without need for strengthening due to unforeseen traffic loads. The absurdity of the rigid application of such thinking was seen at Fingringhoe, Essex, 1924 (perhaps the earliest use of high alumina cement in bridges) where the local country road was too narrow to contemplate driving the standard train along to test load the bridge.[182]

In some areas, notably Liverpool and Glasgow,[183] the local regulations stipulated even more severe loadings to allow for the transport of boilers and other heavy engineering plant. Some engineers, such as W.L. Scott and Owen Williams urged more realistic loadings should be adopted, and in his standard textbook Scott reproduced the French and American regulations as alternative bases for design.[169] The French regulations were used by BRC as the basis for the design of a bridge at Stretford (1925) where a loading of 70 t weight on two axles had been specified.[184]

The Ministry of Transport themselves recognized that some guidance was required on the standard loadings, and their application was discussed at the 1923 Public Works Congress.[185] Williams examined the implications of the loadings on the economic design of beam and slab concrete highway bridges.[186] He concluded that longitudinal beams were unnecessary for short spans (15–20 ft) and if they were required transverse beams should be avoided. Williams was critical of the loadings themselves and called for research on more realistic loadings, which would permit 'more and better engineering' and more attractive bridges.[187]

Influence lines were regularly used as a method of determining live loads.[169,188] For arch bridges with fixed supports Carpenter adopted a method based on an assumption of uniformly distributed live loads.[189] Arch design was also the subject of an extensive series of articles by G.P. Manning (1930) and subsequent textbook.[190]

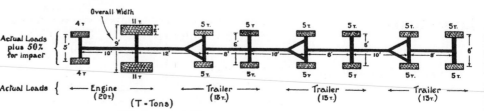

Figure 11.38   Standard loading train (1922).

Ministry of Transport Standard Train.

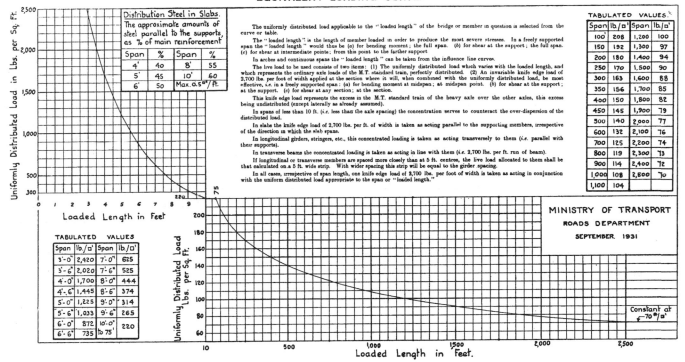

Figure 11.39   Equivalent loading curve (1931).

Scott promulgated Pigeaud's method of analysing concentrated loads in thin slabs.[191,192]

In 1931 the Ministry of Transport produced its equivalent loading curve (Figure 11.39), and introduced knife edge loading, including an allowance for impact.[193] Perhaps in recognition of the rather unhelpful nature of the standard loading train the curve was accompanied by an article illustrating its application by Hargreaves,[194] and given extensive treatment by Chettoe and Adams in their standard textbooks, which were the highway engineers' bible in the 1930s.[92] E.A. Scott discussed the application to slabs.[195] At the same time Westergaard (1930) published his classic paper on wheel loads on slabs.[196]

BS 153,[197] first issued in 1922, was essentially concerned with steel girder bridges. The section dealing with loads (1923) included an impact factor inversely proportional to the span. It was amended in 1925 to include loading trains for railway and highway bridges, the latter being acceptable for MOT purposes. This amendment was incorporated in the 1937 edition; it was not until 1949 that abnormal and normal loads were specified although they were taken into account for the design of major bridges like King George V.

Railway bridge loadings were the subject of extensive research in this period, and previous assumptions were challenged.[198–201]

Load tests

Early load tests were both simplistic and varied as shown in Figures 11.7 and 11.40.

The Kahn Bridges in Northumberland were test loaded with two 15 t traction engines run side by side and in tandem,[44] whereas the York Racecourse Bridge was test loaded with 126 t of earth (Figure 11.7).[45] Farnworth Bridge was test loaded with water enclosed by the parapet girders on each side and clay dams at each end.[44] The Mellor Street bridges in Rochdale were designed for a trailer carrying a Lancashire boiler, estimated load 25–30 t, and test loaded with 25 t of pig

Figure 11.40   Load testing at Ballingdon Bridge, Suffolk, before the First World War.

iron on each pavement, together with a 12 t steam roller and an 18 t bogie of pig iron, and a 16 t traction engine pulling a bogie of cast iron weighing 32 t. The maximum observed deflection was 0.04 in.[37a]

However appropriate test loads might be for steel or iron bridges they penalized concrete bridges. If the bridge were designed for MOT loadings it was necessary for the concrete to have gained sufficient strength to sustain the standard test load after 90 days. This would necessarily involve overdesign as the concrete of that era could be presumed to continue to gain strength after that period, moreover the quite exceptional nature of the standard train loading could impose severe loads, probably unrealistic in terms of service, while the bridge was still gaining strength. In 1931, Chettoe urged that loading and deflection tests should be postponed until the end of the maintenance period, one year after construction.[202]

Service life

Despite the extensive body of literature produced in recent years on bridge assessment and repair, relatively few case studies have been published on the performance of pre-1940 concrete bridges.[203,204] The Institution's Panel for Historical Engineering Works have compiled data on some of the earliest bridges, collated as Table 11.2. Many of the bridges which have been demolished have disappeared because of the redundancy of the structure rather than its unserviceability. Slab vault structures such as those designed by Stead have performed well, certainly better than his beam and slab bridges.[64] The early Mouchel examples in Rochdale have been recently assessed.[37a] Load tests and analysis resulted in both achieving full 40 t assessment capacity. Most early bridges which remain show some evidence of spalling or inadequate cover. In structures of this age this is unsurprising, but some are known to have needed attention after a relatively short time. The Dundee harbour bridge, in a marine environment, was first gunited in 1932 and needed attention again in 1989. Jackfield bridge had a history of repairs dating back to the 1930s. Recently found to have severe carbonation/corrosion problems, it was inadequate for modern traffic loads, and despite its listed status has been

Table 11.2   Reinforced concrete bridges built in the British Isles 1870–1914

System/principal designer	Number of known bridges	Earliest known	Fate known	Still exist	Visual signs of spalling	Repaired pre-1940	Repaired post-1940	Demolished/ derelict
Mouchel	300	1901	96	59	20	2	18	37
Considère	14	1908	3	2			1	1
Coignet	9	c.1910						
Expanded metal	20	[1902–4?]	1	1				
Railway engineers	20	1902						
Kahn/Truscon	12	1906	7	7	2		3	
Local authority	27	1907	24	23			1	1
Other	24	1870	6	4	1		3	2

replaced.[205] The Truscon bridge at Woodbridge (1912) has a weight restriction.[206] Strengthening works made use of the Kahn reinforcement system. Although there was much spalling due to inadequate cover, there was no evidence of chloride attack. The Mizen Head bridge has survived well. Despite chloride ingress and corrosion problems the concrete is generally dense and of high strength, and cover (0.8–1.8 in).[207] Considering the relatively small number of surviving examples of bridges predating 1915 a case could be made for a comprehensive survey of their current condition combined with archival research into the original design specifications, to give a reasoned assessment of their performance. Listed status of selected structures could be combined with regular monitoring to give long term service records of reinforced concrete bridges.

There is more information available on later bridges. Lougher viaduct, a 14 span structure of 1922 had to be extensively strengthened and gunited in 1950, and expansion joints, omitted from the original structure, were introduced at that time. Priory Bridge, Taunton, a Mouchel structure of similar vintage, has recently been condemned and replaced.[208] The Lea Valley viaduct has also been replaced, as part of the North Circular improvements.[209] The majority of its problems related to leaking expansion and construction joints, although there was some honeycombing; the load assessment was generally positive as it was at Wadham Road. There in the absence of waterproofing the camber had provided valuable excess cover. Jordanstown Bridge (1931), a variable depth beam and slab structure of three spans has been strengthened with an overslab and high tensile steel link reinforcement in the deck beams to increase the bending and shear capacity of the structure.[210] Hampton Court Bridge, a three span arch (1931), had local weaknesses in the top slab which spanned between columns and onto the arch, and has recently been strengthened.[210] The more aggressive marine environment at the Royal Tweed Bridge, Berwick, and leakage of deicing salt through joints has caused cracking and spalling, especially where concrete was porous or honeycombed.[211] Montrose Bridge has recently been strengthened. It was suffering from alkali aggregate reaction.[212] Problems generally are said to arise from thin member sections and lack of reinforcement cover, although Maunsell's report revealed that the latter problem could occur with badly built modern structures.[203] Clearly time related phenomena like carbonation depth and chloride penetration can cause problems in older bridges, but data are lacking from which general conclusions can be drawn on the durability of such structures. It is alleged that Considere specified a dryer mix than Mouchel, and it would be of interest to know what impact this has had on performance.

Conclusions

Most early British reinforced concrete bridges were based on foreign engineers' designs and systems. A tremendous debt is owed to a generation of largely forgotten engineers who worked for specialist companies before the Second World

War — T.J. Gueritte of Mouchel, Scott of Considere, Stroyer of D.G. Somerville, Mason of BRC, Legat of F.A. Macdonald. Alert to overseas developments they ensured that there was progress in concrete bridge design in Britain between the wars. British work, however, remained largely derivative, and was generally ignored by foreign commentators. The most innovative British designer of the period, Owen Williams, did not warrant a mention in the New York Metropolitan Museum of Modern Art publication on bridge architecture in 1949; the only British concrete bridges to feature were Berwick and Waterloo.[213]

The design standards and analytic solutions developed in this period in many areas formed the basis of post-war codes. However many of the properties and problems associated with reinforced concrete only became fully understood with the post war work of the Cement and Concrete Association. Until well after the Second World War concrete was believed to be impervious and waterproofing was often not used. Many bridges imitated earlier metal and masonry forms; advantage was not always taken of the potential of the new material. In retrospect detailing was carried out without sufficient consideration of maintenance or of the effect of shape on weathering and deterioration. Nontheless many bridges represented here are worthy of the pride expressed by those concerned with the work.

Space has precluded discussion of abutment and substructure design, consideration of bridge hydraulics, and developments in construction techniques.[92,214-216] These all merit further consideration.

Acknowledgements

The author wishes to acknowledge the help received from R.J.M. Sutherland, W.A. Smyth, David Greenfield and Clive Melbourne in developing this chapter, and Philip Andrews for his original draft.

References

1. There are local exceptions to this pattern. Surrey for example has a relatively large number of older 20th century bridges. Some of these works are described in Robinson, W.P., A few descriptive notes on concrete roads in Surrey. Proc. Instn Munic. Engrs, 1928–29, 55, 229–35.
2. Rings, F., Reinforced Concrete Bridges. Batsford: London, 1913.
3. See Chapter 4.
4. See Chapter 3.
5. See Chapter 7.
6. Ecole [Royale] des Ponts et Chaussees. Collection lithographiques, Paris, 1827, Vol. 2.
6a. PRO Rail 38/16; Oxford Archaeology Unit (1999).
7. Chrimes, M.M., Sir John Fowler — engineer or manager. ICE Proceedings, Civil Engineering, August 97, 135–43.
8. Tests with expanded metal. Concr. Constr. Eng., 1908–1909, 3, 265–67 (these tests were allegedly related to work on the Central Line).
9. Early examples of reinforced concrete. Builders' J., 20 May 1908, 438–39.
10. Bosticco, M., Early concrete bridges in Britain. Concrete, September 1970, 363–66.
*11. Wood-Hill, A., Pain, E.D., On the construction of a concrete railway viaduct. ICE Min. Proc., 1904–1905, 160, 1–61.
12. Wilson, W.S., Some concrete viaducts on the West Highland Railway. ICE Min. Proc., 1906–1907, 170, 304–307. See Newby for illustrations.
13. Coignet's artificial stone. Engineering, 1869, 8, 274, 277 (Coignet's work was widely publicised at the 1867 Paris exhibition. While there is no direct evidence of any influence on Fowler and his colleagues, it is a remarkable coincidence that Fowler's work followed the exhibition in the same year).
14. Coignet, F., Bétons agglomeres appliques a l'art de construire. Paris, 1861 (also British patent 2659, 1855).

*Important general references are indicated with an asterisk

15. Beckwith, L.F., Report on Beton-Coignet, its fabrication and uses. USGPO: Washington, 1868 (report on Paris Universal Exhibition).
16. Wayss, Freytag. Das System Monier. Wayss and Freytag: Berlin and Vienna, 1887.
17. Koenen, M., Fur die Berechnung der Starke der Monierschen Cementsplatts. Centralblatt der Bauverhattung, 1886.
18. Beer, W., The Monier system of construction. ICE Min. Proc., 1898, 133, 376–92.
*19. De Courcy, J.W., The emergence of reinforced concrete 1750–1910. Struct. Engr, 1987, 65a, 315–22.
*20. Hamilton, S.B., A note on the history of reinforced concrete in buildings. National Building Studies, Special Report, 1956, 24.
*21. Haegermann, G. et al., Vom Caementium zum Zcment, Vol. 1. Bauverlag: Wiesbaden, 1964.
*22. Taylor, F.W., Thompson, S.E., Concrete: plain and reinforced, 1st–3rd edn. Wiley: New York, 1905–1917.
23. Labrum, E.A., PHEW Newsletter, 1996, 69, 4.
24. New Civil Engineer, 25 January 1996, 4.
25. Phillips, W and T., Architectural Iron Construction, London, 1870 (gives no clue to its application with concrete).
*26. Institution of Civil Engineers. Preliminary and Interim Report of the Committee on Reinforced Concrete. ICE: London, 1910.
27. Twelvetrees, W.N., Reinforced concrete bridges. Concr. Constr. Eng., 1906–1907, 1, 171–80; 261–68; 341–319; 417–30 (Most examples are foreign, but Reedsmouth Bridge, Northumberland (1904), comprised a concrete slab reinforced with expanded metal and concrete encased RSJs (174–75).)
*28. Christophe, P., Beton arme et ses applications. Beranger: Paris, 1902 (describes many early bridges).
*29. Much of what follows is taken from: L.G. Mouchel and Partners. Mouchel-Hennebique Ferro-Concrete (4th, edn., 1921). Mouchel: London, 1909.
30. Mouchel-Hennebique Ferro-Concrete. List of works 1897–1919. Mouchel: London, 1920 (there is some uncertainty about the dates given, some being date of design rather than completion/opening).
31. Cusack, P., Francois Hennebique: the specialist organisation and success of ferro concrete, 1892–1909. Trans. Newcomen Soc., 1984–85, 56.
32. Hennebique's work is recorded in the house journal 'Beton arme'.
33. Davis, A.J., Historic Shropshire bridges strengthened with Ferro-concrete. Ferro-concrete, 1919, x, 249–62.
34. Otter, R.A., Civil Engineering Heritage: Southern England. TTL: London, 1994: 149–50 (HEW 172) (This was essentially a repair to a brick arch design which had gone horribly wrong).
35. Ferro-concrete bridge over the Sutton Drain, Hull. Engineering, 1903, 75, 14, 16.
36. Ferro-concrete bowstring bridge at Purfleet. Engineering, 1904, 78, 582–83.
37. Pont en beton arme a Surfleet (sic). Genie Civil, 1905, 47, 221.
37a. Platt, S.S., On some of the more recent municipal works of Rochdale. Proc. Instn Munic. Engrs, 1913, 39, 396–408. For a recent analysis see Clapham, P.J., Young, B.K., Supplementary load testing of Mellor Spodden concrete arch bridges. Bridge Management, Vol. 3; Proc. 3rd Int. Conf. Bridge Manage., Guildford, 1996, 675–83.
38. Lowe, W.E., Berw Bridge, Pontypridd. Ferroconcrete, 1910–11, 2, 211–15.
38a. Brear, B., The new ferroconcrete bridge at Ironbridge, Coalbrookdale. Ferro-concrete, 1910, 1, 19–23.
38b. Ferroconcrete bridge at Longtown, Cumberland. Engineering, 1914, 98, 341, plate 23.
38c. Ferroconcrete bridge in Crewe Park. The Engineer, 1908, 105, 346–48.
*39. Emperger, F. von., Handbuch fur Eisenbetonbau, Vol. 3, Part 3: Bruckenbau. Berlin: Ernst, 1908, 216–17 (lists 69 bridges with spans over 30 m the largest being that over the Isar at Grunwald with two 70 m spans). Newby illustrates several continental bridges.
40. Information from Dr R. Cox. HEW 3208.
41. New reinforced concrete viaduct and bridges at Dover. Concr. Constr. Eng., 1922, 17, 399–406.
42. Progress during quarter of a century. Concr. Constr. Eng., 1926, 21, 111–49.

43. Reinforced concrete systems: XII. The Kahn Trussed Bar. Builders' J. Concr. Steel Suppl., 27 March 1907, 38–40.

44. Gould, M. Early Kahn Bridges. PHEW Newsletter (ICE), 61, March 1994, 1–2. (Other early bridges were those at Great Ayton (c. 1909–1911), Woodbridge (Guildford, 1912) and Wergins (1913). The first bridges with which Owen Williams was involved were Aberavon and Carlisle. The Kahn bridge at Farnworth is described in Engineering, 1912, 93, 286, 288.)

45. York Racecourse. Concr. Constr. Eng., 1908, 3, 487–89.

46. Concr. Constr. Eng., 1910, 5, 596–98.

47. Ferroconcrete bridge over the River Lagan. The Engineer, 1913, 115, 493–94.

47a. Ferroconcrete bridge over the River Wansbeck. The Engineer, 1909, 108, 444–45.

48. Steinberg, H.E., Twenty-one years' development in reinforced concrete design. Concr. Constr. Eng., 1926, 21, 93–98.

49. The Considere System of Reinforced Concrete Design. Considere: London [1912] (also List of works, c. 1935).

*50. Considere, A., Experimental Researches on Reinforced Concrete, 2nd edn. McGraw Hill: New York, 1906.

51. Details of this are available on microfilm in the Concrete Archive at ICE. The first contract was awarded in 1910.

52. Other pre-First World War structures were at Dunblane footbridge, Sale, (widening, 1910), Pont Hiw (1910), Bexhill Railway Footbridge (1911), Pontycryft (1911), Holmwood Station (strengthening, 1911), Dorking Station, Debenham (1912), Biel (1912), Stavenston (1912) and Leeming (1913). Some foundation works for bridges were also supplied.

53. Workman, G.C., Some recent works in reinforced concrete. Concr. Inst. Trans., 1912, 4, 18–60.

54. Edmund Coignet and Company. Reinforced concrete construction. London, c. 1913. These included footbridges over railway tracks at Heigham-Hellesden (Norwich) and Erith and two walkways connecting warehouses. A 180 ft long beam and trestle road bridge in 5 spans (max span 40 ft) was erected at Mauld near Inverness, including concrete fenders in the piers. Two 27 ft span beam and slab bridges were erected at Saltley, Birmingham, almost certainly using parapet girders.

55. Reinforced concrete bridge at King's Cross Station. Concr. Constr. Eng., 1912, 7, 780–84.

56. Reinforced concrete bowstring bridge for the Powell Duffryn Coal Company. Concr. Constr. Eng., 1912, 7, 150–52.

56a. Indented Bar Company Bridges. The Company: London, 1913.

57. D.G. Somerville and Company. Handbook of standard steel and reinforced concrete construction. The Company: London, 1908.

58. Reinforced concrete bridge at Bosmere, Suffolk. Concr. Constr. Eng., 1911, 16, 387–88.

*59. Marsh, C.F., Reinforced Concrete. Constable: London, 1904: 205–207, 377–408, 461–94.

59a Meik, C.S., Reinforced concrete in engineering structures. Proc. Instn Munic. Engrs, 1908–1909, 35, 11–46. Meik was the engineer for Purfleet pier.

60. Otter, R.A. (ed.), Civil Engineering Heritage: Southern England. TTSL: London, 1994: 82 [HEW 1087].

61. Expanded steel for reinforced concrete construction. Surveyor, 1912, 41, 864–67.

62. Reinforced concrete bridge at Exmouth. Concr. Constr. Eng., 1910, 5, 845–46.

63. Hutton, S., Exmouth's decade of progress. Proc. Instn Munic. Engrs, 1912, 38, 235–36 I.

64. Information from David Greenfield, MICE. Three were widening schemes.

65. Candidate's circular, ICE archives.

66. Stead, E., Reinforced concrete lining to bridges. Surveyor, 1911, 39, 865.

67. Stone, R.M., Bridge building on Sedgemoor. Surveyor, 1912, 41, 864–67.

68. HEW reports on Donyatt North (HEW 1516) and Bottle Bridge (HEW 1515).

69. Cain, W., Theory of Solid and Braced Elastic Arches. Van Nostrand: New York, 1908 etc.

70. Cain, W., Theory of Steel-Concrete Arches. Van Nostrand: New York. 2nd, edn., 1902; 5th, edn, 1909.

71. Ferro-concrete and the Local Government Board. Engineering, 1910, 89, 207; editorial, Concr. Constr. Eng., 1910, 5, 1.
72. Jeffreys, R., The King's Highway. Batchworth: London, 1949.
73. A Reinforced concrete bridge. Concr. Constr. Eng., 1921, 16, 645–646.
74. Gueritte, T.J., A Study of the views of M. Freyssinet, designer and constructor of the ferro-concrete viaduct at Elorn-Plougastel. Societe des Ingenieurs Civils de France. British Section, 34th ordinary meeting, 1931 (also reported in the Structural Engineer.
75. Knight, H.S.L., Modern trends in road junction design. Roads Road Constr., 1937, 15, 294–99.
*76. Williams work is described in detail in: Cottam, D., Sir Owen Williams. Architectural Association: London, 1986.
*77. Public Works Congress and Exhibition. British Bridges. The Congress: London, 1933: 249 (hereafter British Bridges).
78. British Bridges, 1933, p. 269.
79. Ferro-concrete (special number). Royal Tweed Bridge. Mouchel: London, 1928.
80. Three new Thames bridges. The Engineer, 1933, 156, 17.
81. Chiswick bridge: a new method of jacking open arches. Civil Engineering, June 1932, 36.
82. The New bridge over the Thames at Chiswick. Engineering, 1932, 133, 665.
83. Scott, W.L., Construction of Chiswick Bridge. Concr. Const. Eng., 1931, 26, 39–40; 1933, 29, July (Suppl.), 32–46.
84. Twickenham Bridge. Civil Engineering, March 1933, 88–91.
84a. Scott, W.L., Twickenham Bridge. Concr. Constr. Eng., 1933, 28, July (Suppl.), 14–20.
85. Somers, T.P.M., George the Fifth Bridge, Glasgow. ICE Min. Proc., 1928–29, 227, 155–86.
*86. Road bridges in Great Britain Concrete publications: London, 1939, 93 (reprinted from Concr. Constr. Eng., January–April 1939). (Referred to hereafter as Road bridge in Great Britain, 1939.)
87. Williams, E.O., Montrose bridge. The Engineer, June 1931.
88. British Bridges, p. 435.
89. Mason, A.P., Alveley bridge: cantilevered centering. Concr. Constr. Eng., 1937, 32, 453–59.
90. BRC appear to have begun bridge work just before the First World War, with around 14 bridges built by 1923 (BRC Structures, Manchester; BRC; 1923). Later BRC structures are described in The Concrete Way (1928–33), 5 vols.
91. Road Bridges in Great Britain. 26–29.
*92. Chettoe, C.S., Adams, H.C., Reinforced Concrete Bridge Design. Chapman & Hall: London, 1933. (2nd edn., 1938).
93. Road Bridges in Great Britain. 1939: 45–46.
94. Road Bridges in Great Britain. 1939: 105–107.
95. Road Bridges in Great Britain. 1939: 137–41.
96. Emperger, F. von., Handbuch fur Eisenbetonbau, Vol. 3, Part 3: Bruckenbau. Berlin: Ernst, 1908.
97. Smith, E.C., Influence line diagrams for portal girder bridges. Structural Engineer, 1934, 12, 27–40 (refers to many Mouchel designs). Bottesford Bridge (1910) was a very rigid structure (Ferroconcrete, 1910, 2, 75–76).
98. Cement and Concrete Association. Rigid Frame Bridges. C&CA: London, c. 1937.
99. British Bridges, p. 144.
100. The design and construction of a new canal bridge. Proc. Instn Munic. Engrs, 1930–31, 57, 21–35.
101. Heywood, R., Bridge activities in Lancashire. Proc. Instn Munic. Engrs, 1939–40. 66, 172–200.
102 Reinforced concrete bridge at Wisbech. Concr. Constr. Eng., 1930, 25, 609–611.
103. McIlmoyle, R.L., Reinforced concrete railway bridges. Concr. Constr. Eng., 1930, 25, 37–45.
104. The design of flat slabs was described in Hill, A.W. Reinforced concrete flat slab bridges. Roads Road Constr., 1938, 16, 294–99.
104a. Gould, M.H., The concrete work of the LMS NCC. Innovation in Civil and Structural Engineering, 1997, 41–46.

104b. Hill, W.A., The rigid frame bridge, Civil Engineering, 1932, 27, 40–43.

105. Mason, A.P., Rigid frame bridges in reinforced concrete. Struct. Engr, 1933, 11, 478–502.

106. Leeming, J.J., Some portal frames bridges in Oxfordshire. Struct. Engr, 1937, 15, 146–59; 307–402.

107. Aberuthven Bridge, Perth and Kinross. Road Bridges in Great Britain, 1939, 151.

108. Bridge of Allan. Road Bridges in Great Britain, 1939, 156–59.

109. Aboyne Bridge, Aberdeenshire. Road Bridges in Great Britain, 1939, 88–89.

110. British Bridges, 1933, 204–205 (Stowe Bridge, 1925–26); 437 (Etive Bridge, 1931–32) Concr. Constr. Eng., 1926, 21, 126 (free Bridge, Kings Lynn); 1927, 22, 82 (Welsey Bridge); 416 (Kirkudbright Bridge).

111. Recent bridges by Sir Owen Williams. Concr. Constr. Eng., 1929, 24, 281–88.

112. Bruce, R., The Great North Road over the Grampians. ICE Min. Proc., 1930–31, 232, 113–30.

113. Marsh, C.F., Reinforced concrete (1904). Fig. 468, p. 494.

114. Morsch, E., Der Eisenbetonbau: Der Bruckenbau Eisenbeton, 5th edn. Stuttgart: K Wittner, 1933.

115. Patton, Professor. Pfostenfackwerk, Zentralblatt der Bauwerwaltung, 1907, 558.

116. Emperger, F. von., Handbuch fur Eisenbetonbau, Vol. 3, Part 3: Bruckenbau. Berlin: Ernst, 1908, 255–57.

117. Possible early examples of deck stiffened vault slabs as favoured by Maillart (Fig. 4, Newby) are Considère's design at Dunblane, and the Cruit Island footbridge, County Donegal, designed by Mouchel, both c. 1911. For a later example see Donald, D.A., Glasgow–Edinburgh road (Newbridge Bridge) Proc. Instn Munic. Engrs, 1930–31, 57, 1001–1006.

118. Buckton, E.J., Cuerel, J., The New Waterloo bridge. ICE J., 1943, 20, 145–201.

119. Palmer, P.H., Armoured or reinforced concrete. Proc. Instn Munic. Engrs, 1904–1905, 31, 343–55. A letter to Ferroconcrete 1909, 1, 132 states the first 'ferroconcrete' footbridge was at Copnor near Portsmouth, over the joint L & SWR and LB&SCR line in 1902, but the author has been unable to verify this. It is possible the early bridges were reinforced with expanded metal.[124] (This was certainly used for precast arched segments for three pinned arch footbridges before the First World War. Railway Engineer, July 1914, 35, 209–213.)

120. Reigate Hill bridge. Ferro-concrete, 1912, 3, 30–31. Records in concrete archive at ICE.

121. Railway footbridge: Kew Gardens Station. Ferro-concrete, 1912, 3, 339–41. Records in concrete archive at ICE.

121a. Railway bridge over Enfield Lock. Ferro-concrete, 1909, 1, 102–104.

122. Institution of Civil Engineers. Engineering Conference, 1907. Section I: Railways; Section II Harbours and docks. ICE, London, 1907.

123. Ball, J.B., Tests of reinforced concrete structures on the Great Central Railway. ICE Min. Proc., 1915, 199, 123–32; 145–229.

124. Ball, J.D.W., Reinforced Concrete Railway Structures. Constable: London, 1913.

125. Railway underline bridges. Concr. Constr. Eng., 1929, 24, 91–92.

126. Progress during a quarter of a century. Concr. Constr. Eng., 1926, 21, 135.

127. Reconstruction of Trenance viaduct. Civil Eng., July 1939, 260–62.

128. Ferroconcrete railway structures. Railway Engr, 1908, 29, 257–60.

129. Reinforced concrete viaducts at Valentine's Glen, Northern Ireland. Struct. Engr, 1933, 11, 199–232.

130. McIlmoyle, R.L., Reinforced concrete viaducts near Belfast. Struct. Engr, 1933, 11, 430–43.

131. Follenfant, H.G., A precast reinforced concrete underline railway bridge. ICE J, 1939., 7, 25–34 (see references 133, 136–39 below).

132. Fabricated concrete railway work. Engineering, 1932, 133, 538–40.

133. Railway over-bridge Oxshott. Ferro-concrete, 1909, 1, 99–101. This was apparently not precast, although girders at Enfield (121a) were precast on staging above the line and lowered into position.

133a. Precast reinforced concrete footbridge. Concr. Constr. Eng., 1924, 19, 19–22 (from Railway Gazette 1923, 39, 580–82).

134. Information from Dr R. Cox, Trinity College (HEW 3026).

135. Mizen Head Bridge. Concr. Constr. Eng., 1910, 5, 847–50 (some reinforcement was expanded metal, and part was to Ridley and Cammell patents).

136. Combination pre-cast and *in-situ* reinforced concrete bridge. Concr. Constr. Eng., 1923, 18, 169–175.

137. Wills, E., Some further notes on Chiswick. Proc. Instn Munic. Engrs, 1921–22, 48, 707–709.

138. Limeworks railway bridge. Road Bridges in Great Britain, 1939, 3–7.

139. Dog Lane Bridge, Neasden. Road Bridge in Great Britain, 1939, 60–61.

140. Wolff, W.V., Reinforced concrete bridge at Kirkcudbright. Concr. Constr. Eng., 1927, 414–20.

141. Husband, J., The aesthetic treatment of bridge structures. ICE Min. Proc., 190, 145, 221–43.

142. The Cement Marketing Company's Everyday uses of Portland cement, 5 eds (1909–1930) records the development of its various forms, colours and textures.

143. Stephen's Road Bridge, Bournemouth. Concr. Constr. Eng., 1922, 17, 523–24.

144. New Dorking to Reigate Road. Concr. Const. Eng., 1927, 22, 375–76.

145. Reinforced concrete bridge at Hastings. Concr. Const. Eng., 1922, 17, 537.

146. Colouring a concrete bridge. Concr. Const. Eng., 1925, 20, 666.

147. Bridge in coloured concrete. Concr. Const. Eng., 1927, 28, 36–37.

148. British Bridges, p. 105.

149. British Bridges, pp. 474–75.

150. British Bridges, p. 457.

151. British Bridges, p. 437.

152. Reinforced concrete in 1926. Concr. Constr. Eng., 1927, 22, p. 68.

153. Abbey bridge, Leicester. Concr. Constr. Eng., 1931, 26, 249–50.

154. British Bridges, pp. 464–65.

155. Ministry of Transport. The Design of road bridges. Circular No. 224, Roads. Report on the Road Fund 1924–1925, 43.

156. Report of the Road Fund, 1925–1926, 19–20. Reports on progress on the issue of the quality of bridge design.

157. Concrete in 1926. Concr. Constr. Eng., 1927, 22, 1.

158. British Bridges, pp. 106–107

159. British Bridges, p. 192.

160. Welch, G., Viaduct and bridge at Barking. Concr. Constr. Eng., 1927, 22, 243–51 (this viaduct was designed by Mouchel's, with Ayrton responsible for the architectural treatment).

161. Wadham Road viaduct. Concr. Constr. Eng., 1930, 25, 107–109.

162. Emperger, F. von., Handbuch fur Eisenbetonbau, Vol. 3, Part 3: Bruckenbau, Chapter 1. Other sources suggest Erlanch in 1873.

163. Molitor, D.A., Three-hinged masonry arches; long spans especially considered. ASCE Trans., 1898, 40, 31–85.

164. Marsh, C.F., Reinforced Concrete, 1904, 205–206.

165. New reinforced concrete bridge over the Thames. Concr. Constr. Eng., 1924, 19, 775–78.

166. Stroyer, R.N., Hinges in reinforced concrete structures. Concr. Constr. Eng., 1924, 19, 207–212.

167. Considere, A., Spirally armoured concrete. Engineering, 1910, 89, 578–79.

168. Ferro-concrete bridge at Chateau Thierry. Engineering, 1910, 90, 327, 329.

*169. Scott, W.L., Reinforced Concrete Bridges. Crosby Lockwood: London, 1925–31. (3 eds.)

170. Scott, W.L., Secondary stresses in reinforced concrete arched bridges. Concr. Const. Eng., 1924, 19, 9–11.

171. Chiswick Bridge: lifting the arch. Roads Road Constr., 1934, 12, 22. The idea was taken from Freyssinet.

172. Emperger, F. von., Handbuch fur Eisenbetonbau, Vol. 3, Part 3: Bruckenbau, Chapter 1.

173. Scott, W.L., Reinforced concrete bridges, 1928, 331–33.

174. Chettoe, C.S., Adams, H.G. Reinforced concrete bridge design, 1933, 344.

175. Wolff, C.V., Reinforced concrete bridge at Kirkcudbright. Concr. Constr. Eng., 1927, 22, 415–20.

176. Henderson, W., British highway bridge loading. ICE Proc., 1954, 3, 2, 325–73.
177. Wheel loads and tyre widths. Surveyor, 1911, 39, 556–59.
178. Concrete Institute. Loads on highway bridges; report of a joint committee, London, 1918 (2nd edn., 1926 by Institution of Structural Engineers).
179. Ministry of Transport. Standard load for highway bridges, MoT: London, June 1922.
180. Ministry of Transport. Report on the Road Fund 1921–22. London, 1923.
181. Reinforced concrete road bridges. Concr. Constr. Eng., 1923, 18, 721–33.
182. Weaver, L.T., Fingringhoe bridge, Essex. Aluminuous cement in bridge construction. Concr. Constr. Eng., 1924, 19, 213–16.
183. King George V Bridge, Glasgow. Concr. Constr. Eng., 1928, 23, 2.
184. Warwick Road Bridge, Stretford. Concr. Constr. Eng., 1925, 20, 510–12.
185. Hawkins, J.F., Mitchell, C.G., General construction of bridges. Public Works, Roads and Transport Congress, 1923, 218–35.
186. Williams, O.E., Beam and slab concrete highway bridges. Instn Munic. Engrs J., 1926, report in Concr. Constr. Eng., 1926, 21, 359–363.
187. Williams, O.E., Letter on the Bridge Stress Committee. Concr. Constr. Eng., 1929, 24, 180.
188. Mason, A.P., The Design of reinforced concrete arch ribs: an accurate shortcut method. Concr. Constr. Eng., 1924, 19, 143–50.
189. Carpenter, H., Design of arched bridges with fixed supports. Concr. Constr. Eng., 1930, 25, 612–22; 673–84.
190. Manning, G.P., Reinforced Concrete Arch Design. Pitman: London, 1933 (originally published in Concr. Constr. Eng., 1930–31).
191. Scott, W.L., Design of reinforced concrete slabs. Concr. Constr. Eng., 1930, 24, 167, 221–93.
192. Pigeaud. Recherches sur les plaques rectangulaires appuyees a leur pouvoir. Annales des Ponts et Chaussees, 1921, 5–47.
193. Ministry of Transport. Standard Loading for Highway Bridges. HMSO: London, 1931.
194. Hargreaves, G.H., The application of the equivalent loading curve for bridges. Concr. Constr. Eng., 1931, 26, 661–69.
195. Scott, E.A., Tables for slabs designed to Ministry of Transport load and stress requirements. Struct. Engr, 1934, 12, 382–92.
196. Westergaard, H.M., Computation of stresses in bridge slabs due to wheel loads. Public Roads, 1930, 11, 1.
197. BS 153: British Standard Specification for Girder Bridges: Parts 1 & 2 — 1922: materials and workmanship; Parts 3, 4 & 5 — 1923: loads and stresses, details of construction, erection; Parts 3, 4 & 5 — 1923: appendix no. 1 (1925): tables of British Standard unit loadings; Parts 1 & 2 — 1933; Parts 3, 4 & 5 — 1937
198. DSIR. Bridge Stress Committee. Report. HMSO: London, 1928.
199. Gribble, C., Impact in railway bridges with particular reference to the Report of the Bridge Stress Committee. ICE Min. Proc., 1928–29, 228, 46–153.
200. Inglis, C.E., Impact in railway bridges. ICE Min. Proc., 1931–32, 234, 350–444.
201. Foxlee, R.W., Greet, E.H., Hammer blow impact on the main girders of railway bridges. ICE Min. Proc., 1933–34, 237, 239–418.
202. Chettoe, C.S., Testing concrete bridges. Concr. Constr. Eng., 1931, 26, 5–7.
203. Wallbank, E.J., The Performance of Concrete in Bridges. HMSO: London, 1989.
204. Mallett, G.P., Repair of Concrete Bridges. Thomas Telford: London, 1994.
205. Thomas, W.H. & Partners, Free Bridge Jackfield. Condition report. March 1986. Ref 928/85/SJP/JM
206. Cogswell, G., Herbert, A.P., A25 Woodbridge (old) Guildford Bridge. Highways and Transportation, September 1991, p. 7.
207. MacCraith, S., Performance of an 80-year-old reinforced concrete bridge in an extreme environment. Proceedings of the 3rd International Conference on Corrosion of Reinforcement in Concrete Construction (Society of Chemical Industry) 1990, p. 188.
208. Information from David Greenfield, MICE. The original bridge is described in: New reinforced concrete bridge at Taunton. Concr. Constr. Eng., 1923, 18, 20–24.

209. Elliott, D.W.C., Inspection, load assessment and repair of Lea Valley viaduct and Wadham Road viaduct. Bridge Maintenance 1993, 2, 625–33.

210. Lockwood, S.E., and others. Strengthening concrete bridge decks-increasing the shear capacity. Bridge management 3: Proc 3rd Intl Conf Bridge management, 1996. 173–79.

211. Palmer, J., Cogswell, G., Bridge strengthening in practice. Bridge maintenance 2, 1993, 912–20. This was a Mouchel structure, described in a paper by T.J. Gueritte in Concr. Constr. Eng., 1933, July (Suppl.), 28, 21–31.

212. Wood, J.G.M., Angus, E.C., Montrose bridge: inspection, assessment and remedial work to a 65-year-old bridge with AAR. Structural Faults and Repair 95, Vol. 1, pp. 103–108.

213. Mock, E.B., The Architecture of bridges. Museum of Modern Art, New York, 1949.

214. Adams, H.C., The Design of bridge substructures. Public Works Congress, 1931, 187–241.

*215. Legat, A.W., and others, Design and construction of reinforced concrete bridges. Concrete Publications, 1946 (has much on erection techniques). This was based on articles originally published in Concrete and constructional engineering 1933.

216. Wynn, A.E., Design and Construction of Formwork for Concrete Structures. Concrete Publications: London, 1926 (2nd edn. 1939).

12   UK concrete bridges since 1940

William Smyth

Introduction

This chapter on concrete bridges since 1940, concentrates particularly on the early postwar period and deals with more recent bridges briefly and mainly from an evolutionary point of view. A paper on prestressed highway bridges in the UK, published in the *Proceedings of the Institution of Civil Engineers* in 1989,[1] deals more fully with the more recent bridges as do some of the other references appended to the chapter. As well as the paper referred to above, an indispensible reference is the book *Modern British Bridges*[2] which contains brief descriptions, often with drawings and photographs, of many bridges built or under construction between the end of the war and 1964.

Concrete bridges after the war

The outstanding fact about concrete bridges after the Second World War is the way in which prestressed concrete rapidly became the dominant material for all but the smallest bridges or those which were so large that they had to be built of steel, and even at this end of the scale the spans achievable in concrete have been increasing all the time.

The advent of prestressing meant that construction techniques formerly only used for steel bridges became feasible in concrete and were quickly made use of on the Continent, where large numbers of bridges had to be rebuilt after the land wars which had been fought there. By 1950 bridges by Freyssinet and Magnel had been built by assembling precast segments joined with concrete or mortar, and also Finsterwalder's first *in-situ* concrete box girder bridge built by cantilever construction. During the early 1960s the first incrementally launched concrete bridge was built in Germany, and the first concrete bridge to combine cantilever construction with precast segments jointed with epoxy resin, in France. In each case it took some time before the technique was used for bridges in the UK, possibly because of the lack of opportunity.*

Prestressing techniques underwent significant development during the period and, together with increasingly higher strengths of concrete and steel and further developments of construction techniques, have led to substantial economies in materials and costs and have also produced changes in the character of bridges. Other significant changes which have affected the design of bridges, although in less obvious ways, are a considerable increase in understanding of the behaviour of the ground on which bridges are founded and in the ability to predict its behaviour, and improved understanding of structural behaviour, as well as methods of calculating the way in which a structure will behave. The advent of computers and later of sophisticated electronic calculators increasingly enabled more and more complex calculations to be made with greater and greater ease. Ancillary elements, such as bearings and expansion joints, have also changed over the period. The changes have not generally been dramatic, but a series of evolutionary steps adding up to considerable changes over time.

*Box girders were certainly not new to the UK and a reinforced concrete bridge was built by balanced cantilever construction in 1936.[3]

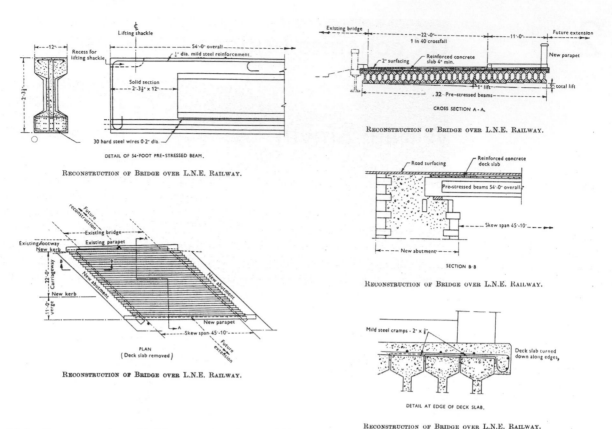

Figure 12.1 *Reconstruction of bridge over London and North Eastern Railway in Yorkshire (1943). The second prestressed concrete bridge deck in Britain.*

Early prestressed bridges in Britain — the 1940s and 1950s

Very few bridges were built in Britain during the war and not many immediately after it, and it was not until the late 1950s that bridges were being built in any numbers. The government had anticipated that large numbers of bridges would be destroyed and in 1940 the Ministry of War Transport had a stock of emergency bridging beams made of prestressed concrete. These beams were either I or box section in various lengths, made by the long-line process. When the military situation improved, it was decided to use some of them for urgent permanent bridgeworks. In 1943 they were used for the reconstruction of two bridges carrying roads over railways, one in Lancashire using 44 ft long beams of box section, and one in North Yorkshire using 54 ft beams of I section (Figure 12.1), and plans for reconstruction of another such bridge were being considered.[4] A stock of these beams remained after the war and at least four other uses are recorded.[5]*

The emergency bridging beams were designed by Dr Mautner of the Prestressed Concrete Company which had been established in the UK in 1938 as a licensee of the Freyssinet organization and a subsidiary of L.G. Mouchel and Partners. Mouchel as consulting engineers and Dr Mautner as designer of the prestressing were the designers of Nunn's Bridge (Figures 12.2 and 12.3), built in 1948 near Boston, the first *in-situ* prestressed road bridge in Britain.[6] When Dr Mautner died in 1949, A.J. Harris, who had been working with Freyssinet in France, succeeded him.

The Freyssinet system and the Hoyer system for long-line pre-tensioning were the first to be used for bridges in England, but other systems were in use by the early 1950s. Walley[7] mentions that ten years earlier the only British one was the Lee-McCall system, and records several others which were in use by 1962.

*According to Sir Alan Harris these beams were so cheap that it was impossible to compete with them, which implies a significant number.

*Figure 12.2   Nunn's Bridge,
Fishtoft, the first in-situ
prestressed concrete road
bridge in Britain.*

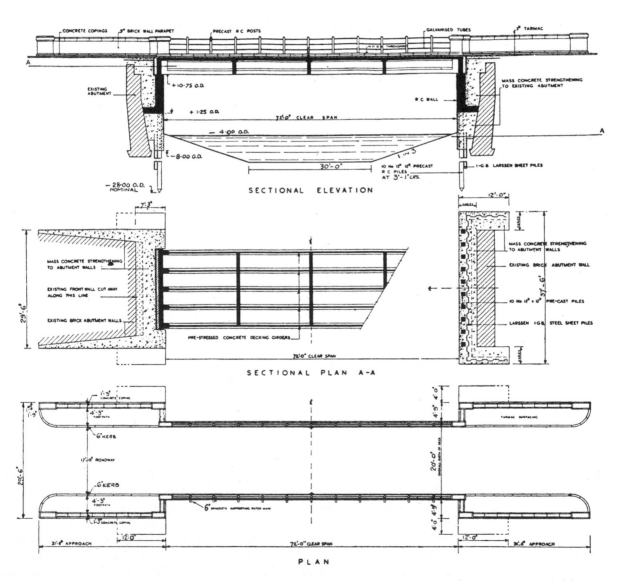

Figure 12.3   Nunn's Bridge, general arrangement. The transverse beams are typical of early post-war bridges.

Figure 12.4   Rhinefield Bridge in the New Forest.

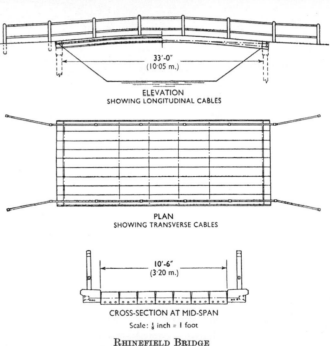

33'-0"
(10·05 m.)

ELEVATION
SHOWING LONGITUDINAL CABLES

PLAN
SHOWING TRANSVERSE CABLES

10'-6"
(3·20 m.)

CROSS-SECTION AT MID-SPAN
Scale: ⅛ inch = 1 foot

RHINEFIELD BRIDGE

Figure 12.5   Rhinefield Bridge, general arrangement and prestressing.

A paper of 1949[8] includes comprehensive lists of prestressed bridges built up to that time and an extensive bibliography. The British references are to the two bridges of 1943, Nunn's Bridge and 'the Adam viaduct, near Wigan ... the first railway under-bridge to be built of pre-stressed concrete' using precast I beams; also to designs for several slab under-bridges using the Freyssinet system, whose construction was to start later in the year for British Railways.

The early prestressed bridges were mostly fairly simple and small. Several interesting ones are mentioned in a paper of 1952 on bridges in Hampshire.[9] The earliest of these is the Rhinefield Bridge (Figures 12.4 and 12.5) carrying a minor road over a stream and the first of a number of bridges in the New Forest constructed of precast, post-tensioned units using the Freyssinet system and stressed together transversely. The slight vertical curvature of the Rhinefield Bridge made the longitudinal cables straighter which reduced friction and produced a very attractive little bridge which still serves its purpose well. Figure 12.6 is from the same paper.

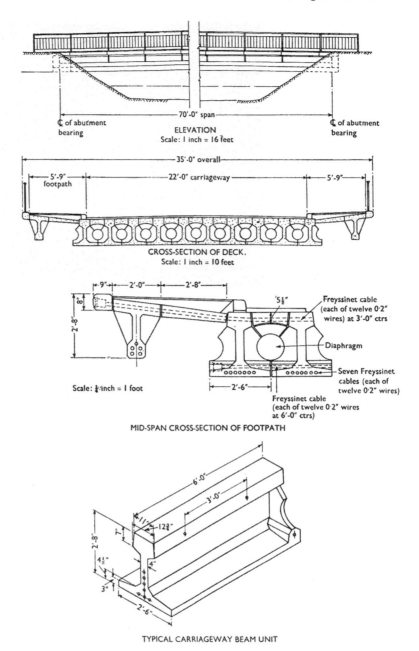

Figure 12.6   Proposed reconstruction of A30 bridge over the Test at Stockbridge. The bridge seems not to have been built to this design.

Because of the shortages of materials after the war and the consequent requirement for licences to use reinforcing steel there was an incentive to use prestressing wherever possible and transverse stressing of bridge decks was quite common. In some of these transversely stressed bridges the surfacing was laid directly on the precast units. Segmental construction was also used, not in large cross sections as in later times but for individual beams or portions of a slab, and there were interesting combinations of precast segments with *in-situ* concrete. A small bridge at Martinhoe in North Devon was destroyed by floods in 1952 and had to be quickly rebuilt on a site not accessible for precast beams. The deck was made from precast trough units side by side and end to end, erected on falsework (Figure 12.7) and stressed together longitudinally with unsheathed cables. Sheathed cables were also provided so that the deck could again be stressed longitudinally and also laterally, after the troughs had been filled with concrete.[10] Freyssinet prestressing was used,

Figure 12.7 Martinhoe Bridge over River Heddon at Hunter's Inn, N. Devon, replaced after the floods of 1952. Tensioning the main cables.

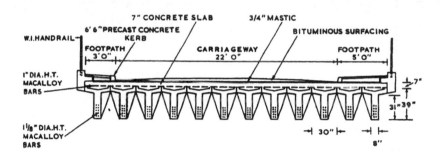

Figure 12.8 Cross-section of Barbrook Bridge, N. Devon. Another bridge which was replaced after the floods of 1952.

but as the deck was designed by Gifford and built by Udalls it may have been a forerunner of the Gifford-Udall system. Other bridges reconstructed because of the floods are described by Criswell (Figure 12.8).[11]

Some interesting footbridges were built. The first prestressed fixed arch in the world was claimed to be a portal frame footbridge across the Cherwell at Oxford (Figures 12.9 and 12.10) in which the only precast elements were the prestressed concrete planks forming the walking surface (since replaced by an *in-situ* concrete slab).[12] Others were a continuous prestressed concrete footbridge, constructed for the Festival of Britain,[13] which was tested to destruction after the Festival ended,[14] and the St James's Park Bridge (Figure 12.11), a three span continuous beam bridge constructed with the end spans initially acting as cantilevers.[15] The Eel Pie Island Footbridge had a mainly precast structure consisting of cross beams and segments of main beams at the same level, stressed together through dry packed mortar joints.[16]

Two service bridges made from precast segments should be mentioned. A pipe bridge at Gunthorpe near Nottingham (since demolished) consisted of an arch made from segments stressed together and tied by a deck, also made from precast units, in the form of a trough; the support at one abutment had Freyssinet hinges top and bottom.[17] The other one is a trough shaped beam bridge consisting of two beams made from precast segments of I section post tensioned by the BBRV system, with precast cross-beams and slabs forming the bottom of the U.[18] A few reinforced concrete bridges were completed during the war or built a few years after it.[19,20] By far the most notable was Waterloo Bridge, completed in 1942.[21]

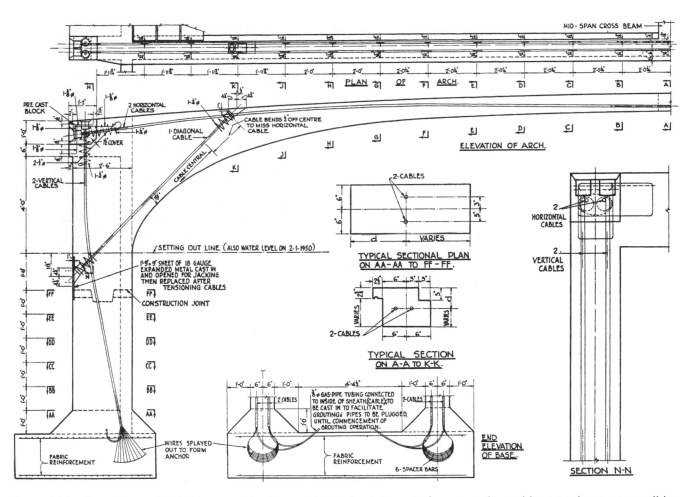

Figure 12.9   Portal frame footbridge over the Cherwell at Oxford. Layout of prestressing cables on what was possibly the first statically indeterminate prestressed bridge in Britain.

Figure 12.10   Footbridge over the Cherwell.

Figure 12.11 St. James's Park Bridge.

Early prestressed railway bridges

Railway civil engineers realized very early the advantages of prestressed concrete. Not only were the first prestressed bridges in Britain railway overbridges, but the first railway underbridge in prestressed concrete, the Adam Viaduct,[22] was built in 1946, and the first partially prestressed bridge ever to be constructed was a railway overbridge built in 1949. Partial prestressing* was a controversial subject. Freyssinet objected to it, and (possibly somewhat later) the UK Ministry of Transport did not allow tensile stresses in prestressed concrete at working loads. Its chief advocate was Dr P.W. Abeles of the Eastern Region of British Railways.

A paper of 1951[23] deals with work being done by British Railways on railway bridges.

Three kinds of railway underbridges were being pursued:

1. Simple constructions comprising precast, prestressed beam and slab units. A range of these was being developed and some had already been built. The design for a bridge at Barmouth[24] using 12-wire cables is illustrated and one for a bridge at Ystalyfera using Macalloy bars is mentioned.
2. Precast, prestressed floor units for use in association with steel main girders, to provide a shallow form of construction, lending itself to rapid site erection, in replacement of existing bridges. Steel end plates were used for anchorages and to allow for connection to the main girders. One of the two designs illustrated has provision for stressing in the longitudinal direction of the bridge (Figure 12.12). A later paper shows developments of this structural type.[25]
3. A proposal for precast, prestressed concrete channel type units, for single track bridges, to be site assembled and post-tensioned to form bridges with completely composite main beams and floor.

Prestressed overbridges designed by British Railways are also covered in the paper. These included slab bridges with pre-tensioned slab units transversely post-tensioned together without any topping (Figure 12.13), and partially prestressed structures.

*That is limited tensile stress allowed under live load and passive reinforcement provided.

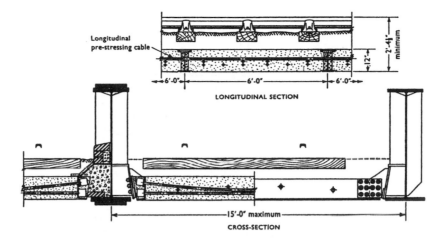

Figure 12.12   Prestressed concrete railway underbridge deck slab for use with welded plate girders.

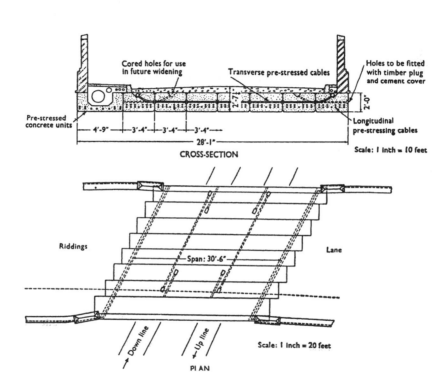

Figure 12.13   Prestressed slab construction for road bridges over railways (1951).

A later paper deals with a bridge at Manchester made from precast box units assembled on falsework with *in-situ* diaphragms and then stressed.[26] Other references are noted.[27–29]

A 1959 paper gives more information on partially prestressed railway bridges.[30] The deck of the first one in 1949 consisted of inverted T beams and the *in-situ* concrete infill contained mild steel reinforcement (Figure 12.14). A similar deck built in 1950 incorporated untensioned wires in the precast beams instead of the bars in the *in-situ* concrete. Somewhat later high-voltage electrification of railways required the depth of overbridge decks to be reduced to the absolute minimum; a new design of deck, partially prestressed and prestressed in two stages was developed which allowed depth to span ratios as low as 1/30 (Figure 12.15). Two such bridges had been built before 1959. In the second and more developed of these groups of precast inverted T beams, including unstressed as well as pretensioned wires, were assembled fully supported near the bridge; duct formers and some transverse reinforcement were placed; *in-situ* concrete infill/topping was

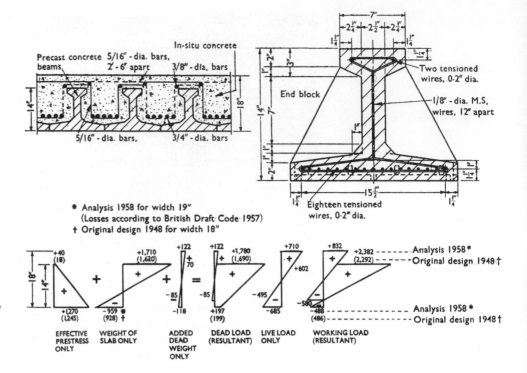

Figure 12.14 Details of road over rail bridge at Buck Lane (1949). This was the first partially prestressed bridge.

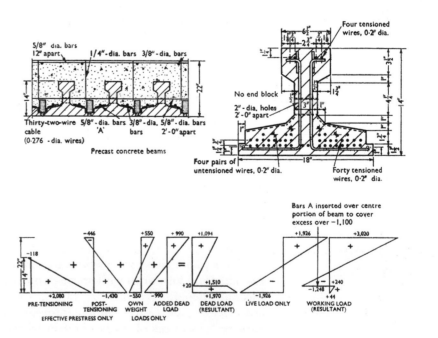

Figure 12.15 Details of road over rail bridge at Colchester about 1958. The 'wafer slab' design was developed to allow exceptional span/depth ratios (up to 1/30) for railway electrification. The precast, partially prestressed inverted T-beams were propped during casting of the in-situ concrete, and the whole was then stressed using the Magnel-Blaton system.

placed to make a slab unit, Magnel-Blaton cables were inserted and tensioned and the slab units were lifted into place by crane. British Railways seems to have lost interest in partial prestressing after 1963.[31]

Railway bridges were also built at industrial sidings. A trough-shaped through bridge built at a Rotherham steelworks in 1952, using the Lee-McCall system, has a skew of 58.5° and a span of 160 ft over the River Don and was thought at the time to be the largest span railway underbridge in the world[32] (Figure 12.16). See also Ref. 33.

Figure 12.16 Railway bridge at Rotherham steelworks, over River Don (1952). At the time this bridge was thought to be the largest span railway bridge anywhere (span 160ft., skew 58.5°. Lee-McCall system.

Railway bridges after 1960

Minimum disruption to train services has always been a major criterion for railway engineers, and one which is particularly difficult with underbridges. One of the main techniques developed is the construction of a new deck alongside the existing track, followed by sliding it into position. A paper of 1961 describes three interesting bridges: one consisting of four separate prestressed box girders rolled in to position; one of transverse prestressed slabs post-tensioned together; the third is a cast *in-situ* box girder whose top flange is a platform, with cantilevers of the bottom flange carrying tracks on either side.[34]

Sliding in of concrete railway bridges and some larger and interesting box girder and arch bridges built in the 1960s are described in other references.[35,36] One of the bridges was recently reassessed.[37]

The first incrementally launched bridge in the UK was a railway overbridge (1977).[38] In the early 1980s a unique prestressed concrete cable stayed railway underbridge was built across the M25 at a considerable skew.[39]

Larger bridges — 1954 onwards

The first substantial prestressed concrete road bridge was the replacement for Northam Bridge over the River Itchen at Southampton (1954, Figures 12.17 and 12.18).[40] This bridge combined up-to-date technology with a style belonging to well before the war. It is typical of its time in that the main deck structure consists of beams rather narrowly spaced and has transverse diaphragms. The beams were precast on site using deflected cables and the deck was made continuous for live and superimposed loads by means of *in-situ* diaphragms between the ends of the beams and precast prestressed slabs clamped between the tops of the beams by transverse stressing over a length where the flanges of the tees were omitted.

After the Northam Bridge prestressed concrete was used for larger and larger bridges, using a variety of forms and construction techniques. Cavendish Bridge (1956), Clifton Bridge[41,42] and Bridstow Bridge[43] were all cantilever and suspended span bridges using precast beams for the suspended span. Clifton Bridge (Figure 12.19) was probably the first postwar concrete bridge in the UK to use box girders and free cantilever construction. The anchor spans and cantilevers

Figure 12.17 Northam Bridge, Southampton, the first major prestressed bridge in Britain.

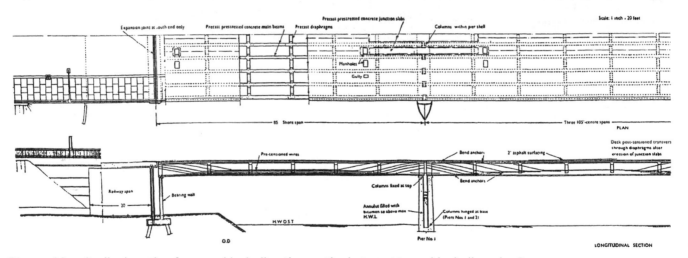

Plan and longitudinal section from and including the south abutment to and including pier 2

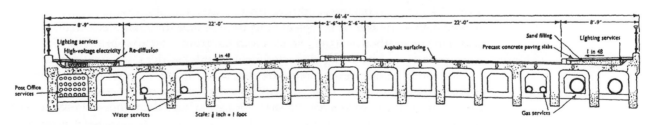

Cross-section at mid-span

Figure 12.18 Northam Bridge. The frequent diaphragms are typical of the period. Near the supports substantial parts of the flanges of the T-beams were removed to allow the junction (continuity) slabs to fit between them. Freyssinet system. Main beams precast, pre-tensioned on site, with deflected tendons. Post-tensioned through diaphragms after junction slabs placed.

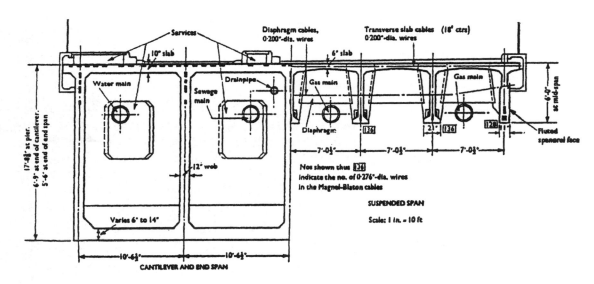

(a)

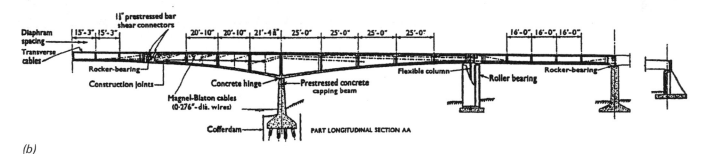

(b)

Figure 12.19   Clifton Bridge, Nottingham. Again the frequent diaphragms can be seen, even in the box-girder. (a) Typical cross-section. (b) Longitudinal section.

are box girders and only the cantilevers were built by free cantilever construction. The anchor spans and cantilevers of the Medway Bridge (1963),[44,45] also a cantilever and suspended span box girder bridge, were built by free balanced cantilevering, and its main span of 152 m was by far the largest concrete span in the UK until the Orwell bridge (1982, 190 m span).[46] It may also have been the first to omit diaphragms except at supports.

Other bridges of interest are: Donnington Bridge,[47] with closely spaced portal frames, partly precast, each leg formed by a tie and a strut. Winthorpe Bridge[48] with three span continuous multiple boxes; Taf Fawr Bridge[49] with precast segmental I beams built by cantilever construction and connected by *in-situ* flanges to form a three-celled box;[50] and Wentbridge Viaduct[51,52] a three-span box girder with raking legs which have concrete hinges at top and bottom.

In 1963 the Ministry of Transport held an open competition for the design of a bridge to carry the M1 over the River Calder, near Leeds. The crossing was on a considerable skew and the bridge had to be capable of coping with mining subsidence. Three of the four prize winning designs had prestressed concrete decks and the other was prestressed concrete except for its composite steel and concrete suspended span.[53,54]

Before the end of the 1960s prestressed concrete had superseded reinforced concrete for all but small bridges and arches, and the box girder had become a dominant structural form for large bridges in concrete and steel, because of its structural efficiency and economy. The scale of bridges increased, not only in span but also in cross section, large single cell box girders with wide edge cantilevers replacing the earlier type with several small cells and smaller edge cantilevers, or none in the earliest bridges. Fewer cantilever and suspended span bridges were built and more of the larger bridges were continuous beam structures.

Reinforced concrete bridges

Particularly in the 1950s some larger bridges were still built wholly or partly of reinforced concrete. For obvious reasons a number of them were large arches, generally with open spandrels and often with twin arch ribs, typified by the Lune Bridge carrying the M6 (1959).[55] The reinforced concrete arches of the Nant Hir and Taf Fechan Bridges[46] were built a few years later by cantilever construction using temporary cables, as described by Hansen.[47] The beam bridge carrying the A1 over the River Wharfe[56] has cantilevers and anchor spans of reinforced concrete although the suspended span uses precast post-tensioned I beams. The long approach viaducts either side of the Queenhill Bridge (1961) were reinforced concrete boxes.[57]

Long motorway-type viaducts

In the 19th century long viaducts were mostly railway viaducts. From the end of the 1950s a number of long road viaducts were built, many in prestressed concrete. Some of the ones built from precast segments are mentioned elsewhere in this chapter, from the Hammersmith Flyover onwards. The Chiswick-Langley Viaduct[58] has tee-headed piers carrying precast pre-tensioned inverted tee beams with an *in-situ* concrete slab, and others have been built from other types of precast beams. Others were constructed *in situ* using span-by-span construction, such as Gateshead Highway (1971),[59] a serious attempt to visually integrate the junctions of the slip road ramps with the main structure. More recently cantilever construction has been used, mostly as post-tensioned box girders of various configurations, some multi-cell and some as groups of single cells.

Larger precast segmental bridges

The Chiswick Flyover (1959) was built from concrete beams one third of the span in length, assembled on temporary supports and post-tensioned.[60] Precast segmental construction gained a new dimension with the construction of the Hammersmith Flyover (1961) (Figure 12.20a, b), the first spine box with edge cantilevers, and a bridge of unique character. Alternating spine segments and diaphragm/cantilever bracket units were assembled on falsework with *in-situ* concrete between them before stressing.[61] Mancunian Way[62] and Westway[63] (Figure 12.20c) which followed used the same construction technique with simpler arrangements of precast segments incorporating the spine box with the edge cantilevers and with diaphragms only at supports. London Bridge[64,65] used a similar technique, but with units suspended from a gantry. Epoxy resin joints were first used on a British bridge, for joining precast segments which were assembled on falsework, at Rawcliffe Bridge (1968).[66] More recent bridges have used match cast segments, with a thin layer of epoxy resin in the joints and built without falsework by cantilever construction. The first two British bridges to use match cast segments in cantilever construction were Byker Viaduct[67], a railway bridge for the Tyne and Wear Metro, which used both balanced and progressive cantilevering, and the M180 bridge across the Trent.[68,69]

Incremental launching

The first incrementally launched bridge in the UK was Shepherd's House Bridge carrying the A4 over a main line railway (1977),[32] and several others have since been completed.[70]

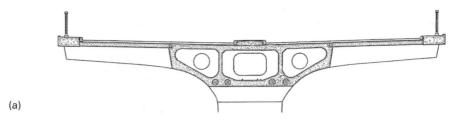

(a)

(b)

(c)

Figure 12.20
(a) Hammersmith flyover:
double cantilver precast units
59 ft long supporting 24 ft
carriageways. (b, c) Westway:
double cantilever precast
units 94 ft long supporting 41
ft carriageways.

Smaller bridges since 1957

Although many quite small bridges were built of prestressed concrete, many were built of reinforced concrete,[71] for example the Kingsgate Footbridge, which integrates its structural form and construction method with its appearance (1963).[72]

Since road construction started seriously in 1957 vast numbers of bridges have been built over grade separated roads. Some of them carry important roads or railways, but many are carrying minor roads, farm accesses, footpaths or bridleways.

On the Preston Bypass, the first section of M6 which was started in 1956 the mix of bridges is shown on Figure 12.21.[73] By 1959 the bypass had been completed and work was under way on the M5 in Worcestershire,[74] the A1 in the West Riding of Yorkshire,[75] the Maidstone Bypass in Kent.[76] The general picture was of a mixture of prestressed and reinforced concrete and sometimes composite steel and concrete bridges. However this was not always the case as can be seen from Figure 12.22 which shows standardized solutions to typical motorway bridging problems on various motorways, but mainly the M1.[77] Eleven types of overbridge are shown in the original figure. One of them (representing four bridges) has universal beams and lightweight concrete. The other ten types (representing 178 bridges) are all of reinforced concrete cast *in situ*.

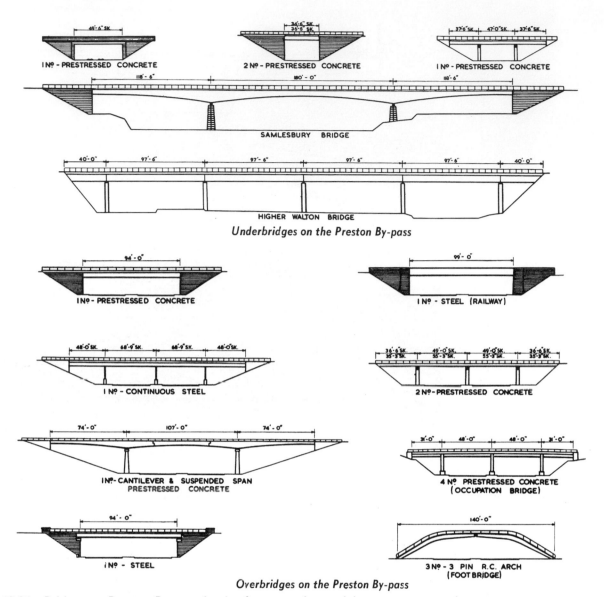

Underbridges on the Preston By-pass

Overbridges on the Preston By-pass

Figure 12.21   *Bridges on Preston Bypass. A mix of types and materials, as on many early motorways.*

Some of these reinforced concrete bridges on the M1 are unusual. The first one in the figure is the anachronistic-looking propped portal frame slab with mass concrete legs/abutments which is such a feature of the southern part of the motorway; the deck thickness is stated to be 1/36 of the span. The following four appear to be a type with hinges in the deck, some of which on the M6 were recently subjected to assessment.[78] On the early motorways a great variety of forms was used, simply supported or continuous beams, portal frames, occasional arches. Cross-sections might be beam and slab, solid slabs or voided slab and various combinations of precast and *in situ*, reinforced and prestressed. Other papers on early motorway bridges may also be of interest[79-81] and *Modern British Bridges* contains a number of early overbridges.[2]

Precast bridges and standard beams

Smaller concrete bridges often had decks consisting of precast concrete slabs, others of *in-situ* slabs supported by and working compositely with concrete beams, either reinforced or, if prestressed, often precast and pre-tensioned. Precast beams were originally designed ad hoc but towards the end of the 1950s and early 1960s

SECTION	TYPE	LOCATION	CONSTRUCTION	SPAN RANGE	NUMBER OR LENGTH BUILT	CLEARANCE	SERVICES	REMARKS
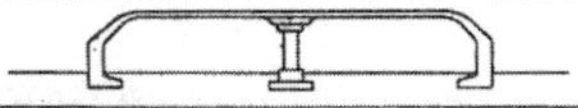	OVER ROAD MAJOR AND MINOR	M1, LUTON TO CRICK AND ST ALBANS BY-PASS	IN SITU R C PORTAL SLAB	2 x 50 FT TO 2 x 125 FT	64	FREQUENTLY CRITICAL	NOTHING EXCEEDING 6" PIPE	LEAST DECK THICKNESS = 1/36 x SPAN
	OVER ROAD 90° TO 60° SKEW	M1, CRICK TO NOTTINGHAM	IN SITU R C SLAB	OVER 3-LANE CARRIAGEWAY	33	NEVER CRITICAL	CONGESTED, UP TO 18" BORE PIPES	
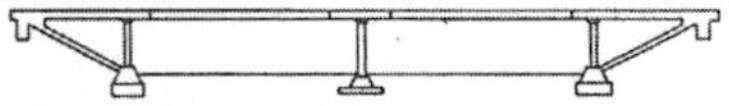	OVER ROAD 60° TO 45° SKEW	M1, CRICK TO NOTTINGHAM	IN SITU R C SLAB		10	NEVER CRITICAL	CONGESTED, UP TO 18" BORE PIPES	
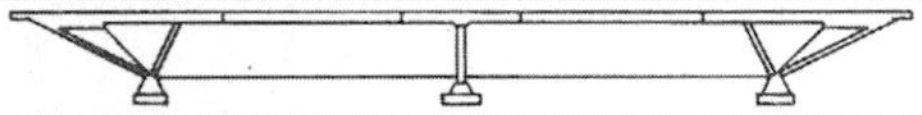	OVER ROAD 45° TO 35° SKEW	M1, CRICK TO NOTTINGHAM	IN SITU R C SLAB		2	NEVER CRITICAL	CONGESTED, UP TO 18" BORE PIPES	
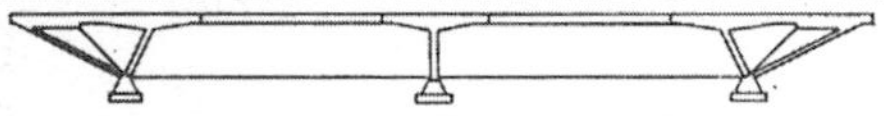	OVER ROAD 35° TO 30° SKEW	M1, CRICK TO NOTTINGHAM	IN SITU R C SLAB		3	NEVER CRITICAL	CONGESTED, UP TO 18" BORE PIPES	
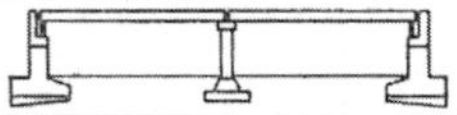	OVER ROAD	M1, NOTTINGHAM TO DONCASTER	IN SITU R C COFFERED SLAB		33		CONGESTED, UP TO 18" BORE PIPES	SUBJECT TO MINING SUBSIDENCE
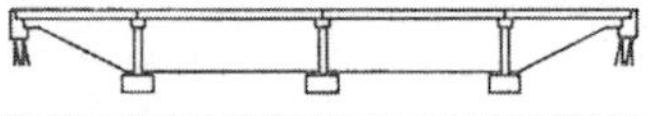	OVER ROAD	M6, BIRMINGHAM – STAFFORD	UNIVERSAL BEAM AND LIGHTWEIGHT CONCRETE		4	HIGH	CONGESTED	DECK THICKNESS 3'-7"
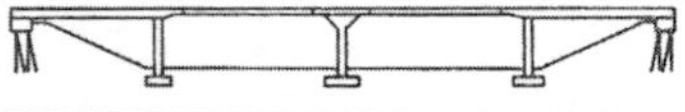	OVER ROAD	M4, M1	IN SITU R C SLAB		6	VERY TIGHT (JUNCTIONS)	NIL	DECK THICKNESS 1'6" OR 2'0"
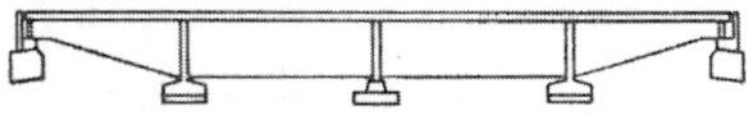	OVER ROAD 90° TO 60° SKEW	M6, CASTLE BROMWICH TO CATTHORPE	IN SITU R C SLAB		12	NOT CRITICAL	CONGESTED	
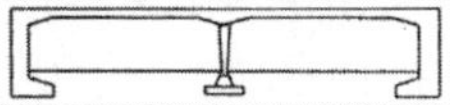	OVER ROAD	M5	IN SITU R C SLAB		8	MINIMUM	CONGESTED	
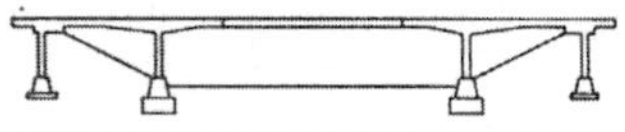	OVER ROAD	NEWPORT (MON) BY-PASS	IN SITU R C SLAB	OVER 2-LANE CARRIAGEWAY	6			IN RETAINED CUTTING

Figure 12.22   Standardized solutions to typical motorway bridging problems. This table shows a number of interesting reinforced concrete bridge types, and indicates that some designers saw no place for prestressing as late as 1971.

the Prestressed Concrete Development Group, under the aegis of the Cement and Concrete Association, prepared standard sections. These and subsequent developments have been well covered by Sriskandan[82] and Taylor.[83]

Estuarial crossings

Most steel bridges incorporate significant concrete elements. Of the four major estuarial road crossings by suspension bridge which were built in Britain after 1940, two have concrete towers. The Tamar Bridge[84] also has a concrete deck slab. The tapering slip-formed concrete towers of the Humber Bridge are themselves major constructions.[85]

Ancillary items

During the postwar period there have been considerable changes in such things as expansion joints, bearings and hinges; involving the use of new materials to produce more effective and often cheaper designs. Laminated rubber bearings were available by 1957 when they were used on Pelham Bridge.[86] Pot bearings were developed in 1959 and sliding bearings using PTFE were first used at the beginning of the 1960s.[87] Older types also were improved, with rollers of increased surface hardness and the use of stainless steel.

Concrete hinges were used on several bridges, and a prototype for Wentbridge Viaduct[48] was tested at the Cement and Concrete Association (C&CA) (Figure 12.23). A type of triple hinge was developed for Wichert Truss bridges in Yorkshire designed to cope with mining subsidence.[88] Mesnager type hinges were also sometimes used, for instance in the decks of some motorway overbridges[75] and between the deck and main piers on the Medway Bridge (Figure 12.23).[41]

Expansion joints with more effective sealing became available. However the problems of leaking joints and the importance of detailing in the vicinity of joints were

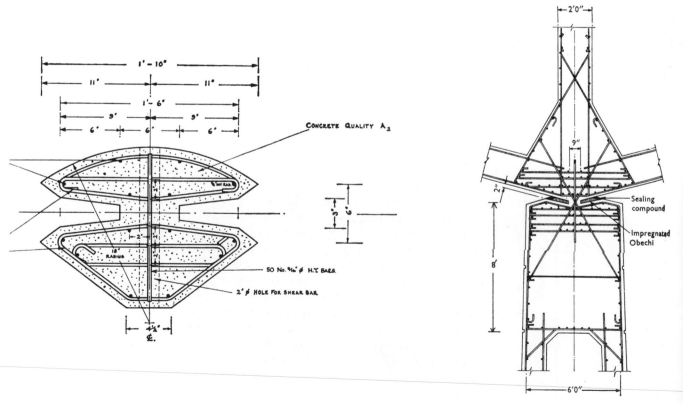

Figure 12.23   Two types of concrete hinge. The ones for Wentbridge Viaduct were precast and those on Medway Bridge were cast in situ.

generally not appreciated as fully as in more recent times. Many bridges consisted of several (sometimes many) spans of simply supported beams with a joint over every support and these have often leaked very badly. Sometimes new materials or methods (e.g. epoxy nosings) seemed promising in trials but were less successful in general use.

Structural analysis

The history of structures since the 18th century has been paralleled by the history of structural analysis, and this was also true in the postwar period. The tools available for analysis also underwent dramatic changes.

Tools

Up to and for some time after the war, the engineer's chief calculating tool was the slide rule which during the 1970s was gradually superseded by increasingly sophisticated electronic calculators. For calculations requiring greater accuracy logarithmic tables and mechanical calculators, often hand operated, were used until the 1960s. Before the advent of computers the solution of simultaneous equations with more than four unknowns was a long winded process, rapidly approaching the impractical as the number increased. By the end of the 1950s computer bureaux could provide rapid solutions to sets of simultaneous equations, although it may be that many engineers were slow to take advantage of them. By the mid-1960s some structural programmes were available and design offices had started to equip themselves with computers. In the late 1970s and early 1980s, programmes became available specifically for analysing prestressed concrete bridges as well as programmes which enabled very demanding construction calculations, such as controlling the alignment in cantilever construction, to be carried out rapidly. The result was to enable engineers who were using innovative methods to do so more easily and to encourage others to follow.

Analysis and the new techniques

New structural techniques needed more sophisticated analyses. Prestressed concrete is more demanding and less forgiving than reinforced concrete, where high local stresses can redistribute themselves without causing failure. Types of structures such as box girders also require more complex analyses to realize their potential. After the advent of computers the proportion of statically indeterminate bridges increased.

Perhaps the most important analytical development of the 1950s was that by the C&CA on methods of load distribution, following on from the work of Guyon and Massonet on the Continent.[89]

Another important development was an increasingly better understanding of how box girders actually worked. The paper on Clifton Bridge[38] refers to aeronautical engineering literature being used to give an idea of the effects of skew. The work of Vlasov,[90] based on thin shell theory, became available in the early 1960s, introducing the idea of the 'bimoment', self-equilibrating moments associated with warping and distortion of the cross-section. For British engineers the papers of Richmond[91] from the mid 1960s made the subject more comprehensible, and computers made the calculations feasible. Also during this period thermal stresses began to be calculated as a matter of routine and the ability to carry out more complex calculations was generally followed by demands for them to be carried out.

For bridges which were difficult to analyse, physical models, of perspex and sometimes of micro-concrete, were tested, many at the laboratories of the C&CA.

Clifton Bridge was possibly the first, others were Mancunian Way and Gateshead Highway. Following model tests at the C&CA the Medway Bridge was built without diaphragms in the spans, showing the way for later concrete box girders.

Problems with prestressed bridges

Concrete bridges have not been immune to the problems experienced by concrete structures generally due to alkali silicate reaction and the use of calcium chloride. Bridges are particularly exposed to the elements and the salting of road bridges has produced more corrosive environments. Some of the earliest prestressed bridges still seem to be in good condition; however problems have been experienced with others, particularly corroding tendons where bridges have been salted and ducts were not completely filled with grout. It was a number of years after the war before suitable ducts were found and satisfactory techniques for grouting them. Cables external to the concrete sections were used as early as Clifton Bridge (completed 1958, detailed design 1953) and in a number of subsequent bridges. However corrosion of tendons was experienced on some of these and as a result in the early 1980s the Department of Transport banned the use of external cables on its bridges. In 1985, a segmental bridge with grouted tendons failed, due to corrosion at the joints between segments,[92] and corrosion has been found in some other bridges with grouted cables. In 1992 the Department placed a ban on grouted cables in new bridges being constructed for it, and for some years only external cables were acceptable. A Concrete Society working party was set up to find means of producing grouted cables with acceptable reliability[93–95] and after some four years the ban was lifted.

The changing character of bridges

The changes in bridges since 1940 were not driven by style, but by the search for more effective and economical methods of construction. The changes in the characteristic appearance of bridges may have a cultural component, but it is hard to isolate it, particularly as the architecture of the early postwar period emphasized function as the basis for style. Some of the larger bridges were designed with advice from architects and may have had an influence on others. Bridges of this period are simpler in appearance than their predecessors, larger in scale, lacking in detail. This is just as true of steel bridges as it is of concrete ones. Large modern concrete and steel bridges are often surprisingly similar in appearance, particularly where the concrete has a smooth surface.

Typical bridges of the 1950s may be simpler in appearance than those of the 1930s, but they still do not have the distinctively modern character which started to appear in the 1960s. A typical 1950s beam bridge often has a number of beams very close together without any cantilever over the edge other than a slightly projecting fascia. It also has cross beams or diaphragms to stiffen the cross section and distribute loads. These are not present in modern bridges which have had more sophisticated analyses and a different economic climate for construction. Even box girders had frequent internal diaphragms, and Medway Bridge must have been one of the first to do without them. A typical modern road bridge has few main members, possibly only one or two box girders, often with edge cantilevers which are often quite large, and diaphragms only at the supports. Bridges are still evolving, as always.

Postscript

In February/March 1996 English Heritage put forward, for public consultation, 65 postwar buildings recommended for listing. Eleven of these were bridges (nine of concrete) and the only 'buildings' recommended Grade I were the Severn Bridge and the Kingsgate Footbridge. In 1998 these two were listed as Grade I, Winthorpe

Bridge as II* and the following as Grade II:

- English and Welsh sections of the Wye bridge;
- Rhinefield Bridge, near Brockenhurst;
- Footbridge over the River Cherwell at Parson's Pleasure Punt Rollers, Oxford;
- Garret Hostel Bridge, Cambridge;
- Wentbridge Viaduct, near Selby;
- Swanscombe Footbridge, A2 near Ashford;
- West Footbridge, London Zoo.

All but the Severn and Wye Bridges are concrete. The Adam Viaduct was listed Grade II in 2001.

Acknowledgements

The starting point for this chapter was a study of postwar bridges for English Heritage. It could not have been written without the help of Michael Chrimes and the libraries of the Institution of Civil Engineers, the British Cement Association and Ove Arup and Partners; and would have been even more imperfect than it is without discussions with Dr George Somerville, Sir Alan Harris, Dr E.W.H. Gifford, Alfred Goldstein and Dr Francis Walley, and help from too many others to mention by name. The help of John Henry was invaluable in finding out how to get to bridges considered for listing.

References

1. Sriskandan, K., Prestressed concrete road bridges in Great Britain: a historical survey. Proc. ICE, 1989, 86, 269–302.
2. Henry, D., Jerome, J.A., Modern British Bridges. CR Books: London, 1965.
3. Mason, A.P., Alveley Bridge — cantilevered centering. Concr. Constr. Eng., 1937, 32, 453–59.
4. Paul, A.A., The use of pre-stressed concrete beams in bridge deck construction. J. ICE, 1943, 21, 19–30.
5. Anon., An exhibition of prestressed concrete. Concr. Constr. Eng., 1949, 44, 129.
6. Anon., Reconstruction of Nunn's Bridge, Lincolnshire. Concr. Constr. Eng., 1948, 43, 239–44.
7. Walley, F., The progress of prestressed concrete in the United Kingdom. Lecture given at the Institution of Civil Engineers, 1962.
8. Thomas, F.G., Pre-stressed concrete. Proceedings of Conference on Prestressed Concrete held at Inst. Civ. Engrs, 1949.
9. Gifford, E.W.H., Recent developments in highway bridge design in Hampshire. Proc. ICE, Part 2, 1952, 1, 461–85.
10. Anon., A prestressed bridge in North Devon. Concr. Constr. Eng., 1954, 49, 169–70.
11. Criswell, H., Prestressed concrete bridges in North Devon. J. Inst. Highw. E., 1956, 3, 60–66.
12. Goldstein, A., Design and construction of a prestressed concrete arch footbridge at Oxford. Concr. Constr. Eng., 1950, 45, 347–57.
13. Anon., Concrete structures at the festival of Britain. Concr. Constr. Eng., 1951, 46, 199–206.
14. Anon., Test of a prestressed concrete footbridge. Concr. Constr. Eng., 1952, 47, 185–88.
15. Walley, F., St. James's Park Footbridge. Proc. ICE, 1959, 12, 217–21.
16. Prestressed Concrete Development Group, Prestressed Concrete Footbridges, Prestressed Concrete Development Group, 1962.
17. Anon., A prestressed pipe bridge. Concr. Constr. Eng., 1955, 50, 229–32.
18. Anon., A Swiss system of prestressing. Concr. Constr. Eng., 1958, 53, 431–39.
19. Anon., Two reinforced concrete bridges at Boroughbridge. Concr. Constr. Eng., 1947, 42, 217–22.
20. Anon., Boonshill Bridge, Rye, Sussex. Concr. Constr. Eng., 1947, 42, 271–72.
21. Buckton, E.J., Cuerel, J., The New Waterloo Bridge. J. ICE, 1943, 20, 145–201.
22. Anon., A pre-stressed concrete railway bridge near Wigan. Concr. Constr. Eng., 1947, 42, 305.

23. Dean, A., Pre-stressed concrete applied to the construction of railway bridges and other works, Proc. ICE, Rly. Div., 1951, 44, 14–33.
24. Anon., A railway viaduct in pre-stressed concrete. Concr. Constr. Eng., 1952, 47, 385–87.
25. Berridge, P.S.A., Prestressed concrete slabs for railway bridges. Concr. Constr. Eng., 1954, 49, 283–88.
26. Anon., A prestressed bridge at Manchester. Concr. Constr. Eng., 1958, 53, 395–98.
27. Note in Concr. Constr. Eng., 1949, 44, 37.
28. Anon., Prestressed concrete bridges in Yorkshire. Concr. Constr. Eng., 1950, 45, 99–100.
29. Anon., A new type of prestressed bridge deck. Concr. Constr. Eng., 1958, 53, 145–46.
30. Sadler, R.E., Development in overhead electrification of railways as it affects the civil engineer. Proc. ICE, 1959, 12, 125–51.
31. Abeles, P.W., Partial prestressing and its suitability to limit state design. Struct. Engr, 1971, 49, 67–86. Discussion 529–41.
32. Anon., Prestressed concrete railway bridge at Rotherham. Concr. Constr. Eng., 1953, 48, 19–21.
33. Anon., Test of a prestressed concrete railway bridge. Concr. Constr. Eng., 1951, 46, 186–87.
34. Turton, F., Three prestressed concrete railway bridges. Proc. ICE, 1961, 20, September, 1–18, discussion 1962, 22, July, 317–330.
35. Mann, F.A.W., Developments in the construction of concrete railway bridges. Concrete, 1968, 2, 373–79, 426–29.
36. Mann, F.A.W., Railway Bridge Construction: Some Recent Developments. Hutchinson Educational, 1972.
37. Blackler, M.J., Cooke, R.S., Besses O' Th' Barn Bridge: inspection and testing of a segmental post-tensioned railway bridge. Proc. ICE Structs Bldgs, 1995, 110, 19–27.
38. Best, K.H. et al., Incremental launching at Shepherd's House Bridge. Proc. ICE, Part 1, 1978, 64, 83–102.
39. Kretsis, K., Lyne railway underbridge. Proc. ICE, Part 1, 1982, 72, 585–610.
40. Wooldridge, F.L. et al., The New Northam Bridge, Southampton. Proc. ICE, Part 1, 1955, 4, 269–89.
41. Finch, R.M., Goldstein, A., Clifton Bridge, Nottingham: Initial Design Studies and Model Test. Proc. ICE, Part 1, 1959, 289–316.
42. Finch, R.M., Goldstein, A., Clifton Bridge, Design and Construction. Proc. ICE, Part 1, 1959, 4, 317–52
43. Henry, D., Jerome, J.A., Modern British Bridges. CR Books: London, 1965.
44. Kerensky, O.A. et al., Medway Bridge: design. Proc. ICE, 1964, 29, 19–52.
45. Kier, M. et al., Medway Bridge: construction. Proc. ICE, 1964, 29, 53–100.
46. Lewis, C.D. et al., Orwell Bridge — design. Proc. ICE, Part 1, 1983, 74, 765–78.
47. Henry, D., Jerome, J.A., Modern British Bridges. CR Books: London, 1965, 144–45.
48. Anon., Bridges on the Newark Bypass Road. Concr. Constr. Eng., 1964, 59, 351–55.
49. Coombs, A.S., Hinch L.W., The Heads of the Valleys Road. Proc. ICE, 1969, 44, 89–118.
50. Hansen, F., A contractor's view of design and its influence on construction. Proc. Meet. Design Concr. Bridge Struct., Concrete Society: London, 1967.
51. Sims, F.A., The design of the Wentbridge Viaduct. Struct. Concr., 1963, 1, 573–80.
52. Markham, R.B., The Construction of the Wentbridge Viaduct. Struct. Concr., 1963, 1, 553–65.
53. Anon., Prize-winning designs for a motorway bridge. Concr. Constr. Eng., 1964, 59, 113–18.
54. Gifford, E.W.H. et al., The design and construction of the Calder Bridge on the M1 motorway. Proc. ICE, 1969, 43, 527–52. Discussion 1970, 46, 355–71.
55. Henry, D., Jerome, J.A., Modern British Bridges. CR Books: London, 1965: 97.
56. Henry, D., Jerome, J.A., Modern British Bridges. CR Books: London, 1965: 176–77.
57. Gibb, M.E., Tansley, F.J., Queenhill Bridge over the River Severn. Proc. ICE, 1962, 23, 545–63.
58. Henry, D., Jerome, J.A., Modern British Bridges. CR Books: London, 1965, 160–61.
59. Brown, P.A., Prestressed concrete bridges 1970–74. Concrete, 1974, 8 (April) 45–49.
60. Henry, D., Jerome, J.A., Modern British Bridges. CR Books: London, 1965, 108–109.
61. Rawlinson, J., Stott, P.F., The Hammersmith Flyover. Proc. ICE, 1962, 23, 565–600.
62. Bingham, T.G., Lee, D.J., The Mancunian Way elevated road structure. Proc. ICE, 1969, 42, 459–92.
63. Baxter, J.W. et al., Design of Western Avenue Extension (Westway). Proc. ICE, 1972, 51, 177–218. Also paper by Nundy on construction 219–50.

64. Brown, C.D., London Bridge: planning, design and supervision. Proc. ICE, 1973, 54, 25–46.
65. Mead, P.F., London Bridge: demolition and construction. Proc. ICE, 1973, 54, 47–69.
66. Sims, F.A., Applications of resins in bridge and structural engineering. Int. J. Cement Composites Lightweight Concrete, November 1985.
67. Smyth, W.J.R. et al., Tyne and Wear Metro: Byker Viaduct. Proc. Inst. Civ. Engrs, Part 1, 1980, 68, 689–700.
68. Sims, F.A., FIP 78, UK National Report on Bridges. Concrete, 1978, May, 16–25.
69. Sims, F.A., Applications of resins in bridge and structural engineering. Int. J. Cement Composites Lightweight Concrete, 1985.
70. Rowley, F.N., Incremental launch bridges: UK practice and some foreign comparisons. Struct. Engr, 1993, 71.
71. Henry, D., Jerome, J.A., Modern British Bridges. CR Books: London, 1965, 86, 134 and others.
72. Anon., An elegant footbridge at Durham with an original method of construction. Concrete Quarterly, 1964, 60, January–March.
73. Anon., The Preston by-pass, road and bridge works. Roads & Road Constr., 1957, 35, 200–207.
74. Thomson, W.R., Motorway design and construction in Worcestershire. Roads & Road Constr., 1959, 57, 45–50.
75. Anon., The Great N. Road in the W. Riding of Yorkshire. Roads Road Constr., 1959, 57, 55–56.
76. Anon., The Maidstone bypass road: bridges for the eastern section. Concr. Constr. Eng., 1958, 53, 195–200.
77. Williams, O.T., The consulting engineer's view. Some particular considerations. Proc. ICE Conf. Motorways Br. Today Tomor., London, 1971: 48.
78. Wilson, C.B., Assessment of the reinforced concrete hinges on five M6 overbridges in Staffordshire Proc. ICE Structs. Bldgs., 1995, 110, 4–10.
79. Anon., The London-Birmingham Motorway. Concr. Constr. Eng., 1959, 54, 337–44, 413–14.
80. Anon., Bridges on the Birmingham-Preston Motorway (Cheshire Section). Concr. Constr. Eng., 1962, 57, 31–32.
81. Anon., Concrete Bridges on the New Motorways. Concr. Constr. Eng., 1964, 59, 389–400, 437–43.
82. Sriskandan, K., Prestressed concrete road bridges in Great Britain: a historical, survey. Proc. ICE, 1989, 86, 274–77.
83. Taylor, H.P.J., The precast concrete bridge beam — the last 50 years. Struct. Engr, 1998, 76, 407–14.
84. Anon., Concrete piers for the Tamar Road Bridge. Concr. Constr. Eng., 1961, 56, 130–34.
85. Sims, F.A., UK national reports to FIP78 — Bridges. Concrete, 1978, 12, May, 25.
86. Anon., Rubber bearings for bridges. Concr. Constr. Eng., 1959, 54, 350.
87. Lee, D.J., The theory and practice of bearings and expansion joints for bridges, 1971. Cement and Concrete Association, London.
88. Sims, F.A., Bridle, R.J., The design of concrete hinges. Concr. Constr. Eng., 1964, 59, 277–286.
89. Rowe, R.E., Concrete Bridge Design. CR Books: London, 1962.
90. Vlasov, V.Z., Thin walled elastic beams, Israel Program for Scientific Translations, Jerusalem, 1961.
91. Richmond, B., Twisting of thin-walled box girders. Proc. ICE, 1966, 33, 659–75.
92. Woodward, R.J., Williams, F.W., Collapse of Ynys-y-Gwas Bridge, West Glamorgan. Proc. ICE, Part 1, 1988, 84, 635–69.
93. Woolley, M.V., Clark, G.M., Post-tensioned concrete bridges. Struct. Engr, 1993, 71, 409–11.
94. Porter, M.G., Repair of Post-tensioned Concrete Structures. Concrete Society, Concrete Bridges — Investigation, Maintenance and Repair, London, 1985.
95. Raiss, M., Lasting Effect, Concrete Engineering, supplement to New Civil Engineer, 1995, 46–48.

13 Reinforced and prestressed concrete in maritime structures

Brian Sharp

Introduction

In this chapter, the development of reinforced concrete and the later innovation of prestressed concrete in maritime structures is reviewed, from its first such introduction in France in 1896 until the era of North Sea Oil Platforms in the 1970s. Most of the applications are for quay structures but related works include coastal structures, offshore structures and lighthouses. The review is divided into broad periods. Starting with the early applications of reinforced concrete from the turn of the century to 1920, it traces the extensive developments of the 1920s and 1930s, then the Second World War followed by the activity of post-war construction, which saw the increasing size of maritime facilities. Although focused on the UK, the review includes related overseas activity.

The aim of the chapter is threefold:

- To understand and appreciate what was achieved in these past periods and what was done, how and why.
- Thereby to enable present-day engineers to assess, repair and renovate where appropriate, by providing an insight into what system and details may lie beneath the surface. For this reason, more details are given of earlier structures as it is hoped that later ones should be better recorded and be more recognizable in structural form and detailing.
- To learn from the past and interpret the reasons for success or failure in the performance, especially relating to structural form, detailing and durability, to improve the design of future structures.

Engineering related to the sea is one of the oldest branches of civil engineering, the sea being a major highway for exploration, demographic expansion, fishing and trade. Loading and exposure conditions for works in the sea are usually exceptionally severe, due to the sheer mass of ships and the loads they carry, the natural forces of winds and waves and the corrosive properties of seawater.

To combat these conditions, plain Portland cement concrete and then reinforced concrete were readily employed as soon as the materials and systems became available. Consequently, the seawater environment has served as a proving ground and accelerated test-bed for methods and materials.

By virtue of its strength in bulk and versatility in both *in situ* and precast form, concrete has obvious advantages in maritime structures. Portland cement was mistrusted for use in seawater, being reported as first used in blocks for the original Admiralty Pier at Dover in 1849.[1] Previously, mass (unreinforced) concrete was made with lime or pozzolan as the binder. In France, prior to 1914, the use of Portland cement in port works was negligible.[2]

The medium of reinforced concrete by definition excludes plain unreinforced concrete, termed 'mass' concrete in UK but not in the USA. However, as the introduction of steel reinforcement on structural grounds causes its own problems for durability, and as unreinforced concrete predates reinforced concrete and can still remain the optimum solution when structurally and economically feasible, its historic and modern use is summarized as a preview. Gravity structures in the sea using masonry blockwork and rubble concrete infills have been used from early times and precast concrete blocks were and remain one of the most convenient means of gravity wall construction, such as used at Folkestone between 1897 and 1905,[1] as shown in Figure 13.1. Blocks of 1:6 concrete, average mass 16 t, were laid in bonded courses.

A modern example is the 'Tema' type wall[3,4] as shown in Figure 13.2. This innovatory form was dubbed the 'upside-down' wall. The heavy lines on the figure illustrate how the classical trapezium shaped wall with the widest section at the base is inverted to provide the widest section at the top. This inversion, initiated in fact by Alex Leggatt when at Halcrow, provides a most efficient restoring mass, the capping block providing an enhanced lever arm to the toe, and maximum mass in air with minimum deduction for 'submerged weight'.

As mentioned above, both plain concrete and even more so, and for good reason, reinforced concrete was mistrusted by maritime engineers at the turn of the century, and some of the questions related to its use, as raised in the reports reviewed below, sound remarkably familiar. A comparable spur to the replacement of timber in buildings by reinforced concrete because of its fire resistance was the scourge of sea worms in timber-piled sea structures.

Maritime structures figure prominently amongst the earliest applications of reinforced concrete from the end of the 19th century. The early days of building construction were dominated by the patented systems, but this was not so for maritime structures. Although the Hennebique system figures prominently in early maritime work, both this and several of the examples in the ICE Reports of 1913[5] all employed ordinary straight rods.

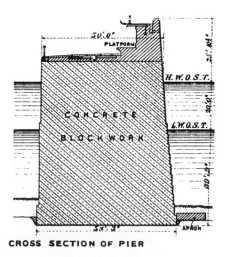

CROSS SECTION OF PIER

Figure 13.1   Folkestome pier, 1897–1905.

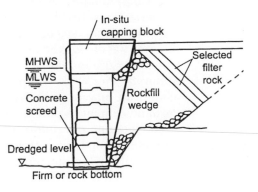

Figure 13.2   'Tema'-type blockwork wall, 1952–1965.

Francois Hennebique is the name most associated with the development of reinforced concrete in port works, beginning with his 8.25 m cantilever quay at Nantes[6,7] in 1896. Mouchel[7] introduced the Hennebique Ferro-Concrete system into Great Britain in 1897. As explained in Chapter 4 and illustrated in its Figure 4.7 and 4.34, the Hennebique system used normal round bars and resisted diagonal tension due to shear by inclined bars and open-topped stirrups made from flat strip. The stirrups embraced single bars. Patent links were used in columns and piles (Figure 4.2 of Chapter 4). Mouchel patented a hollow circular pile, and also the construction of massive cylindrical piers by driving one or more square-section piles in a cluster, which were then enclosed with a reinforced concrete cylinder, which was filled with concrete and steel (Figure 13.3).

The main problems for earlier port engineers were the stability of the cement compounds in seawater and the corrosion of reinforcement. Wentworth-Shields[8,9] explained that mass (Portland cement) concrete had been freely and successfully used in railway and dock construction in the last 50 years of the 19th century. When failures occurred, they were due to 'unsound' cement, that is cement containing uncombined lime. Concrete made with such cement swelled and disintegrated in seawater. Port engineers were constantly on the watch for unsound

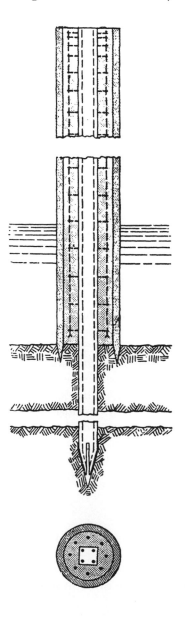

Figure 13.3   Mouchel piled cylinder piers (from Mouchel-Hennebique[7]).

cement and required cement to be turned over by shovel at weekly intervals. Pats of neat cement about 75 mm diameter were immersed in cold water; if the cement contained an excess of free lime the edges would crack. Cement for Folkestone pier[1] was checked by plunging broken tensile briquettes into boiling water for 3 h, to detect signs of swelling. Disasters had been experienced at Belfast, Aberdeen and in France in which the concrete had been reduced to gravel and sand. Fortunately, improved methods of cement manufacture removed these risks quite early on.

An uninformed observer may, these days, gain the impression that corrosion of reinforcement is new. It was observed as a major problem from the earliest stages,[2,5,6] but, mostly because of the time it can take to show in better-made structures in cold and temperate climates, was never really adequately understood. The mechanisms of the corrosion of reinforcement in concrete are only now beginning to be fully analysed, and it is not surprising that electrolysis from the effects of stray currents from electric cranes and 'moist' air were strongly suspected.

One of the major problems of analysing the performance of maritime structures is that success is as much a matter of detailing as it is of materials and is highly sensitive to the particular 'microclimate', that is the location of specific members in relation to the fluctuating sea level, in addition to the geographical 'macro-climate'.[10,11] Many of the earlier applications of reinforced concrete in quay construction involved a reinforced concrete deck spanning between timber, iron, steel or massive piles. Such an application, often being out of the range of seawater action and washed by rain water, was inherently durable. In other earlier examples, reinforced concrete was used in the substructure, and surmounted by conventional mass concrete or masonry in the upper tidal and splash zones, again affording ideal conditions for longevity.

This chapter concerns the details likely to be met in earlier construction as opposed to performance, but the two are mutually related. Where possible, feedback is given from later chapters' references to the same works.

The Institution of Civil Engineers Committees

Maritime works figured prominently in the Institution of Civil Engineers (ICE) Committee convened to review reinforced concrete practice in 1910.[6] Port works in the UK, New Zealand and Australia were included in the second report of 1913,[5] some details of which are given below.

One of the most valuable and longest contributions to the use of concrete in maritime work is that of the ICE Sea Action Committee, initiated in 1916 and lasting until 1960. The programme of research into the deterioration of structures in seawater included the use of timber, metal and concrete and observations in some 40 ports of the UK and in the Dominions. The first report,[12] published in 1920, had identified the main problems and recognized that, as the deterioration mechanisms were not rapid, a long-term programme was required. The 15th report[13] in 1935 gave an overview of the work to that time.

The conclusions of the final 20th report on the durability of reinforced concrete in seawater,[14] published in 1960, were that the primary cause of deterioration was corrosion of reinforcement and not disintegration of the concrete matrix itself. In hindsight, the conclusions, as were those of programmes carried out in other countries, were somewhat understated in comparison with the observations of 1920.[12]

The early years (1900–20)

According to Wentworth-Shields,[8] the first reinforced concrete maritime structure in the UK was built at a Southampton shipyard on the River Itchin in 1899, in imitation of a timber jetty. It consisted of a reinforced concrete deck 30.5 m by 12 m on reinforced concrete piles, designed and supervised by Hennebique, and was reported[8] to be in excellent condition in 1956. This was the Woolston jetty,[7,15] (Figure 13.4), which was later incorporated into an extensive jetty complex for the Ministry of Defence. A 1-t piece of this structure is kept in the Science Museum Annexe at Wroughton, Wiltshire.

A coaling jetty was built at Southampton, from 1901 to 1903, for transhipping coal from South Wales colliers to barges, and carried six large electric cranes (Figure 13.5).[7] The reinforced concrete beam and slab deck was carried on trestles of reinforced concrete piles which were stiffened with longitudinal, transverse

Figure 13.4   Woolston Jetty at Southampton, 1899 (from Mouchel-Hennebique[7]).

Figure 13.5   Coal barge jetty, Southampton, 1903 (from Mouchel-Hennebique[7]).

and diagonal bracing. The details, drawn from the ICE report[5] of 1913, are given in Table 13.1. (Principal dimensions will usually be given in metric units, but details in the tables are given in the original units, except where otherwise stated.) Within a few years this jetty was suffering considerably from reinforcement corrosion, above the high water level.[5] The flat stirrups had contributed to the trouble, and cover had been reduced by bars sagging and stirrups dropping in the forms. The reinforced concrete Town Quay at Southampton[7,8] also dates from this time. Widening of 427 m of quay by 13.7 m at Southampton docks between 1902 and 1905 is recorded in the 1913 report.[5] This was also a beam and slab deck on braced piles. The concrete was 1:3 and 1:4, with a maximum aggregate size of 22 mm (⅞ in) ring with minimum cover of 38 mm (1½ in). The flat stirrups were made of steel. Extensive damage had occurred by the 1913 report, very similar to the case above. Unlike the first reported jetty, the concrete was mixed 'wet'. Until the universal acceptance of the role of water-to-cement ratio, argument centred on whether the concrete mix should be 'wet' or 'dry'.

Details of the various stages of extension of the jetty up to 1915 were described in a paper[16] on repairs carried out by the Gunite process in 1936–40. A number of practical expedients had been used, including building onto earlier cast iron piles. Some decking incorporated Kahn bar reinforcement. The Associated Portland Cement Manufacturers' Swanscombe Jetty[12] on the Thames, built in 1906, was also detailed with Kahn bars, and had straight bars in the beams and 'Johnsons Lattice' in the decks (Figure 13.6; see also Figure 4.1 and Table 4.1 of Chapter 4). The heavy piers were built up of large diameter circular concrete blocks with holes through which rails were threaded and grouted up.

The jetty head extension at Gladstone[5,12] in Australia had a deck slab supported by continuous longitudinal beams on braced piles and was built between 1907 and 1908 (Table 13.1). A similar braced jetty, the Clyde Wharf[5,12] built at Wellington, New Zealand in 1908–1909 was also, in 1913, described as being in good condition (Figure 13.7 Table 13.1). The pile extensions and bracings above low water level were all cast in situ.

The early development of reinforced concrete piles was reported by Walmisley[15] in 1906–1907. Piled construction has obvious benefits for working over water and the precast production of piles has many advantages over in-situ work. The idea was suggested by Coignet in 1869. The various types of pile are described in Chapter 7. The reinforcement consisted of longitudinal bars tied together by transverse bars, spiral winding or expanded metal. Rolled joists and sections connected

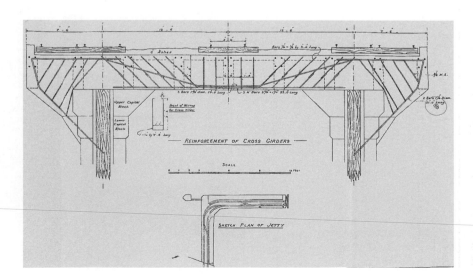

Figure 13.6 Swanscombe Jetty, 1906 (from Institution of Civil Engineers[12]).

Table 13.1 Project details: 1903–1909 (imperial units unless otherwise stated)

Details	Coal barge jetty, Southampton, 1903[5]	Jetty head at Gladstone, Australia, 1908[5,12]	Clyde Wharf, New Zealand, 1909[5,12]
Overall dimensions	110 m long × 6 m wide. Deck 10.5 m above seabed Tidal rise 4 m Deck supported on beams supported on three-pile braced trestles	61 m long Depth alongside at low water 7.3 m Tidal rise 3.7 m	Deck 3.7 m above ordinary low water springs Bottom edge of walings 0.6 m above OLWS (Ordinary low water springs) Tidal rise 1.2 m
Basis of calculation working stresses	Not available in 1913	Reinforcement designed by contractor Designed to carry a test load of 7 cwt/ft^2 (3.8 t/m^2)	Mode of calculating stresses as laid down in Marsh and Dunn *Reinforced concrete*, London, 1906 Limiting (working stress) stresses column concrete: 400 psi (2.8 N/mm^2) compression; beams, concrete in compression: 500 psi (3.4 N/mm^2); steel in tension: 15,000 psi (103 N/mm^2)
Piles and struts	Piles: 16 in × 12 in. 4 No. 1⅝ in bars Struts: 12 in × 12 in. 4 No. 1 in bars	15 in × 15 in. Each tier of six piles braced diagonally (original piles were hollow, 18 in square with gravel aggregate but failed in vessel impact). Braces and wales 14 in × 19 in mortized into piles	Piles 18 in square, spaced 20 ft c/c in longitudinal direction and 9 ft transverse. Piles carrying decking only reinforced with 4 No. 1 in ms rods at 12 in c/c. Piles carrying crane rails had 8 No. 1½ in ms rods spaced around the sides of a 12 in square
Beams	Under railway rails: 10 in wide × 12 in deep. 4 No. bottom bars and 2 No. top, all 13/16 in Under crane rails: 12 in wide × 15 in deep. 6 No. bottom bars and 3 No. top, all 13/16 in	Longitudinal beams: 24 in × 12 in. 10 ft apart, 20 ft span Transverse beams: 9 in × 12 in. 6–8 ft apart, 10 ft span	Main beams (longitudinal): spans reduced by means of struts to the piles. Beams 13½ in wide by 10½ in deep below deck slab. Outer crane beam had 3 No. 1⅛ in rods top and bottom. Other beams 3 No. 1 in top and bottom Secondary beams (transverse): 9 in wide by 7½ in below deck slab. 2 No. 1⅛ in rods in bottom face at 13 in below upper surface. 2 No. in top, 2¼ in below surface Wales: reinforced with 4 No. ⅞ in rods
Stirrups to piles and struts	Round steel 9/16 in loops at 10 in c/c		
Stirrups to beams	Flat wrought iron stirrups 2½ in × 12 BWG (Birmingham wire gauge) on the lower bars. Inverted stirrups 1⅜ in × 15 BWG on the upper bars		Round rod shear bars in the forms of loops Main beams: ⅜ in at 4 in c/c Secondary: 5/16 in at 3½ in c/c
Slabs	6 in thick. Maximum span 5 ft	6 in slab	7½ in thick. Panels 9 ft long by 3 ft–4 in wide between main and secondary beams. Covered with 4½ in wood blocking. ¾ in rods at 3½ in c/c at 6¼ in depth from upper surface, bent up over supports. Rods of adjacent spans overlap so that steel area is double over the supports
Concrete	1 : 4 cement: natural river gravel with sand. Crushed to ¾ in ring	Piles: 3 cement : 4 sand : 8 blue metal chips Rest: 3 cement : 5 sand : 8 gravel, mixed wet	1 : 2 : 3. ¾ in maximum aggregate, mixed 'wet', almost sloppy Cement locally made, to British Standard of 1904. Average tensile test briquettes with neat cement, 701 lb at 28 days
Reinforcement	Steel ratio: Railway rail beams 1¾% Crane beams 2½% Struts 2¼% Piles 4¼% Cover 1 in	Plain round bars. Siemens Martin steel. UTS (ultimate tensile strength) 28–30 t/in^2 (432–463 N/mm^2)	UTS (ultimate tensile strength) 60,000 psi (414 N/mm^2), elastic limit between 50% and 60% slung by wires from battens)

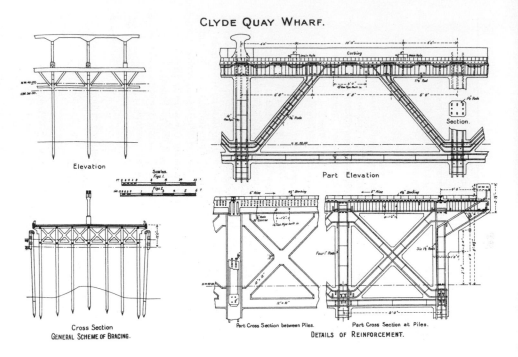

Figure 13.7   Clyde Wharf, Wellington, New Zealand, 1909 (from Institution of Civil Engineers[5]).

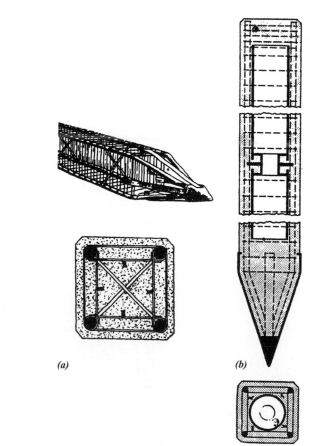

Figure 13.8 (a) Hennebique piles (from Walmisley[16]). (b) Mouchel Piles[7].

together by flat bars were also employed. Considère employed hooped concrete in which a thin rod was wound spirally around vertical rods. A concrete stress of 13.8 N/mm^2 could be taken in hooped concrete, as opposed to 2.8 N/mm^2 for concrete in unrestrained compression.[15]

Most piles were made according to individual patents. Coignet made hollow piles without any steel. Hennebique piles (Figure 13.8a) were used on jetties at Woolston near Southampton;[15] Mouchel's type of Hennebique pile had a hollow cavity (Figure 13.8b)

Other applications of braced piled jetties of this type include Purfleet Pier,[12] a coal handling structure on the Thames designed by C.S. Meik employing the Hennebique system, built in 1904 and extended in 1911. Due to the flat stirrups and detailing as in the structures at Southampton, cracks were soon observed in the high water regions. Through annual inspections and regular repairs the structure remained in service, until requiring major repairs in 1950. It was converted for use by oil tankers in 1961 and, after later repairs in 1979, was in use until recently. The bowstring Girder Bridge illustrated in Figure 7.3 of Chapter 7 still appears to be in place.

Extensive works for Parkeston Quay at the continental station of the Great Eastern Railway Co., Harwich, figured in reports of 1908,[17] 1933,[18] 1934[19] and 1935.[20] A common feature of this type of quay was that large areas of water were decked over on piles. The new braced piled quay at Harwich,[17] which used the Hennebique system, was 329 m by 16 m (Figure 13.9). The dredged depth was 6 m at low tide and 9.8 m at high tide. The piers supporting the main beams along

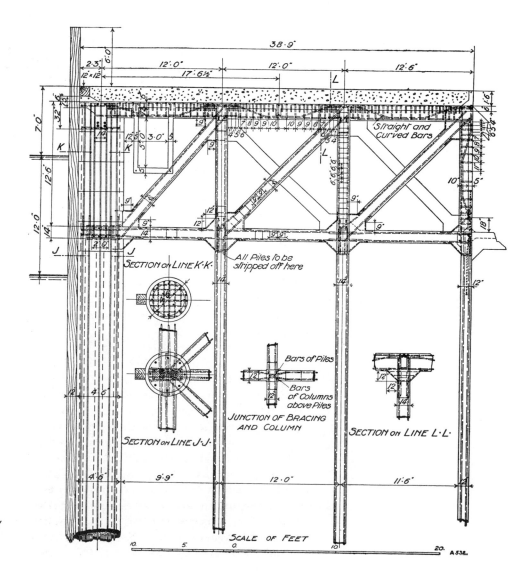

Figure 13.9 Parkeston Quay extension, Harwich, 1908 (from Twelvetrees[17]).

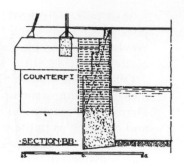

Figure 13.10   Coventry Ordnance Dock on the Clyde, 1908 (from Twelvetrees[22]).

the front were 1.40 m diameter cylinders, containing pairs of foundation piles driven into the clay beneath. All the independent piles were 355 mm square. The deck slab was 150 mm thick. Repairs by the Gunite process were carried out in the 1930s, 1963 and 1977.[21]

A very different use of reinforcement is illustrated by the dock for the Coventry Ordnance Works on the Clyde at Scotstown in 1908.[22] The tidal dock wall was 15 m high, to accommodate a tidal range of about 6 m from a lowest level below half way up the wall. The gravity wall was economic in its use of concrete and was achieved by a stepped back to the wall, and buttressing by wide counterforts 8 m apart (Figure 13.10). The reinforcement did not take bending tension, but was placed in a horizontal direction such 'as to give ample resistance to tension wherever developed, to guard against temperature cracks, and to tie the counterforts securely to the walls where connexion of the kind is desirable'. The bars were patent indented bars of the type which had been used throughout the sea wall at Galveston in the USA.

A similar jetty to Parkeston was built at Dunston, Newcastle-upon-Tyne, in 1907–1908, to serve the reinforced concrete flour mills, granary and grain cleaning tower of the Cooperative Wholesale Society, all built using the Hennebique system, and described in Chapter 4.[68]

The 1920s and 1930s

Extensions of Parkeston Quay were always in the news and 'one of the most important contracts of its kind' was reported in 1933–35.[18–20] The existing quay was extended by 345 m by 44.5 m, with two curved approach viaducts 150 and 170 m long. Its construction, undertaken by the Yorkshire Hennebique Contracting Co., was similar to that described above. At the front, the piles were driven in pairs and sleeved with cylinders 1.8 m in diameter. Repairs in the 1930s, 1960s and 1970s are reported by Dyton.[21]

The berthing arm at Clacton Pier,[23] reported in 1934, was designed to resist berthing forces by opposing rakers forming trestles of four piles (Figure 13.11). To reduce the tension loads in the raker piles, the dead weight of the berth was increased by constructing the deck as a stone-filled box.

By this time the rational advantages of reinforced concrete for resisting berthing forces as an elastic structure with designed fender systems were being exploited. The earlier reinforced concrete jetties were braced and strutted together in all directions as low as the tide would allow, as had been done for timber jetties.[8] Later, bracings were discarded in favour of raking piles and, even later, monolithic portal bents, which had increased strength for less cost and removed the need to cast members in the tidal zone. Berthing forces, originally allowed for by experience, were now subject to analysis of the kinetic energy to be absorbed. Then Chief Engineer of Christiani and Nielsen, Ove Arup, contributed interesting articles in 1934 and 1935[24] on this subject.

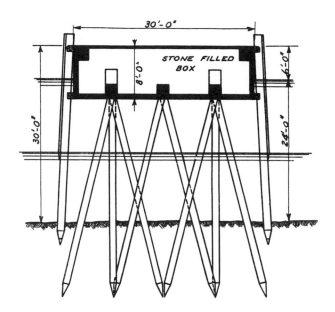

Figure 13.11   Berthing arm at Clacton Pier, 1934.[23]

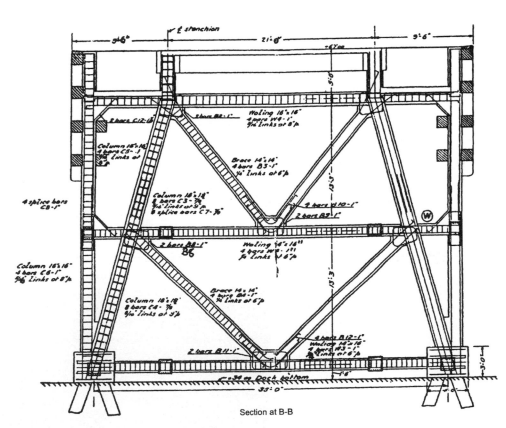

Section at B-B

Figure 13.12   New fish dock at Grimsby: coaling appliances for two ships at once, 1930 (from Comrie[25]).

Nevertheless, extensive works for the new impounded Fish Dock Complex at Grimsby[25] commenced in 1930 and were mostly built to the piled and braced format (Figure 13.12 and Table 13.2).

The extension to Southend Pier[26] in 1929 (Figure 13.13) had two decks, the lower deck being at mid-tide (range 5 m) with a stiff Vierendeel-type truss without bracings in order to obviate obstruction. The lower deck was made of narrow precast slabs 159 mm (6¼ in) deep, including a 19 mm (¾ in) granolithic top finish, supported on beams built *in situ*. The slabs were 229 mm (9 in) thick at

Table 13.2   Project details: 1930s (imperial units unless otherwise stated)

Details	Fish Dock at Grimsby, 1930 (South Quay)[25]	Quay extension at Newcastle-upon-Tyne, 1930[27]	War Department jetty at Deptford, 1934[28]
Working stresses	Not stated	Not stated	Concrete in compression — 600 psi (4.1 N/mm^2) Steel in tension — 16,000 psi (110 N/mm^2)
Piles	14 in square Piles stripped and capped by pile caps 1 ft 8 in × 1 ft 8 in × 3 ft 0 in	Front piers: Groups of 3 No. octagonal piles encased in PC (Portland Cement) cylinders 7 ft 3 in diameter cast in 10 ft lengths General piles: octagonal section at 12 ft c/c	295 No. 16 in square, up to 72 ft long, in raker bents at 16 ft c/c Raked at 1 in 3¾
Struts and columns	Struts: 10 in × 12 in Longitudinal walings: 12 in × 14 in Columns surmounting piles: 14 in × 14 in	Struts: 12 in × 12 in precast Superstructure columns surmounting piles: 18 in × 18 in	Diagonal braces at 6 ft 3 in above LWOST (low water ordinary spring tides) 18 in × 12 in
Beams	Main beams: 12 in × 18 in Secondary beams: 10 in × 14 in	Crane beams: 20 in wide × 33 in deep Railway beams: 12 in wide by 20 in deep	Main beams: 3 ft deep Secondary beams: 19 in deep × 12 in wide at 5 ft c/c
Deck	7 in thick, reinforced with ⅝ in straight bars at 12 in c/c and ⅝ in bent up bars at 12 in c/c, span 6 ft 3 in Overlaid by 2 in granolithic not included in design strength Roadway at rear, 8 in thick with 2½ in topping	9 in thick, overlaid by 6:1 concrete laid to a fall and 2 in granolithic topping	Vaulted deck on main and secondary beams: minimum thickness 6 in and 1½ in thick granolithic finish
Cement			Below high water level: rapid hardening Elsewhere: ordinary
Concrete	Piles: 204 lb cement, 3¾ ft^3 fine, 7½ ft^3 coarse (i.e. 1 : 1.7 : 3.3) All other: 204 lb cement, 5 ft^3 fine, 10 ft^3 coarse (i.e. 1 : 2.2 : 4.4)		Piles and work below high water: 1 : 1½ : 3 Superstructure: 1 : 2 : 4 Surface treatment: three coats of sodium silicate applied to piles before driving and superstructure after completion

Figure 13.13   Extension to Southend Pier, 1929 (from Moller[26]).

the top and 203 mm (8 in) at the bottom, to facilitate washing away of seaweed. Dyton[21] reported that no maintenance was carried out until 1976. The lower deck and piles were 'reasonably intact' but the upper parts of the frames were severely decayed.

The 1930 quay extension at Newcastle upon Tyne[27] was designed by Mouchel with braced piles, sleeved piles at the front and bracing (Figure 13.14 and Table 13.2). The jetty was designed to accommodate very heavy loading of $5\,t/m^2$, and octagonal section piles were chosen, at 3.7 m centres. The bracing and struts were precast and the only *in-situ* concrete was in the junctions of the precast intertidal members, with a thickened section in order to maintain larger cover.

The War Department jetty at Deptford[28] (Figure 13.15 and Table 13.2) was designed and completed in 1934, with raked piles connected by a triangular system of low-water bracing to provide efficient strength against lateral forces and low-level support to the fender timbers.

Meanwhile, on the continent and elsewhere, more 'modern' use of reinforced concrete was being made, as reported in the Permanent International Association of Navigation Congresses (PIANC) sessions in London in 1923.[2] The general report by Humphries[29] the Chief Engineer to the London County Council, reviewed 14 reports from international sources. There was discussion of the advantages of slag cement in sea water in Belgium. Examples were given[30] of reinforced concrete caissons constructed at Rotterdam in 1908 with $1:4\frac{1}{2}$ concrete ($340\,kg/m^3$). Reinforced concrete caissons had been used in the Dutch East Indies, including

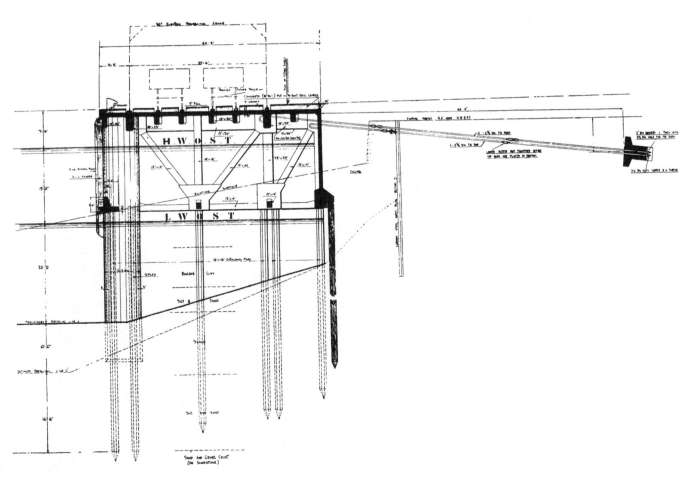

Figure 13.14   Cross section of the quay extension at Newcastle upon Tyne, 1930.[27]

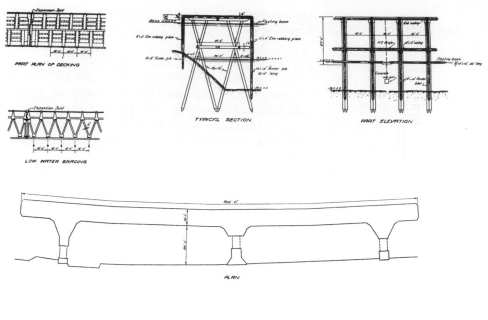

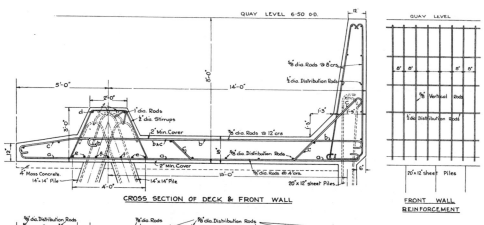

Figure 13.15   War Department jetty at Deptford, 1934.[28]

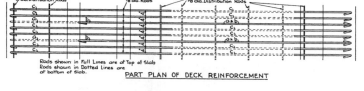

Figure 13.16   Hamworthy Wharf, Dorset, mid 1930s (from Wentworth-Shields and Gray[9]).

Belawan, since 1910. A number of types of 'relieving platform' design, with reinforced concrete decks on reinforced concrete sheet and bearing piles, had been developed in Amsterdam. A UK example of this type of wall is the Hamworthy Wharf (Figure 13.16).[9] Reinforced concrete caissons were used at Yokahama[31] from before 1920 and also in Spain. Previously, 'caissons' had been steel or iron structures filled with concrete. The cement contents used on the continent were contrasted with much lower figures associated with a 1 : 2 : 4 mix (referred to as $260\,kg/m^3$) which were 'usual in the USA'; $260\,kg/m^3$ appears to be a very low figure for 1 : 2 : 4 concrete.

War time

A reinforced concrete 'screwcrete' pile had been developed for founding in silt or sand free from boulders, and was used for a slipway at Southend[32] in 1934. The piles were formed by light metal cylinders attached to the screw shoes, which were

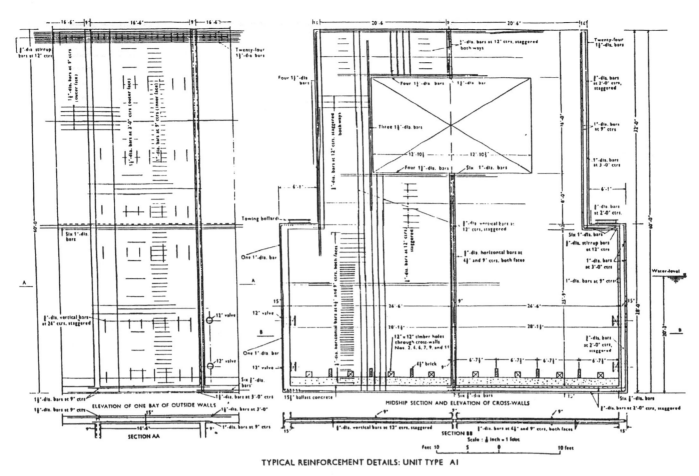

Figure 13.17   Reinforcement of Phoenix units (from Wood[34]).

screwed into the sea bed using a mandrel within the casing. The casing was maintained watertight and a reinforced concrete column formed within it. This system was used for the piles at the deepwater wharf for No. 1 Military Port built in Faslane Bay[33] in 1940. The beams and deck over the screwcrete piles were reinforced concrete.

The floating breakwaters and piers built for the invasion of Europe incorporated many types of components, including steel ship-like units, blockships, precast concrete boat-like units, bridge units and concrete caissons. Six different sizes of concrete 'Phoenix' caisson units were produced, varying in depth from 7.6 to 18 m, but mostly of the same length, 62 m.[34] Engineers are unlikely to meet examples in service and the archive interest is therefore mostly as a record of what was done and how it compares with what we would do now. A survey of two surviving units was carried out in 1980 under the 'Concrete in the Oceans Project,'[35] aimed at providing additional knowledge to improve the design, construction and performance of concrete oil production platforms.

The total of 213 'Phoenix' cellular caissons[34] were built with a $1 : 1\frac{1}{2} : 3$ mix with 19 mm maximum aggregate size. The working compressive stress for concrete was 5.86 N/mm^2 (850 psi). In order to save time and materials, only straight bars were used and no splays were employed at the cell corners. The floors varied from 300 to 380 mm thick, the external walls from 350 to 380 mm thick and internal walls 228 mm. Reinforcement details are shown in Figure 13.17. A leftover Phoenix caisson was in fact used to close the largest gap at Schelphoek in Holland, after the storm surge damage of 1953.

Feedback from the 1980[35] survey was not particularly illuminating as to performance. The estimated cube strength from cores was, at 37 N/mm², not perhaps as high as might be expected, and the cover was between 10 and 30 mm. During the installation of the units in 1944, problems were experienced due to lack of anchorage of the walls to each other, caused by the lack of splays and bent bars and guniting was used to repair miscast sections.[36]

Post-war until the 1970s

This period was, arguably, the golden age for building reinforced concrete maritime structures. Information is available in many ICE papers, but sometimes the overview gives scant information on the details of the concrete and reinforcement. Immediately post-war, there was a vast amount of work to be done and no lack of confidence in methods and materials. The principles of reinforced concrete had been developed from the 1934 DSIR code into CP114 in 1948 (see Chapter 5), there was a confident but relatively small library of concrete publications, and prestressed methods were established. Steel was in short supply and concrete had obvious advantages in making use of indigenous bulk materials.

The increased size of structures to accommodate the rapidly growing size of vessels, and the transfer from the outdated enclosed and river docks to the open sea, led to the need to build more for less and hence to re-examine the techniques and materials previously used, mostly in other structures, such as precasting and prestressing. Confidence was also boosted by the developments in concrete technology, with 'improvements' in mix design and quality control. In hindsight, not all of these developments were beneficial.

The dry dock built at South Shields for Brigham and Cowan,[37] between 1953 and 1956, is an example of reinforced concrete replacing mass concrete for graving docks. Dry docks had previously been mostly built to accommodate passenger liners, but the increased size of oil tankers led to the need for this dock for the repair of vessels up to 38,000 t deadweight. The dock walls were built by placing precast reinforced concrete cantilever buttresses at 1.5–2.3 m centres against a near vertical clay face. The buttresses were only 300 mm wide (Figure 13.18), and were subsequently anchored into the floor slab. Between the buttresses, concrete panels 600 mm thick were cast directly against the clay face. The reinforced concrete was generally 1:6.5 by weight, 19 mm aggregate and 28 N/mm² at 28 days.

Development of the oil industry also led to the deep water jetties for the BP Oil Terminal at Angle Bay, Milford Haven in 1958–60,[38] constructed mainly of precast units carried on Rendhex steel box piles. A paper in the contractor's house magazine[39] gives a deeper insight into the mix design than in normal construction papers. What may not now be obvious to some engineers is that the nominal volumetric proportions 1:1:2, 1:1½:3 and 1:2:4 covered a range of mixes with different richness and strength. The common 1:2:4 mix as a 'normal' grade to CP114 had a specified works crushing strength of 20.7 N/mm², but higher and lower grades of this mix could be used by control of the cement content and water to cement ratio to give stresses 25% above or below the nominal level. The mixes for Milford Haven were specified as in Table 13.3. The development of the practical mixes from this specification is given in Table 13.4. The average strength of the 'A' mix was some 51 N/mm² and the slump only 17 mm.

Erith Jetty,[40] built on the tidal Thames between 1955 and 1957, is a typical example of the use of high quality precast and prestressed concrete. The jetty (Figure 13.19) accommodated vessels up to 14,000 t displacement and carried three electric cranes. The construction consisted of rigid portal frames at 7.6 m centres for the jetty head and 15.2 m on the approach, with no other bracing. The portal columns were cylinders consisting of prestressed shells, themselves capable of carrying the imposed loads without assistance from the colloidal concrete plug core.

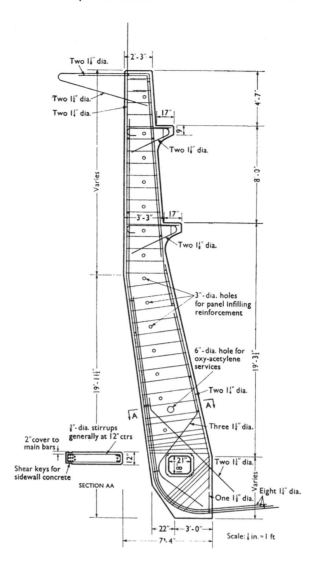

Figure 13.18 Dock walls for Brigham and Cowan, 1956 (from Stott and Ramage[37]).

Table 13.3   Mixes specified for Angle Bay, Milford Haven

Quality	Nominal proportions by volume	Aggregate size (mm)	Maximum water-to-cement ratio	Works cube (N/mm^2)	cf. CP 114 range (N/mm^2)
A	1 : 1½ : 3	20	0.45	38	32.3 25.9 20.7
B	1 : 2 : 4	20	0.50	31	25.9 20.7 16.5
B-1	1 : 2 : 4	20	0.55	24	25.9 20.7 16.5

The cylinder piers were made up of units 1.5 m long by 1.8 m outside diameter and 159 mm thick walls. The units had 12 mm spiral reinforcement outside the prestressing tendons, which consisted of sixteen 28 mm Macalloy bars, post-tensioned to give a uniform stress of 7.24 N/mm^2 after allowing for losses due to shrinkage and creep. The minimum cover of 38 mm was monitored by frequent

Table 13.4   Development of practical mixes* for Angle Bay, from specifications in Table 13.3

Nominal mix by volume	Trial mixes in laboratory						Final works mix	
	Cement	Sand (0–5 mm)	Stone		Water-to-cement ratio	Proportion by weight	Cement	Water-to-cement ratio
			(5–10 mm)	(10–19 mm)				
A 1 : 1½ : 3	391	590	353	901	0.4	1 : 1.5 : 3.2	365	0.46
B 1 : 2 : 4	302	630	368	747	0.5	1 : 2.1 : 4.4	<302	?
B-1 1 : 2 : 4	286	640	445	913	0.55	1 : 2.3 : 4.5	<286	?

*Mix proportions in kg/m^3.

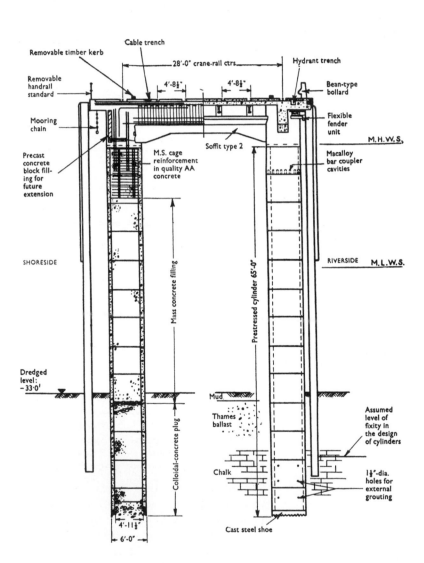

Figure 13.19   Erith Jetty, 1957: cross section on jetty head (from Carey and Cumming[40]).

checks with a covermeter, then a cumbersome article requiring a separate car battery.

To maintain the weight of individual units to within 8 t, only the soffits of the heavier beams were precast, the webs and, of course, the deck being completed *in situ*. The beams were also post-tensioned with Macalloy bars (Figure 13.20). The cover to ordinary reinforcement was 25 mm and that to the 'Kopex' tubing used to form the prestressing bar ducts was 38 mm. The paper[40] includes sample calculations with prestressing diagrams. The concrete mix and cube record details are reproduced in Table 13.5.

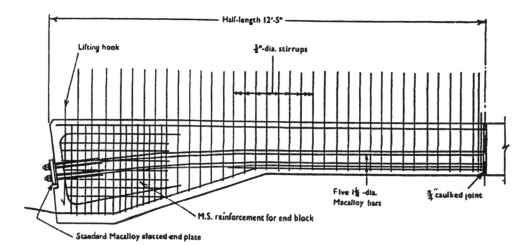

Figure 13.20   Erith Jetty, 1957: half-section of soffit type 2 (from Carey and Cumming[40]).

Table 13.5   Erith Jetty 1957: concrete records

Class and location	Type of cement	Mix proportions (kg/m³)				28-day strength (N/mm²)			
		Cement	Fine	Coarse (19 mm)	Water-to-cement ratio	Specified (minimum)	Maximum	Minimum	Average
PS									
Precast cylinders	Rapid hardening	411	550	1302	0.36	51.7	64.3	47.6	54.7
Precast beams and soffits	Rapid hardening	481	498	1232	0.36	51.7	60.9	51.9	54.5
AA									
Cylinder caps, *in-situ* beams and deck	Ordinary Portland	427	525	1230	0.42	25.9	54.1	31.6	45.6

Notes: 1. Mix proportions in kg/m³ were re-calculated from original figures given in lb/112 lb cement.
2. Note that the second and third mixes have almost a 'standard' water content of 170–180 kg/m³ for such a mix. The first mix, for the same water-to-cement ratio as the second, has a low water content of 147 kg/m³.

Tees Dock No. 1 Quay,[41–43] built between 1959 and 1962, is similar in that the substructure consisted of prestressed concrete cylinders in bents of three at 9.1 m centres, carrying a prestressed beam and slab deck. The cylinders were precast hollow sections 1.8 m long, 1.9 m in diameter, with a wall thickness of 457 mm. The front cylinders were stressed together with eighteen 22 mm Macalloy bars and the other cylinders with six bars. The deck consisted of twin main transverse prestressed I beams spanning between the cylinders and supporting prestressed deck units. The deck units were novel, being 432 mm thick prestressed concrete dovetail interlocking units, transversely post-tensioned together with CCL cables. In later years, by 1982, there were some random failures in the transverse post-tensioning tendons associated with some incompletely grouted ducts, as has been experienced in some bridges.

A landmark description of the development of reinforced concrete and steel piled walls at Rotterdam was presented in 1966.[44] Twelve years' post-war experience had given the Dutch engineers considerable confidence in the use of reinforced concrete in decks and piles in conjunction with heavy section steel sheet pile walls to form a typical relieving platform quay of the type shown in Figures 13.16 and 13.21. Examples of reinforced concrete caisson walls were given which, of course, included those quoted earlier.[30] In the discussion of this Chapter,[45] the possible corrosion problems arising from the use of steel piles and the use of higher

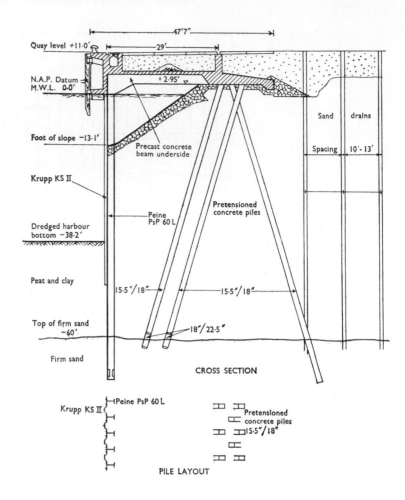

Figure 13.21   Dutch relieving platform, 1966 (from Bokhoven[44]).

tensile steel reinforcement, which had previously been strictly avoided in maritime works in order to reduce flexural cracking, were debated. At this time, the limitation of crack width according to the hostility of the environment was being more widely introduced. The crack width limit suggested was 0.1 mm. The concrete strength was lower than might be the case in the UK, being $22.5 \, N/mm^2$, and the cover was only 35 mm. The salinity of the water was low, of course, and, as had been clearly stated in the 1923 papers,[30] the cement used in Holland was likely to have been a high slag blend.

It was now common for quays and jetties to consist of either beam and slab or even thick plain reinforced concrete slabs seated on the growing range of steel box or tubular piles, including the hexagonal Rendhex piles and composite sheet piles. The steel piling was easier and quicker to drive and extend, and was capable of absorbing larger berthing and superstructure loads.

Port Talbot Harbour[46,47] was the first tidal harbour to be built in the open sea in the UK since the extension of Dover in 1909. It was built between 1966 and 1969 to accommodate iron ore carriers of 100,000 t deadweight and above. The approach structure consisted of precast concrete beams carried on Rendhex steel piles, with precast concrete slabs spanning between the beams (Figure 13.22). The head of the berth consisted of a conventional concrete slab deck supported on 760 mm diameter steel tubular piles.

Two very large dry docks, one to accommodate tankers and bulk carriers up to 200,000 t deadweight and one to build ships up to 1 million tons deadweight, were built in Belfast in 1965 and 1969, respectively.[48,49] The more conventional repair dock had a reinforced concrete floor, but walls of Peine steel sheet piling anchored

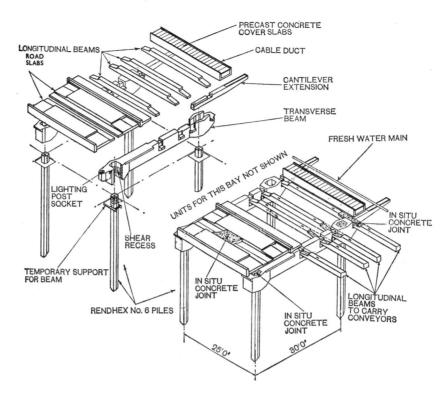

Figure 13.22   Port Talbot, 1969 (from McGarey and Fraenkel[46]).

by prestressed concrete raking piles. The shallower building dock was made of normal reinforced concrete.

The last example of this period is the Brighton Marina,[50,51] built in 1971 on an exposed site with circular precast vertical breakwater caisson units 12.5 m diameter. The characteristic strength of the concrete was 31.5 N/mm^2 and reinforcement was generally with high yield deformed bars. The walls were designed to CP114, with 65 mm cover to reinforcement and surface crack widths were limited to 5% exceedance of 0.26 mm.

Offshore and North Sea oil

The offshore scene began with light-houses. There is firstly the curious Nab Tower, built in the First World War with a reinforced concrete base. Kish Bank Lighthouse[52] was built for the Commissioners of Irish Lights, 8 miles out of Dublin Bay, from 1963 to 1965, using a patented Swedish method of telescopic caisson construction. The solid structure was founded directly on the seabed, and was designed for a wave height of 13.7 m. The telescopic form of construction improved the stability during towing and sinking. The outer caisson consisted of three concentric cylinders standing on a 19 m thick base slab and interlocked by 12 radial walls (Figure 13.23). The design required a 28-day cube strength of 31 N/mm^2, and 4% entrained air. The mix was 1 : 4.43 by weight with a water to cement ratio of 0.425. Ordinary Portland cement was used except for the outside wall extending from below mean sea level to the top, where sulphate-resisting cement was used. The cover to the base slab and external wall face was 50 and 38 mm to the inside faces of the outer wall. A condition survey commissioned after some 25 years, in 1990, showed the structure to be in very good condition. Although difficult conditions restricted access, the average cover to reinforcement recorded by covermeter was very close to the specified figure. Measured equivalent cube strengths from cores at the first extended landing were consistently high, being between 80 and 90 N/mm^2.

The Royal Sovereign Light Tower[53,54] was built a few years later, 11 km offshore from Eastbourne, using the telescopic principle in a different way. The tower was prestressed. Ordinary reinforced concrete was specified as 41 N/mm^2 and

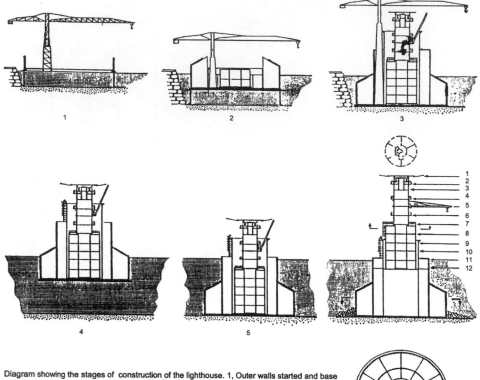

Diagram showing the stages of construction of the lighthouse. 1, Outer walls started and base slab constructed on platform. 2, Further construction on deeper bed. 3, Platform removed and construction completed. 4, Tower crane removed and structure towed to Kish Bank. 5, Sunk on to prepared bed. 6, Tower raised and fixed. Caisson filled with sand. stone protection placed.
Details of lighthouse: 1, Helicopter platform. 2, Lantern floor. 3, Lantern room. 4, Watch room. 5, Radio room. 6, Rad/beacon room. 7, Entrance deck. 8 and 9, Living quarters. 10, Storage. 11, Generator and heat plant. 12, Storage and magazines.

Figure 13.23   Kish Bank Lighthouse, 1965 (from Hansen[52]).

prestressed as 55 N/mm^2. Cover to reinforcement was 50 mm in ordinary reinforced areas and 75 mm in prestressed areas. Sulphate-resisting Portland cement was used in the wave zone. A limited condition survey[55] was also conducted for this structure some 10 years later in 1980 under the 'Concrete in the Oceans Project'.

Oil and gas in the North Sea lie some 3000 m below the waves, in depths of water up to 150 m with wave heights of 30 m or more. The mass of a concrete structure in these conditions is likely to be of the order of 250,000 t. A new generation of concrete sea structures began in 1971 with the Ekofisk artificial island, 270 km from the North Sea coast.[56] The structure consisted of a nine-cell crude oil reservoir, similar to a grain silo, measuring 50 m by 50 m in plan and 90 m high. This was surrounded by a 'Jarlan' perforated breakwater screen with a mean diameter of 95 m (Figure 13.24). The prestressed structure was constructed by slipforming and precasting, the whole construction upwards from the cellular foundation being carried out afloat in deep but protected water. The concrete was specified at 50 N/mm^2 cube strength, using plain Portland cement. As a result of an anticipated subsidence of about 6 m caused by oil extraction, an additional protective barrier, which forms a complete ring around the earlier structure, was completed in 1990.[57]

The Condeep structures for the Beryl A and Brent B production platforms were ordered in 1973 and completed in 1975. They consisted of a base of 19 cylindrical cells providing oil storage capacity and three tapering concrete shafts supporting a steel deck.[58] The plan dimension across five interconnected cells was 100 m and the total height to the top of the shafts was 147 and 173 m for Beryl A and Brent B, respectively, producing a truly impressive and beautiful structure (Figure 13.25). Like Ekofisk, they were built partly in dry dock and partly afloat

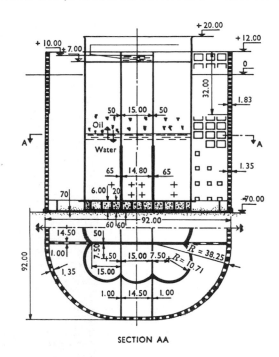

SECTION AA

Figure 13.24 Ekofisk artificial island, 1973 (from Marion and Mahfouz[56]).

Figure 13.25 Condeep platform under construction, 1975.[58]

in a deep water fiord. The concrete technology for the prestressed structure was described by Moksnes[59] in 1975. Platforms of this nature were ordered and constructed in a similar way to ships, that is under certification and insurance provisions. A British example is Cormorant 'A', 300,000 t, at Ardyne Point. (Figure 13.26).

In the period 1970–80 a considerable number of papers were written about the requirements for concrete in what were seen as the onerous conditions of the

Figure 13.26 Cormorant 'A' at Ardyne Point, 1974–1978 (Sir Robert McAlpine).

North Sea. A number of these papers[60,61] by Browne and others reported the current state of the art and established a confidence. In reality, however, the exposure conditions of the cold North Sea as regards the performance of concrete and reinforcement corrosion are not quite as aggressive as one might initially think, and the nature of construction in prestressed concrete led almost automatically to high grade concrete and high quality construction which withstood the conditions for the required commercial service life, which is likely to be between 25 and 50 years.

Overseas and the Middle East

British engineers were very active in the design and construction of port works overseas in the post-war reconstruction era. Starting with the emerging African states and then Libya, Aden and the new oil states of Kuwait and then the rest of the Gulf states, Saudi Arabia and Iran, there was a strong accent on maritime facilities. The speed of construction was usually of the essence, often with a shortage of inland transport and hence locally available cement and aggregates. Although mass concrete was preferred in a number of cases for a number of reasons, with the demand for speed and economy and the increasing size of many of the facilities, it is not surprising that steel-piled and reinforced concrete structures tended to predominate.

There were, naturally, fears expressed concerning the use of structural steel in the sea, which, in time, have largely been seen to be over-cautious. Renewed confidence in the application of reinforced concrete, however, hit a number of snags which were not so obvious or avoidable as is sometimes assumed, and which are only now being more fully understood. The environment, in fact, provided accelerated exposure conditions and served to broaden understanding, owing to experience of conditions not met anywhere else worldwide, except perhaps in bridge decks and tunnels. A number of examples are quoted by Gerwick.[62]

Some earlier rapid reinforced concrete failures in the Middle East had been dismissed as failures of materials and workmanship and rapidly demolished. What had really been demonstrated was the limitation of standard design and details to European and American practice which worked well enough in very different exposure conditions elsewhere in the world. The comparison of environmental conditions is quantified by Fookes.[10]

A graphic illustration of the significance of the water-to-cement ratio to the discontinuity of capillary pores, drawing from the classic work of Powers and Brownyards,[63] amongst others, was given by Torben Hansen[64] in 1989, who described the deterioration of the concrete superstructure of a pier at Brega Industrial Port in Libya after less than 10 years' service. The high absorbency of surface concrete of good quality under these conditions is explained in other papers,[11] and it is in this context that the differential performance of reinforced concrete structures in various parts of the world needs to be analysed.

Other major examples of reinforced concrete include the harbours at Port Rashid, Dubai and Mina Zaed, Abu Dhabi, built from 1968 and the supertanker dry docks at Bahrain (1973/74) and Dubai[65,66] (1973/80). Feedback on these projects is available in papers by John[11,67] and others.

Conclusion

In the maritime field, design and construction details are often more significant than the choice of materials themselves. Reinforced concrete in the sea has a relatively long and successful history, but many of the questions raised and discussed almost a century ago are sometimes revived as though they were novel. Our predecessors, naturally, linked air and moisture with corrosion, but drew close parallels with the performance of bare metal, masonry and timber. It is only recently that the very different deterioration mechanism of reinforced concrete in some specific maritime conditions has been better understood, and can explain the very early observations[12] that, depending on the magnitude of cover, it is the infrequently wetted areas above the intertidal zone which can be most at risk, because drying of the surface zone enables the chloride content to concentrate and permits the flow of oxygen. Wetness itself, and regular 'wetting and drying', can be an advantage. Concrete does not dry out below the surface layer, and the depth of this layer is critical in arid conditions.[67,68]

Design for durability of reinforced concrete subjected to severe marine exposure or deicing salts has not matched advances in structural computation.[69,70] Unlike land-based concrete, both the practice and expectations of concrete in the sea remain controversial. The next generation of European (CEN) codes will include a much improved classification of exposure conditions specific to the various deterioration mechanisms, which is explained with reference to sea structures by Leeming,[71] Slater and Sharp,[72] and is reflected in the revision of BS 6349, Pt. 1: 2000.[73] Detailing practice and the specification of achievable tolerances for fixing reinforcement are likely to improve. Materials, particularly cement, continually change, and so one cannot simply compare specifi-cations and achievements of different dates, and codes of practice can take too long to adapt and recognize some basic principles in relation to durability.

The above review of structures has been mainly drawn from the literature and personal experience. Wherever possible, subsequent reports of earlier works have been traced. As it is essential that future designs are based on the performance of structures in the field and not on academic experiment, a scientific follow-up of structures such as those that have been identified, with the interest and support of their current owners, would provide an ideal sequel and benchmark of performance. The author would welcome any feedback on the structures identified, or others.

Finally, an illustration of changing properties is provided by a recent case when a contractor was surprised to meet 'high' concrete strengths when demolishing a

1950s jetty on the Thames. He had expected strengths akin to the standard 20–25 N/mm^2 specified during the 1960s. However, in those days, the mean strength to meet 25 N/mm^2 was likely to be 40 N/mm^2. The concrete most likely contained some 400 kg/m^3 cement and had a water to cement ratio of about 0.42. The mean 28-day strength, according to Road Note 4,[74] would be some 40 N/mm^2, increasing to 65 N/mm^2 after 1 year. Unlike modern cements, strength gain was likely to continue, to be in the region of 80 N/mm^2 today. Hopefully, the information given in this book may reduce the risk of such a misunderstanding.

The author acknowledges the encouragement given by the associated authors in this book, and James Sutherland (his first section leader in the design office) and Mike Chrimes in particular. Also to his senior mentors in port engineering, particularly Peter Scott,[3] Harry Ridehalgh, Bob Daniels and Eric Loewy[4] who, in fact, recommended and reviewed the original paper.

References

1. Ker, H.J., The extension, widening and strengthening of Folkestone Pier. Discussion comment by Sir William Mathews KCMG (President). Min. Proc. Instn Civ. Engrs, 1908, 171, 106.
2. Permanent International Association of Navigation Congresses. Proc. 13th International Congress on Navigation, London, 1923, Section 11, Ocean navigation communications.
3. Scott, P.A., Port of Tema. Proc. Instn Civ. Engrs, 1965, 32, 211–53.
4. Loewy, E.I., Pannet, R.J., Collinge, H.S., Blockwork quay walls. Design construction and utilisation in developing countries. PIANC 27th Int. Navigation Congress, Osaka, May 1990, Section II — Subject 3, 141–46.
5. Institution of Civil Engineers. Second Report of the Committee on Reinforced Concrete. Institution of Civil Engineers, William Clowes & Son, 1913.
6. Institution of Civil Engineers. Preliminary and Interim Report of the Committee on Reinforced Concrete. Institution of Civil Engineers, 1910.
7. Mouchel-Hennebique, Ferro-Concrete, 4th edn. L.G. Mouchel and Partners Ltd.: Westminster, 1921.
8. Wentworth-Shields, F.E., Early marine structures. Concr. Constr. Eng. 1956, January, 25–29.
9. Wentworth-Shields, F.E., Gray, W.S., Reinforced Concrete Piling. Concrete Publications Limited: London, 1938.
10. Fookes, P.G., A simple guide to risk assessment for concrete in hot dry salty environments. Proc. 4th International Conference on Deterioration and Repair of Reinforced Concrete in the Gulf Region. Bahrain Society of Civil Engineers, 1993.
11. John, D.G., Corrosion deterioration of reinforced concrete structures in the Middle East. Conference on Structural Improvement Through Corrosion Protection. The Institute of Corrosion: London, 1992.
12. Institution of Civil Engineers. First Report of the Committee Appointed to Investigate the Deterioration of Structures of Timber, Metal and Concrete Exposed to the Action of Sea Water. Institution of Civil Engineers, London, 1920.
13. Department of Scientific and Industrial Research. Deterioration of Structures of Timber, Metal and Concrete Exposed to the Action of Sea Water. Fifteenth Report of the Committee of the Institution of Civil Engineers, London, 1935.
14. National Building Studies. Research Paper No. 30. The Durability of Reinforced Concrete in Sea Water. Twentieth report of the sea action committee of the Institution of Civil Engineers. HMSO: London, 1960.
15. Walmsiley, A.T., Reinforced concrete piles in tidal waters. Concr. Constr. Eng. 1906–1907, 1, 88–100.
16. Pannel, J.P.M., Cement-gun repairs to maritime reinforced concrete structures with special reference to the Town Quay, Southampton. Institution of Civil Engineers, Maritime and Waterways Engineering Division Meeting, Maritime paper No. 2, December 1945, 3–37.
17. Twelvetrees, W.N., Parkeston quay extension, Harwich. Concr. Constr. Eng. 1907–1908, 2, 166–94.
18. Parkeston Quay extension: report. Concr. Constr. Eng. 1933, 28, 63.
19. Parkeston Quay, Harwich: report. Concr. Constr. Eng. 1934, 29, 64.

20. Extension to Parkeston Quay, Harwich: report. Concr. Constr. Eng. 1935, 30, 56–57.
21. Dyton, F.J., Case histories of repairs of maritime structures. Maintenance of Maritime Structures. Institution of Civil Engineers, 1978, 89–103.
22. Twelvetrees, W.N., Reinforced concrete wharves and quays. Article III. Docks and quays on the Clyde. Concr. Constr. Eng. 1907–1908, 2, 273–78.
23. Berthing arm at Clacton Pier: report. Concr. Constr. Eng. 1935, 30, 58–59.
24. Ove Arup., Design of piled jetties and piers. Concr. Constr. Eng. 1934, 29, 37–41; 1935, 30, 41–48.
25. Comrie, J., The new fish dock, Grimsby. Concr. Constr. Eng. 1934, 29, 545–58.
26. Moller, A., Extension to Southend Pier. Concr. Constr. Eng. 1930, 25, 57–61.
27. Quay extension at Newcastle-upon-Tyne: report. Concr. Constr. Eng. 1934, 29, 65–69.
28. War Department jetty at Deptford: report. Concr. Constr. Eng. 1934, 29, 86–87; 677–80.
29. Humphries, G.W., General Report. Concrete and reinforced concrete. Their applications to hydraulic works; means to insure their preservation and their watertightness. Proc. 13th International Congress on Navigation, London, 1923, Section 11, Ocean navigation communications. PIANC, Brussels.
30. Ringers, J.A., Tellegen, G., Report (Dutch experience), Holland. Proc. 13th International Congress on Navigation, London, 1923, Section 11, Ocean navigation communications. PIANC, Brussels.
31. Aki, K., Okabe, S., Reinforced concrete caisson for the Yokohama Harbour new extension works. Proc. 13th International Congress on Navigation, London, 1923, Section 11, Ocean navigation communications. PIANC, Brussels.
32. Reinforced concrete screw piles, slipway at Southend-on-Sea: report. Concr. Constr. Eng. 1929, 29, 273–77.
33. Knight, C.W., Stork, S.G., Nos 1 & 2 Military Ports. The Engineer in War, Vol. 2. Docks and Harbours. Institution of Civil Engineers, 1948, 3–35.
34. Wood, C.R.J., Phoenix (Mulberry units). The Engineer in War, Vol. 2. Docks and Harbours. Institution of Civil Engineers, 1948, 336–68.
35. Marine Durability of the Mulberry Harbour Units at Portland. Concrete in the Oceans Project, Phase 2, Project 6. Taylor Woodrow Construction Ltd., Research Laboratories, 1981.
36. Pannel, J.P.M., Cement-gun repairs to maritime reinforced concrete structures with special reference to the Town Quay, Southampton. Institution of Civil Engineers, Maritime and Waterways Engineering Division Meeting, Maritime paper No. 2, Observations by Mr. G.M. Trehorne Rees, December 1945, 25.
37. Stott, F.P., Ramage, L.M., The design and construction of a dry dock at South Shields for Messrs Brigham and Cowan Ltd. Proc. Instn Civ. Engrs, 1957, 8, October, 161–92.
38. Grove, G.C., Siting, design and construction of two terminals for large oil tankers. Proc. Instn Civ. Engrs, 1964, 27, January, 99–152.
39. Michelson, T., Concrete design and control — some information from the execution of the deep water jetties for BP Oil Terminal, Angle Bay, Milford Haven. CN (Christiani and Nielsen) Post, 1961, February, 24–27.
40. Carey, R., Cumming, C.G., The design and construction of Erith jetty. Proc. Instn Civ. Engrs, 1961, 18, January, 15–42.
41. A prestressed concrete quay at Tees dock: the substructure. Concr. Constr. Eng. 1961, December, 413–20.
42. A prestressed concrete quay at Tees dock: the superstructure. Concr. Constr. Eng. 1962, February, 71–77.
43. Tees Dock No. 1 Quay. Cement and Concrete Association: London, 1961.
44. Bokhoven, W., Recent quay wall construction at Rotterdam Harbour. Proc. Instn Civ. Engrs, 1966, 35, December, 593–613.
45. Bokhoven, W., [Discussion on above reference.] Proc. Instn Civ. Engrs, 1967, 37, 685–700.
46. McGarey, D.G., Fraenkel, P.M., Port Talbot Harbour: planning and design. Proc. Instn Civ. Engrs, 1970, 45, April, 561–92.
47. Ridgeway, R.J., Kier, M., Hill, L.P., Low, D.W., Port Talbot Harbour: construction. Proc. Instn Civ. Engrs, 1970, 45, April, 593–626.
48. Geddes, W.G.N., Sturrock, K.R., Kinder, G., New shipbuilding dock at Belfast for Harland and Wolff Ltd. Proc. Instn Civ. Engrs, 1972, 51, January, 17–47.

49. Ross, K., Rennie, W.J.H., Cox, P.A., The new dry dock at Belfast. Proc. Instn Civ. Engrs, 1972, 51, February, 269–94.
50. Terrett, F.L., Ganly, P., Stubbs, S.B., Harbour works at Brighton Marina: investigations and design. Proc. Instn Civ. Engrs, 1979, 66, May, 191–208.
51. Llewellyn, T.J., Murray, W.T., Harbour works at Brighton Marina: construction. Proc. Instn Civ. Engrs, 1979, 66, May, 209–26.
52. Hansen, F., Design and construction of Kish Bank Lighthouse. Trans. Instn Civ. Engrs Ireland, 1965–66; 92, 248–99.
53. Antonakis, C.J., A problem of designing and building for a structure at sea (Royal Sovereign Lighthouse). Proc. Instn Civ. Engrs, 1972, December, 95–126.
54. Thorskov, S., The Royal Sovereign Light Tower. CN (Christiani and Neilsen) Post, 1962, February, 12–22.
55. P66 final report, Concrete in the Oceans, Phase 11. Marine durability study of the Royal Sovereign Lighthouse. Taylor Woodrow Research Laboratories, December 1981.
56. Marion, H., Mahfouz, G., Design and construction of the Ekofisk artificial island. Proc. Instn Civ. Engrs, 1974, 56, November, 497–511.
57. Broughton, P., Waagaand, K., The Ekofisk protective barrier. Proc. Instn Civ. Engrs Wat. Marit. Energy, 1992, 96, June, 103–19.
58. Norwegian Contractors. Promotional literature. Oslo, 1975.
59. Moksnes, J., Condeep platforms for the North Sea — some aspects of concrete technology. Offshore Technology Conference, Houston, Texas, May 1975, 339–50.
60. Browne, R.D., Domone, P.L., Concrete for surface and underwater structures. Symposium on Materials for Underwater Technology, Admiralty Materials Laboratory, Poole, 1973, Society for Underwater Technology.
61. Browne, R.D., Domone, P.L., Geoghegan, M.P., Deterioration of concrete structures under marine conditions — their inspection and repair. Institution of Civil Engineers Conference on Maintenance of Marine Structures, London, October 1977.
62. Gerwick Jr., B.C., International experience in the performance of marine concrete. Concrete International, American Concrete Institute, May 1990, 47–53.
63. Powers, T.C., Brownyands, T.C., Studies of the physical properties of hardened Portland cement paste. Research Laboratories of the Portland Cement Association, Chicago, Bulletin No. 22, March 1948.
64. Hansen, C.T., Marine concrete in hot climates — designed to fail. Mater. Structs, RILEM, 1989, 22, 344–46.
65. Daniels, R.J., Sharp, B.N., Dubai Dry Dock: planning, direction and design considerations. Proc. Instn Civ. Engrs, 1979, Part 1, 66, 75–92.
66. Cochrane, G.H., Chetwin, D.J.L., Hogbin, W., Dubai Dry Dock: design and construction. Proc. Instn Civ. Engrs, 1979, Part 1, 66, 93–114.
67. John, D.G., Leppard, N., Wyatt, B.S., Cathodic protection repair applied to reinforced concrete deck support beams for Mina Zayed Port, Abu Dhabi. Proc. 4th International Conference on Deterioration and Repair of Reinforced Concrete in the Gulf Region, Bahrain, 1993.
68. Bijen, J.M., Durability aspects of the King Fahd Causeway. RILEM Conference on Concrete in Hot Climates, Torquay, September 1992.
69. Aitcin, P.C., Durable Concrete. In: S. Nagataki, T. Nireki, F. Tomasawa (eds), Durability of Buildings and Components, Vol. 6, E & F.N. Spon, 1993.
70. Rostam, S., Schiessl, P., Next generation design concepts for durability and performance of concrete structures. In: S. Nagataki, T. Nireki, F. Tomasawa, (eds), Durability of Buildings and Components, Vol. 6. E & F.N. Spon, 1993.
71. Leeming, M.B., Durability of concrete in and near the sea. In: R.T.L. Allen (ed.), Concrete in Coastal Structures. Thomas Telford, 1998, Chapter 3.
72. Slater, D., Sharp, B.N., Design guides, specifications and the design of coastal structures. In: R.T.L. Allen (ed.), Concrete in Coastal Structures. Thomas Telford, 1998, Chapter 4–6.
73. British Standards Institution. Maritime Structures, Part 1, Code of Practice for general criteria. BS 6349: Pt. 1: 2000.
74. Road Research Road Note No. 4. Design of concrete mixes. Department of Scientific Research, Road Research Laboratory, HMSO, 1950.

14 The Concrete Institute 1908–23, precursor of the Institution of Structural Engineers

Anita Witten

Synopsis

The reasons for the foundation of the Concrete Institute are briefly outlined. Its organization and scope are described, and its achievements, particularly its role in drafting the London County Council Reinforced Concrete Regulations and in establishing a standard notation for reinforced concrete, are discussed. A short analysis of its membership is given.

Introduction

There are subtle variations of emphasis in the accounts of the reasons for the founding of the Concrete Institute (from 1923 the Institution of Structural Engineers), but essentially there was concern amongst building control officers, architects, engineers and others that concrete and designs using it should be better understood, not least by those ultimately responsible if failure occurred. The consensus of the Institution of Civil Engineers (ICE), whose members[1] had long experience of concrete for foundations and were certainly using it for buildings, was that regulations for reinforced concrete construction would unduly fetter its members while there was still so much about the material which was imperfectly understood.[2] Others saw regulations as a necessary step forward.[3] The early history of the adoption of reinforced concrete as a building material in the UK is described in papers by Hurst[57] and Bussell[58] elsewhere in this issue and in works such as Hamilton.[5]

It seems likely that the number of people across several professions involved in construction using concrete and the number wishing to know more about it and its uses, none of whose needs were being adequately met by the existing institutions, was sufficient to encourage the formation of another body. Concrete was the first new construction material with such great potential for all sorts of uses for several generations, and it would be surprising if it did not cause a ferment in the industry as people began to become familiar with it.

The driving force behind the Concrete Institute was Edwin O. Sachs (Figure 14.1), a dynamic architect with much experience of theatre design and the effects of fire upon buildings. In 1897 he established the British Fire Prevention Committee and as its Chairman initiated the first independent fire testing station in the world,[4] and in 1906 he founded the excellent journal *Concrete and Constructional Engineering* (although his involvement with this was probably not public knowledge).

Figure 14.1   Edwin O. Sachs.

Events preceding the foundation

At this distance we must rely on published information and one or two published histories such as that by Hamilton.[5] Table 14.1 lists the important publications which preceded the founding of the Institute. Sachs was 'induced' to look for a suitable basis for forming a scientific society that appreciated and welcomed the cooperation and advice of all those commercially concerned in concrete

Table 14.1   Important publications preceding the founding of the Concrete Institute

Date	Content
8 April 1905[6]	Letter from William Dunn summarizing the current position and urging the RIBA, with the ICE, to appoint a committee to prepare a standard specification for reinforced concrete work (including calculations) to help architects to assess designs submitted by the specialists
March 1906	First issue of the journal *Concrete and Constructional Engineering*. Initially bi-monthly, recent works in reinforced concrete at home and abroad were described and illustrated, extensive reviews of the position of reinforced concrete in other countries given, and developments in theory and practice noted. Edited and partially funded by Edwin O. Sachs[7]
28 April 1906[8]	Report on the first meeting of the Joint Reinforced Concrete Committee, chaired by Sir Henry Tanner and comprising representatives of RIBA, the District Surveyors Association, the Institute of Builders, the Incorporated Association of Municipal and County Engineers, the War Office, the Admiralty and 'distinguished scientists'
December 1906[9]	Announcement of the formation by the British Fire Prevention Committee of a Special Commission on Concrete Aggregates, chaired by Sir William Preece, and comprising engineers, architects, surveyors, firemen, cement manufacturers and representatives of the Admiralty and War Office for the purpose of reporting on and defining 'the aggregates suitable for concrete floors intended to be fire-resisting, having due regard to the question of strength, expansion and the chemical constituents and changes of the aggregates'
27 May 1907[10]	First report of the RIBA Joint Reinforced Concrete Committee adopted at RIBA General Meeting
Autumn 1907	Circular issued by a member of one of the reinforced concrete firms, proposing the formation of a society made up of those firms
First half of 1908[11]	Reports in the industry press of the formation of the Concrete Institute

whilst having the standing of a technical institution of the first order and the active guiance of the leaders of the technical professions affected.[12]

The new Institute

The first meeting of the Council of the new Institute was held on the 21st of July, 1908 at the Ritz Hotel, with Sachs in the chair. The Earl of Plymouth, First Commissioner of Works 1902–1905, was appointed as President.[13] One hundred founder members had signed up[14] at a subscription of 1 guinea (£1.05). Tables 14.2 and 14.3 show the employment of Council members and the overlaps in membership between the influential bodies, in collaboration not competition, and Figure 14.2 shows the growth in membership of the new body. The Council members must all have been known personally by Sachs; this shows the breadth of his interests and contacts in the industry and his ability to inspire hard work and enthusiasm in others. The objects of the Institute were agreed by the meeting to be:

(a) to advance the knowledge of concrete and reinforced concrete, and direct attention to the uses to which these materials can be best applied;

(b) to afford the means of communication between persons engaged in the design, supervision and execution of works in which concrete and reinforced concrete are employed (excluding all questions connected with wages and trade regulation);

(c) to arrange periodical meetings for the purpose of discussing practical and scientific subjects bearing upon the application of concrete and reinforced

Table 14.2   Council members in 1908 also sitting on other committees

Council member and employment	ICE Council member at some time	RIBA 1907 Joint Committee on RC	BFPC Special Commission on Concrete Aggregates	RIBA 1911 Joint Committee on RC	ICE 1908 RC Committee	ICE 1911 RC Committee
Earl of Plymouth						
Sir Douglas Fox, President ICE 1899–1900	x					
Sir William Mather						
Sir William Preece, President ICE 1989–99, Chief Engineer, GPO	x		x			
Sir Henry Tanner, HM Office of Works		x		x		x
Edwin O. Sachs			x			
H.H.D. Anderson, manufacturer of Portland cement						
B. Blount					x	x
A.E. Collins, City Engineer, Norwich		x	x	x		
C.H. Colson, Admiralty			x	x		
William Dunn, Architect		x		x		
B. Hannen, Cubitt & Co.						
W.T. Hatch, Chief Engineer, Metropolitan Asylums Board			x			
W.H. Hunter, Chief Engineer, Manchester Ship Canal			x			
W.H. Johnson, Chairman, Johnson, Clapham & Morris						
C.F. Marsh, Metropolitan Water Board		x		x		
F. May, Chairman, Trussed Concrete Steel Co.		x		x		
J. Munro, Director, Stuart's Granolithic Co						
F. Purton, Manager, New Expanded Metal Co.						
A. Ross, President ICE 1915–16, Chief Engineer, Great Northern Railway	x		x			
L. Seraillier, Manager, Patent Indented Steel Bar Co.						
J.S. de Vesian, Mouchel & Partners						
J. Winn, School of Military Engineering			x			
G.C. Workman, Coignet						
E.P. Wells						

BFPC: British Fire Prevention Committee; RC: Reinforced Concrete.

Table 14.3   Composition of the 'concrete bodies'

	Concrete Institute Council 1908 (25 in total)	RIBA Joint Committee on Reinforcced Concrete 1907 (15 in total)	BFPC Special Commission on Concrete Aggregates (23 in total)	RIBA Joint Committee on Reinforced Concrete 1911 (18 in total)
Architects	2	5	4	6
Engineers	9	2	14	8
Contractors or producers of proprietary systems	8	1	1	0
A public service	6	4	8	5
Unascribed	6	5	5	3

Excluding those unascribed, some people fall into two categories.

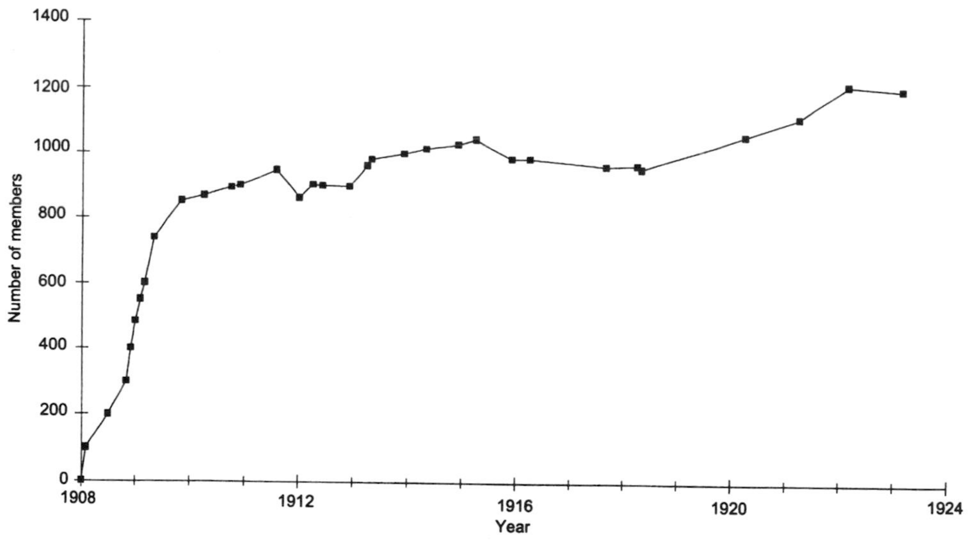

Figure 14.2   Number of members, derived from reports in the Concrete Institute Transactions and annual reports (excludes 'special' subscribers and honorary members).

concrete, and to conduct such investigations and to issue such publications as may be deemed desirable.

The qualifications for membership were:

(a) persons professionally or practically engaged in the application of concrete or reinforced concrete and the production of their constituents;
(b) persons of scientific, technical or literary attainments specially connected with the application of concrete, reinforced concrete and their constituents.[15]

All members of the Council were to be British subjects.

It seems likely that many of the founder members were nearing the end of their careers and enrolled to encourage the new body; there are quite a number of deaths recorded in the first volumes of *Transactions* and others doubtless retired. By April 1910, when the first list of members was published, only 89 founder members remained; this was reduced to 78 by February 1912.

Four committees were established immediately (science, parliamentary, reinforced concrete practice and tests), and their terms of reference approved in March 1909.[16]

Administration and organization

Initially the administrative work was done by an Executive of six members, chaired by Sachs, which included the Honorary Secretary, A.E. Collins, who was city engineer of Norwich (and resigned at the end of 1909 due to pressure of work) and the Honorary Treasurer, E.P. Wells. Sachs did a great deal of the day-to-day work himself — the Institute's offices were at 1 Waterloo Place, Pall Mall, London, SW1, while his own address at the time as 7 Waterloo Place — and it proved difficult to get the Executive together as often as necessary.[17] The Institute got off to a good start, but there was a hiatus when Sachs was taken ill in November 1909 and was still convalescing at the first Annual General Meeting (AGM) on the 17th of February, 1910.[18] Only 10 days' notice of the date of the AGM was given (contrary to the *Articles of Association*, which stipulated 30) and there were one or two discussions at the meeting which show clearly the lack of anyone present with a sufficient working knowledge of the *Articles*. Sachs would undoubtedly have had this knowledge.

As a result of this obvious over-reliance on one person, the *Articles* were changed so that Council had the responsibility and the management of the Institute's affairs. Sachs seconded the resolution and received an enthusiastic vote of thanks at the end of the meeting which approved the changes.[19] A permanent secretary was appointed to assist Council.[20]

Possibly because Sachs was out of action for quite a long period,[21] the Institute floundered slightly in its direction forward. This may have been a natural occurrence because the initial momentum had worn off and much time needed to be spent on the London County Council (LCC) regulations, and because Sachs' undoubted vision and energy were missing from Council meetings. However, around the end of 1910 the Earl of Plymouth retired from the office of President and Sir Henry Tanner was appointed in his place. He was Chief Architect to HM Office of Works from 1898 to 1913 and promoted the use of reinforced concrete in the UK, overseeing the first official use in the UK of reinforced concrete for a complete building in the King Edward Building, Newgate Street, London, for the Post Office, and he capably and carefully steered the Institute through the next 15 months,[22] and was succeeded by E.P. Wells. Sachs was appointed as fifth Vice-President.

During 1911 the way forward was considered, and this resulted in approval of a proposal before Council for extending the scope of the Institute. A subcommittee of ten Council members[23] considered how to put it into practice, and its report was adopted at the third AGM on the 9th of May, 1912. The revised objects were:

(a) to advance the knowledge of concrete and reinforced concrete, and other materials employed in structural engineering, and to direct attention to the uses to which these materials can be best applied;

(b) to afford the means of communication between persons engaged in the design, supervision and execution of structural engineering works (excluding all questions connected with wages and trade regulation);

(c) to arrange periodical meetings for the purpose of discussing practical and scientific questions bearing upon the application and use of concrete and reinforced concrete and other materials employed in structural engineering for any purpose whatsoever.[24]

Structural engineering was defined as 'that branch of Engineering which deals with the scientific design, the construction, and the erection of structures of all kinds in any materials', and structures as 'those constructions which are subject principally to the laws of Statics, as opposed to those constructions which are subject principally to the laws of Dynamics and Kinematics, such as engines and machines'.[25]

The full title was changed to 'The Concrete Institute: an institution for structural engineers, architects, etc.', the breadth of the Institute's membership being considered one of its great strengths, allowing all points of view on a subject under investigation to be taken into account.[26]

An annual course of technical lectures on 'some branch of structural engineering' was to be started, and examinations in structural engineering were to be held annually, 'to test the scientific or technical attainments of applications for Studentship'.[24] Other small changes to the *Transactions* and meeting arrangements were made, and a Bronze Medal was introduced for the best paper presented each session (Table 14.4).

This new scope had a beneficial effect on the number of members,[27] which continued when additional classes of membership were added[28] and the first Honorary Members elected (Table 14.5).[29]

During the 1913–14 session, informal meetings of junior members were instituted, and an Examination Board was established in December 1913, with Professor Henry Adams as Chief Examiner, to run entrance examinations for the Graduate and the Associate Member (now Member) grades of membership. A syllabus was compiled[30] and it was proposed to hold the first examination in 1915.

The First World War, consequent increases in the cost of paper and printing, and some controversy over policies[31] brought things to a relative standstill for a few years. The publication of the *Transactions* was postponed and the content reduced[32] (volume 11 was never published at all) and from 1920 onwards announcements, very brief meeting reports, etc., were published in *Concrete and Constructional Engineering*.[33] Work and discussion on the various reports in progress continued through the war, but publication was greatly delayed.

The first examinations were deferred, and were eventually held on the 13th and 14th of May, 1920. By 1920 educational lectures had resumed, and visits of

Table 14.4   *Bronze medals awarded for the best paper presented in each session*

Session	Title and author
1910–11	The aesthetic treatment of concrete, by Beresford Pite
1911–12	Fireproofing, by Richard L. Humphrey
1912–13	Steel-frame buildings in London, by S. Bylander
1913–14	Calculations and details for steel-frame buildings from the draughtsman's standpoint, by W.C. Cocking
1914–15	The design of quay walls, by F.E. Wentworth-Shields
1915–16	Shearing resistance of reinforced concrete beams, by Oscar Faber
1916–17	Southampton Docks: re-modelling of an old dry-dock, by Robert N. Sinclair
1917–18	No records available
1918–19	The geology of aggregates and sands, by P.G.H. Boswell
1919–20	The attrition of concrete surfaces exposed to sea action, by J.S. Owens
1920–21	No records available
1921–22	What is the use of the modular ratio? by H. Kempton Dyson

Table 14.5   *Honorary Members elected during the 1912–13 session*

C. Bach	Director of the Materialprüfanstalt, Stuttgart
N. Belelubsky	Director of the Technical Laboratories, St Petersburg
H. Le Chatelier	Inspector General of Mines, Paris
Paul Christophe	Principal Engineer, Office of Works, Brussels
Edmond Coignet	Director of Edmond Coignet Ltd.
Armand Considère	Inspector General of Public Works, Paris
Fritz von Emperger	Editor of *Beton und Eisen*, Vienna
R. Feret	Chief of the Technical Laboratories, Boulogne
F. Hennebique	Director of the Hennebique Ferro-Concrete Company
A. Martens	Principal of the Prussian Testing Laboratories, Berlin
A. Mesnager	Director of the Technical Laboratories, Paris
Emil Morsch	Chief Engineer, Wayss & Freitag AG
C. Rabut	Engineer to the Office of Works, France
F. Schule	Director of Materialprüfungsanstalt, Zurich
Arthur N. Talbot	Professor of Engineering, University of Illinois, USA
W.C. Unwin	President ICE 1911–12

inspection (including one to the works for the new Selfridge store) were being arranged again.

The number of members, the finances and activities continued to be healthy after the war, and it seems that during 1921 the idea of changing the title to The Institution of Structural Engineers was considered.[34] Very little background information is available on this as the Council papers are not available at the time of writing and published sources say nothing, but as no changes were made to the Objects of the Institute established in 1912, it seems likely that the new name generally reflected the facts. Some concern at the new title was reflected at the ICE Council meeting on the 26th of September, 1922, but it was decided to take no action.

The title change was approved on the 28th of September, 1922 and The *Structural Engineer* issued monthly from January 1923. Branches, first mooted some years earlier, were now permitted, and the Lancashire and Cheshire Branch was inaugurated in November 1922, followed by the Western Counties Branch in January 1923. The first Gold Medal of the Institution of Structural Engineers was awarded to Henry Adams in recognition of his services to the Institution and his long and distinguished service as Chairman of the Board of Examiners.[35]

All the signs are of a professional association flinging itself confidently, with energy and purpose, towards the future.

Membership

The Institute attracted members from all professions with an interest in reinforced concrete. Figures 14.3 and 14.4 show the remarkably consistent professional and geographical spread over time. Most of those abroad were working in the British Empire or on railway construction. The subscription was initially 1 guinea (£1.05), raised to 2 guineas (£2.10, or £105.50 at 1993 prices) in 1913 when the membership reached 1000.

Achievements

Meetings

Papers on a wide range of subjects were presented and discussed at meetings of the Institute (Table 14.6). There are useful remarks about various aspects of the industry in the verbatim discussions published in the *Transactions* (e.g. Seraillier[36]).

Building regulations

In December 1908 the Institute was asked by the LCC to consider proposed amendments to the London Building Acts included in the LCC (General Powers) Bill to be presented to Parliament in 1909, and to respond almost immediately. In January, the LCC requested delegates from 'the leading societies connected with building works'[37] to consider the proposals in the Bill 'regulating the erection of buildings of iron or steel skeleton construction'.[38] William Dunn (Chairman of the Science Standing Committee and member of the RIBA Joint Committee on Reinforced Concrete) and E.P. Wells (who had much experience of testing work and research) represented the Concrete Institute. In addition, the Parliamentary Standing Committee put in considerable effort petitioning against the Bill.[39] As a result, when the Act came into force, under Section 23 the Concrete Institute, the Institution of Civil Engineers, the RIBA and the Surveyors Institution (now the RICS) were to consider proposed regulations to do with the construction of buildings wholly or partly of reinforced concrete. During 1910, the Concrete Institute committees spent 19 h discussing these draft regulations and recommending amendments in addition to those put forward by the four societies in conference.[40] The annual reports summarize the progress of the consultation;[41] an examination of the original LCC proposals and of the amendments recommended by all the various

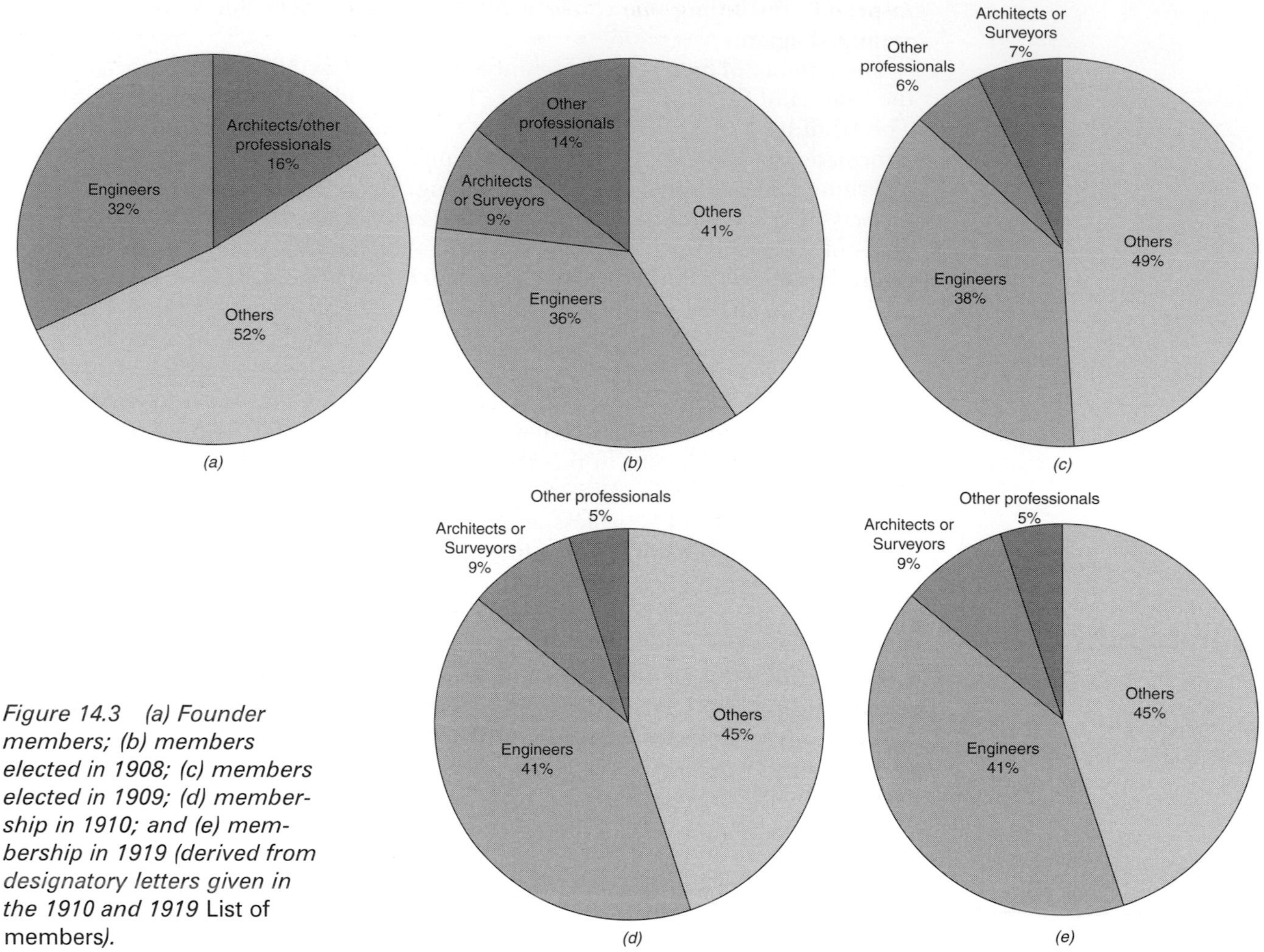

Figure 14.3 (a) Founder members; (b) members elected in 1908; (c) members elected in 1909; (d) membership in 1910; and (e) membership in 1919 (derived from designatory letters given in the 1910 and 1919 List of members).

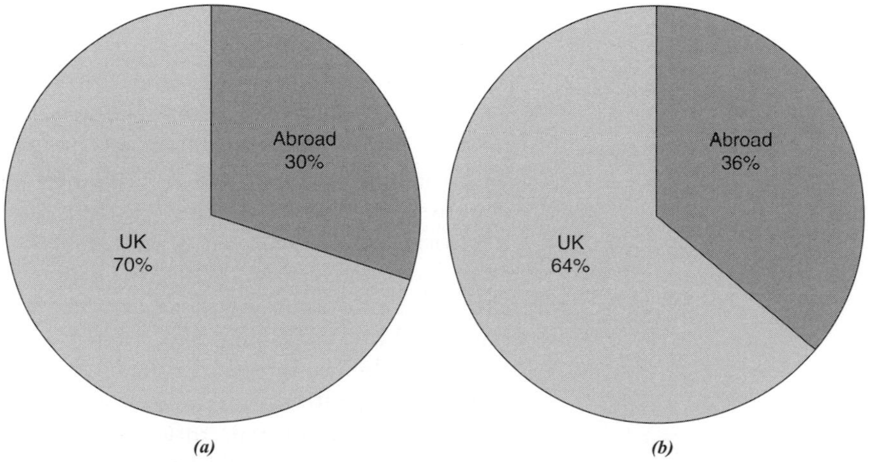

Figure 14.4 (a) Location of members in 1910; and (b) location of members in 1919 (derived from the 1910 and 1919 List of members).

bodies would be interesting in charting the development of reinforced concrete, but unfortunately space does not permit it here. The draft of the first set of Regulations occupied 12 h of Institute committee time in 1911–12 and that of the second set 47 h in the 1913–14 session.[42] The RIBA and Surveyors Institution supported most of these recommendations, and the ICE did not make any detailed

Table 14.6   Papers presented to the Concrete Institute

1908	The composition and uses of plain and reinforced concrete, by Charles F. Marsh
	The examination of designs for reinforced concrete work, by William Dunn
1909	The commercial aspect of reinforced concrete, by Lucien Seraillier
	Concrete in arched bridge construction, by E.P. Wells
	Some notes relating to the setting of Portland cement with description of method adopted for regulating the same, by H.K.G. Bamber
	Some points relating to reinforced concrete as applied in the United States, by R.L. Humphrey
1910	Reinforced concrete chimney construction, by Ernest R. Matthews
	Notes on the Le Chatelier boiling test of Portland cement, by D.B. Butler
	The effects of sewage and sewage and sewage gases on Portland cement concrete, by Sidney H. Chambers
	Reinforced concrete bins, by H. Kempton Dyson
	The British Aluminium Company's works at Kinlochleven, by A. Alban H. Scott
	The manufacture of Portland cement, by A.C. Davis
	General concrete practice, by Thomas Potter
1911	The dissociation of competitive designs and tenders, by R.W. Vawdrey
	Swanscombe reinforced concrete pier, by C. Percy Taylor
	The aesthetic treatment of concrete, by Beresford Pite
	The YMCA building, Manchester, by Alfred E. Corbett
	Some recent works in reinforced concrete, by G.C. Workman

amendments.[43] The Concrete Institute's suggestions were largely embodied in revised draft regulations[44] which came into force on the 1st of January, 1916,[45] followed in 1916 by the first explanatory handbook.[46] Debate in the industry, and the Institute's involvement continued. The development of design theory and practice during the 1920s and 1930s is described in Bussell's[58] paper elsewhere in this issue.

International standard algebraical notation

In response to a letter to members asking for suggestions for matters which the new Concrete Institute should consider, J. Sherwood Todd suggested the consideration of the standardization of methods and symbols in calculations for reinforced concrete work, and E. Fiander Etchells (President 1920–23) suggested the publication of a standard notation for reinforced concrete formulae. Both suggestions were discussed by the Science Standing Committee, together with a draft international notation proposed by the International Commission on Reinforced Concrete (an offshoot of the International Association for Testing Materials) which had been forwarded to Sachs who was the UK representative on the Commission. The principles adopted were:

(a)   the use of initial letters;
(b)   the use of significant subscripts;
(c)   discrimination between the use of small and capital letters;
(d)   sparing use of the Greek alphabet.[47]

A draft report was issued on the 28th of July, 1909, but the principle of using initial letters was found to be a barrier to its adoption in Europe[48] (different languages preferring different initial letters) and the Continental use of Greek letters for intensity of stresses would not be readily adopted in the UK and USA. The draft was despatched to the American Joint Committee on Concrete and Reinforced Concrete for consideration. It was adopted for use in textbooks, for the second RIBA report[49,50] and by the LCC for its proposed reinforced concrete regulations. Revisions were made periodically and a version with explanatory notes by E. Fiander Etchells was published by E. and F.N. Spon in 1918.[51]

Other activities

A library was established as soon as space was available, and members were asked to donate books. The first list of additions appeared in the *Concrete Institute Transactions*, Vol. 3, page xiii.

The Institute was invited to appoint representatives to various committees and to participate in joint efforts, including:

(a)   in 1908, the Special Commission on Concrete Aggregates (set up by the British Fire Prevention Committee);

(b)   in 1910, the Engineering Standards Association (now the British Standards Institution) Sectional Committee on Portland Cement and the Sectional Committee on Bridges and Building Construction;

(c)   in 1910, the RIBA Joint Committee on Reinforced Concrete;

(d)   in 1913–14, the International Association for Testing Materials, for reporting of accidents to reinforced concrete buildings;

(e)   in 1916–17, the conference convened by the District Surveyors Association to consider the interpretation of the LCC (General Powers) Act 1909 with reference to steel-frame buildings.[52] The report was complete and awaiting publication in May 1918.[53]

The Institute also initiated various investigations by its committees, although too much was attempted for the time and finance available (the LCC regulations took up a great deal of time) and nothing of great impact resulted, although publication in the *Transactions* (and discussions) of several interim reports must have been helpful in disseminating current thinking and are indicative of the state of knowledge in the UK. Some of the subjects investigated were:

(a)   a standard specification for reinforced concrete work;

(b)   adhesion of and friction between concrete and steel;

(c)   reinforced concrete piles;

(d)   the effects of oils and fats on concrete;

(e)   standardization of methods of taking off quantities;

(f)   advice to clerks of works and others on methods of properly executing concrete work;

(g)   standard concrete mixtures;

(h)   the use of cinder, ash and breeze in concrete;

(i)   methods of making concrete watertight.

Several cooperative efforts came to fruition. In 1910 the Joint Committee for Loads on Highway Bridges was set up, with representatives from the Institution of Municipal and County Engineers and the Institution of Municipal Engineers. Their report was published in 1918. In 1913–14, jointly with the Quantity Surveyors Association (now part of the RICS), drafting began on a Standard Method of Measurement for Reinforced Concrete. It was also considered by the National Federation of Building Trades Employers of Great Britain and Ireland, and by the Institute of Builders. The final report, *Measurement of reinforced concrete in building works*, was approved during the 1914–15 session[54] and published by the Quantity Surveyors Association. The Concrete Institute wished to extend it with a second part on engineering works and intended to publish both parts at a later date, but this seems never to have happened.

In 1915–16 an advisory council appointed by the Committee of the Privy Council for Scientific and Industrial Research requested information on specific problems of concrete industry which required 'scientific investigation'.[55] The Institute offered further help and later applied for a research grant, intending to coordinate research in various different universities and laboratories. In the following

Table 14.7   Publications

1918	Mnemonic notation for engineering formulae Report of Joint Committee on Loads on Highway Bridges Recommendations to clerks of works and foreman concerning the execution of reinforced concrete works
1920	A standard specification for reinforced concrete work
1921	Reports to the Department of Scientific and Industrial Research on the research work on concrete carried out under the direction of the Concrete Institute during the years 1917–19
1924	International Cement Congress 1924: report on papers and discussions
1925	Report of Joint Committee on Loads on Highway Bridges (2nd edn)
1926	Scale of charges for consulting structural engineers Report on aluminous, rapid hardening, Portland and other cements of a special character
1927	Report on steelwork for buildings. Part I: loads and stresses Report on steelwork for buildings. Part II: details of design and construction Report on loads and stresses for gantry girders
1928	Report on reinforced concrete for buildings and structures. Part III: materials and workmanship

session (1916–17) a programme was drawn up for testing the properties of various sands, aggregates and their concretes, and work started at various universities.[56]

After the war, some publications started to appear and once the Institution of Structural Engineers got going several more were published (Table 14.7) within a few years of its name change which were particularly important; the valuable experience of the early years had not been wasted.

It would be interesting to compare the initial aims and the influence and effectiveness of the Concrete Institute with those of other bodies formed later: the Reinforced Concrete Association (founded c. 1930), the Cement and Concrete Association (founded 1935) and the Concrete Society (founded 1966). Unfortunately, that must remain a subject for future study.

References

1. Harrison, C.A., Reinforced concrete for railway structures. 1907 Engineering Conference. Institution of Civil Engineers: London, 1907: 24–25.
2. Read, R.D.G., Discussion on reinforced concrete structures. 1907 Engineering Conference. Institution of Civil Engineers: London, 1907: 56–57. (Unpublished letter from the Institution of Civil Engineers to London County Council, 21 November 1910.)
3. Anon., The Concrete Institute. Builders J. Arch. Eng., 1908, February, 197.
4. Welch, C., London at the Opening of the 20th Century. W.T. Pike & Co: London, 1905; Anon., Our Fiftieth Anniversary. Concr. Constr. Eng., 1956, 51, 1–3; Anon., Obituary Notice. Concr. Constr. Eng., 1919, 14, 556.
5. Hamilton, S.B., A note on the history of reinforced concrete in buildings. National Building Studies Special Report 24. HMSO: London, 1956.
6. Dunn, W., Reinforced concrete floors. RIBA J., 1905, 8 April, 373.
7. Anon., Our Fiftieth Anniversary. Concr. Constr. Eng., 1956, 51, 1–3.
8. Anon., Chronicle. RIBA J., 1906, 28 April.
9. Concr. Constr. Eng., 1, 408.
10. RIBA J., 1907, 1 June, 515–41.
11. Anon., A Concrete Institute. Concr. Constr. Eng., 1908, 2, 423; Anon., A Concrete Institute. Builders J. Arch. Eng., 1908, 29 January, 95; Anon., A Concrete Institute. 26 February 1908, 197; Anon., The Concrete Institute. RIBA J., 1908, 9 May, 412.
12. Sachs, E.O., Memorandum. Concr. Inst. Trans., 1, v.
13. Anon., The Righ Hon. the Earl of Plymouth, GBE, CB etc. Struct. Engr, 1923, 1, 76–77.
14. Anon., The Concrete Institute. Builders J. Arch. Eng., 26 February 1908, 197.
15. Anon., Notes: constitution. Concr. Inst. Trans., 1, vii.

16. Anon., References to Standing Committees. Concr. Inst. Trans., 1, xxiii.
17. Concr. Inst. Trans., 3, 12.
18. Anon., Discussion on annual report and accounts. Concr. Inst. Trans., 2, 75.
19. Anon., Extraordinary general meeting. Concr. Inst. Trans., 3, 1–3.
20. Second annual report, Concr. Inst. Trans., 3, 224.
21. Anon., Our Fiftieth Anniversary. Concr. Constr. Eng., 1956, 51, 3.
22. Concr. Inst. Trans., 4, 275, 285.
23. Concr. Inst. Trans., 4, 163–64.
24. Anon., Objects of the Institute. Concr. Inst. Trans., 4, iv.
25. Concr. Inst. Trans., 4, 272.
26. Concr. Inst. Trans., 4, 278–79.
27. Anon., Report of Council for 1912–13 session. Concr. Inst. Trans., 5, 407; Concr. Inst. Trans., 6, 2.
28. Concr. Inst. Trans., 5, 414.
29. Concr. Inst. Trans., 5, 411.
30. Anon., Syllabus of the examination. Concr. Inst. Trans., 6, 21–24.
31. Anon., Award of the gold medal to Professor Henry Adams. Struct. Engr, 1923, 1, 33.
32. Concr. Inst. Trans., 9, Diii.
33. Anon., The Concrete Institute. Concr. Constr. Eng., 1920, 15, 444.
34. Anon., Annual general meeting and annual report. Concr. Constr. Eng., 1922, 17, 407.
35. Anon., Award of the gold medal to Professor Henry Adams. Struct. Engr, 1923, 1, 33.
36. Serraillier, L., Discussion at Fourteenth Meeting of the Concrete Institute. Concr. Inst. Trans., 3, 115–16.
37. Concr. Inst. Trans., 1, xxi.
38. Anon., Amendments to the London Building Acts (London County Council General Powers Bill, session 1909). Concr. Inst. Trans., 1, xii.
39. Concr. Inst. Trans., 1, xxv–xxvii.
40. Concr. Inst. Trans., 3, 220–21.
41. Concr. Inst. Trans., 5, 409; Concr. Inst. Trans., 6, 9; Concr. Inst. Trans., 6, 156.
42. Concr. Inst. Trans., 6, 10.
43. Concr. Inst. Trans., 6, 9–10.
44. Concr. Inst. Trans., 6, 156.
45. Regulations made under the provision of section 23 of the London County Council (General Powers) Act 1909 with respect to the construction of buildings wholly or partly of reinforced concrete. LCC: London, 1915.
46. Andrews, E.S., The Reinforced Concrete Regulations of the London County Council, July 1915 (under the London County Council General Powers Act 1909 and now in force: a handy guide containing the full text, with explanatory notes, diagrams and worked examples). Batsford: London, 1916.
47. Anon., The report of the Concrete Institute on standard notation. Concr. Constr. Eng., 1910, 5, 2.
48. Anon., Standard notation for reinforced concrete. Concr. Inst. Trans., 2, xii.
49. RIBA Joint Committee on Reinforced Concrete: Second Report. RIBA: London, 1911.
50. Concr. Inst. Trans., 4, 269.
51. Concrete Institute. Mnemonic notation for engineering formulae: Report of the Science Committee of the Concrete Institute with explanatory notes by E.F. Etchells. Etchells Concrete Institute: London: Spon, 1918.
52. Concr. Inst. Trans., 9, Dvii.
53. Concr. Inst. Trans., 9, Miv.
54. Concr. Inst. Trans., 6, 163.
55. Concr. Inst. Trans., 7, 162.
56. Concr. Inst. Trans., 9, Dv.
57. Hurst, B.L., Concrete and the structural use of cements in England before 1890. Proc. Instn Civ. Engrs Structs & Bldgs, 1996, 116, 283–95.
58. Bussell, M.N., The development of reinforced concrete: design theory and practice. Proc. Instn Civ. Engrs Structs & Bldgs, 1996, 116, 317–34.

15 Concrete in tunnels

Alan Muir Wood

Introduction

Concrete in tunnels serves usually in a simple functional capacity, as an arch or continuous ring providing support to the ground, acting in simple compression. So far as is practicable, reinforcement is avoided, since it complicates the process of emplacement of concrete *in situ* and may need special protection against corrosion, especially where the concrete is exposed to aggressive water under pressure.

Generally, aesthetic attractions of concrete in a tunnel would be otiose where the lining is hidden from view or only passed in semi-darkness at high speed. The exception to this rule is predominantly for underground railway (predominantly metro) stations or caverns for recreational or similar use. The severe geometrical artistry of features of sewer tunnels have a special fascination from their hermetic nature, exploited for example by Carol Reed's film of The Third Man.

The adoption of concrete for a tunnel lining needs to be considered essentially as part of a system, intimately dependent on the overall logistics of the other processes of the system: excavation, spoil removal, primary support, preparation, transport and placing the concrete into position. A particularly direct example of this feature of the system explains the timing of concrete as successor to masonry (in this context including brickwork as well as natural stone) — or, particularly in North America, the use of timber — as a 'permanent' lining. In consequence, this brief account relates the developments in the use of concrete in tunnels to developments of other aspects of the total tunnelling process. Much of the development is seen, on this account, to have occurred incrementally from simple beginnings, as special plant was developed in response to the specific economic demands for placing concrete in a tunnel.

Applications of concrete

Concrete may be used structurally in tunnels in three different forms:

(i) as concrete placed *in situ* behind formwork (in the section Concrete placed *in situ*), of which a sub-set entails the grouting of previously placed aggregate;
(ii) as linings composed of precast units, normally assembled in rings, forming a complete, usually circular, ring or a partial arch (in the section Precast segmental linings);
(iii) as sprayed concrete (shotcrete) which requires no formwork (in the section Shotcrete).

The functional objectives of a concrete lining may satisfy one or more of these requirements:

(a) to support the ground or, more precisely, to provide an adequate degree of support to allow the ground to support itself;
(b) to exclude water or other fluid or gas in the ground;
(c) to contain water or other fluid or gas under pressure;
(d) to provide a smooth intrados to the tunnel, especially to reduce resistance to the flow of water or gas.

The lining may serve secondary purposes, for example for the attachment of an internal facing or for support of service pipes, cables or equipment.

Concrete placed *in situ*

The traditional form of primary support for tunnels entailed heavy timbering developed on different principles in different geographical areas. The forms were appropriate to the conditions for tunnelling and installed in accordance with traditional rules of the pattern of the system. All such systems required the partial or full removal of the timbering to be undertaken in short lengths as the secondary support (i.e. the permanent lining) was advanced, either in headings occupying a limited fraction of the cross-sectional area of the tunnel or for the full-area section, the choice depending on the system of timbering and upon the nature and special problems presented by the ground. A masonry (in this text including brickwork) lining was suitable for such a system; concrete placed *in situ* was not, since the work had to be undertaken in short and possibly variable lengths and sections. Success often depended on early use of the lining for support.

In consequence, the earliest uses of concrete placed behind formwork occurred either where the ground required virtually no support (e.g. where protection against rock falls in service remained a consideration) or where steel supports, initially in the form of arches packed off the rock surface, had replaced timbering as primary support. These examples are therefore to be found where competent relatively unfractured rock coexists with areas of population growth or early industrial development.

Early examples of concrete linings included tunnels of the early 1880s on the now disused Glenfarg line on the North British Railway between Kinross and Perth, for which W.R. Galbraith was responsible,[1,2] constructed contemporaneously with the Forth (rail) Bridge. Concrete on this occasion was so novel a material to the Government (railway) Inspectorate that holes needed to be made for the purposes of their examination, who pronounced the work to be 'infinitely better than brickwork'.[3] Another example in Scotland, demonstrating that absence of timbered tunnel support hastened progress, was in the provision of concrete lining to the previously unlined sections of the 30-km length of the water supply rock tunnels to Glasgow from Loch Katrine in 1886–88, for the purpose of increasing flow (by 35% according to Simms[3]). The concrete for tunnels at this time was hand-mixed using as aggregate stone broken on site and local sand.

For the Perkasie and Muscanetong Tunnels of the Philadelphia and Reading Railroad company in 1886–88, concrete sections of arch lining were cast in sections 20–75-ft long on centring placed and removed by the railroad company[4,5] (Figure 15.1).

There are a number of accounts from North America of the use of *in-situ* concrete for water and rail tunnels in the period 1890–1900. One of these was the Cascade Tunnel for the Great Northern Railway in Columbia (1891–92) for which the arch was cast in 12-ft sections after side-walls had been built in concrete shovelled from a platform, at the level of the wall-plate, built in 500-ft lengths (Stevens[25]), above the level of the muck trains.

For the Tremont Street Tunnel for the Boston Subway (~1900), the side-walls were constructed in rectangular drifts, with the arch constructed behind a roof shield travelling on track along the 2-ft 9-in wide side-walls, the completed tunnel being 20-ft 5-in high × 23-ft 3-in wide. This is probably the first instance of an *in-situ* concrete lining being placed behind a shield. The concrete was placed manually into the crown along curved cast-iron troughs and contained by timber stop-ends. Vertical pipes were cast in for the purpose of filling the crown with grout after concreting. The 16 thrust rams of the shield each shoved against a 3.25-in dia. iron 'push bar' set end to end against previous push-bars cast into the

Figure 15.1   Concrete placed in situ *c.1890.*

concrete lining (see also Bonnin[23]). A similar system had been attempted for the Siphon de l'Oise in Paris but had to be abandoned.[6]

Lee[7] describes concrete placed for the double-track Peakshill Tunnel, using timber side forms and arch forms in 12-ft lengths, with the final key in 4-ft lengths, the concrete placed by shovels and tampers, with external waterproofing in tarpaper treated with hot tar (elsewhere at this time bitumen felt was being used for this purpose).

As steel arch supports came into more general use at this period, concrete arch formwork initially comprised timber laggings on steel arches themselves slung from the tunnel supports. An early use of travelling formwork is described for the Gallitzin Tunnel, Harrisburg, Penn. The roof was timbered with the arch lined behind a 20-ft long travelling shutter of timber laggings on steel frames, the shutter travelling on rails and lowered 9-in when being advanced.[8]

During the next 25 years or so, collapsible steel formwork, hinged to each side of the crown, was being developed by the Blaw Steel Centering Co. and others, initially requiring diagonal tendons to hold the formwork to shape during concreting. During the same period developments in all aspects of batching, mixing, transporting and placing concrete achieved great economies and accompanied a general acceptance of the superiority of concrete as a tunnel lining material.[9] Concrete was being conveyed down vertical pipes from the ground surface with provisions to prevent segregation, and satisfactory pneumatic placers were developed after years of trials. Subsequently, collapsible forms allowed the placing of continuous lengths of arch lining (Figure 15.2), while slip-forming was later exploited for the tunnel invert.[10]

Subsequent developments of note included the use of plasticizers and waterproofing agents, developed particularly by Kaspar-Winkler, later Sika, from around 1920,[11] with the earliest uses of mortars to control leakages of the Alpine Tunnels.[12]

The most celebrated use of a tunnel lined by prepacked aggregate subsequently grouted is that of the 11-ft dia. Kemano penstock tunnel for Alcan's Kitimat

Figure 15.2   The development of collapsible formwork between 1929 and the present day (courtesy of CIFA, Milan).

hydropower project, in British Columbia, to ensure that a high proportion of internal pressure (up to about 2400 ft head) would be transmitted from a 5-in thick steel lining to the surrounding rock (the proportion measured as around 75%). The aggregate was placed by 'rock-blower', the grout pumped over horizontal distances up to 3000 ft and up to 1500 ft vertically. The project was completed in 1954. Jaeger[13] describes several experiences in prestressing pressure tunnels, by external grouting and by the use of circumferential prestressing cables, of the concrete surrounding a steel lining, together with accounts of several failed attempts.

One particular development of *in-situ* concrete concerns the placing of concrete immediately behind a tunnel boring machine (TBM), of which the Boston Subway described above, was an early forerunner. The Press Beton TBM was based on successful experiences in constructing metro tunnels for the Moscow metro in 1972 (Roisin — personal account). The Press Beton TBM would excavate the ground by means of a rotating cutter-head (Figure 15.3) with the thrust developed against two sets of hydraulic rams, one set bearing directly against special shoes which compacted the concrete lining, the other set against (say) 300-mm wide rings of flanged steel segments providing the internal shutter for the newly placed concrete. As the lining advanced, so was the 'trailing' ring — of (say) six

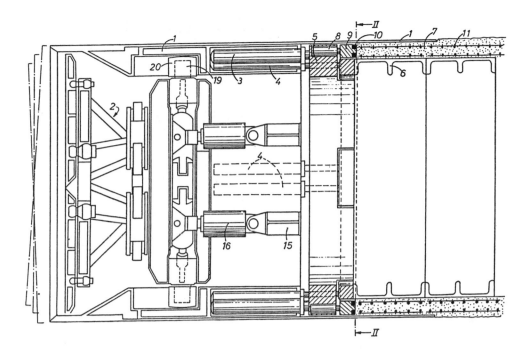

Figure 15.3   Concrete in situ behind shield (from Brevet d'Invention [patent application] of Bade, CFE, Ed. François & Fils, Frankignoul 29 Jan 1976).

rings of formwork — progressively dismantled and re-erected at the leading position. The system was developed and adapted by Hochtief in Hamburg in 1978 using steel-fibre reinforcement for the concrete, and later for a metro tunnel in Lyon.

Precast segmental linings

One of the earliest systems of segmental lining of a complete ring was that introduced by McAlpine[14] in 1903, first used for the City of Glasgow Main Drainage. The rings of concrete segments had tongue-and-groove circumferential joints, in each of which was embedded a steel bar hoop as reinforcement. Such linings were then used in different sizes for many years.

A widely used form of precast concrete lining, resulting from the general progress in the manufacture of precast concrete, comprised a series of bolted flanged segments, following the general form of cast-iron segments, the traditional lining for shield-driven tunnels since Barlow's Tower Hill Subway of 1869.[15] Reinforced concrete segments were used for a New York subway in 1930 but their wider adoption followed from the 2.75 mile extension of the Central Line of London's Underground in 1936.[16] Linings of this type were most commonly erected within the protection of the skirt of a shield with the annular void filled with cement grout after each advance of the shield.

Speed, economy and control of ground settlement were much enhanced by the development of linings in concrete segments directly expanded against the ground. These linings have been developed for shield-driven tunnels in stiff clay, particularly the London clay. The Don-Seg lining (Figure 15.4) was first used for the Lee valley — Hampton water tunnels, following a successful use in an experimental tunnel in 1951.[17] This lining, of 21-in rings of 10 identical tapered segments, was expanded by thrusting alternate segments into place by the shove rams of the shield. Subsequent linings of similar type but using fewer tapered segments are less readily adaptable than the Don-Seg to slight variations in the cut profile of the ground (on account of corrections for line, for curves or for increased 'coming-on' of the ground prior to building the lining).[15] For tunnels of 8–10 m dia, the preference has been, on account of ratio of diameter to ring width, to expand the ring by means of jacks inserted in recesses between selected segments,

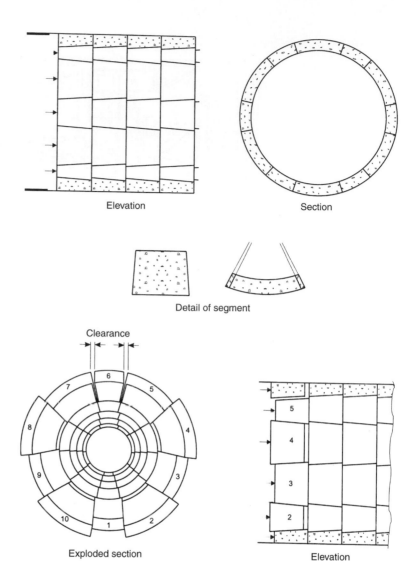

Elevation

Section

Detail of segment

Clearance

Exploded section

Elevation

Figure 15.4   Don-Seg.

especially high economy being achieved for the Cargo Tunnel at Heathrow Airport in 1968[22] illustrated by Figure 15.5. To ensure no significant imposed bending stresses in the lining, longitudinal joints between segments were shaped convex — convex, requiring high precision (~0.2 mm) in casting and in overall dimensional controls.

There have been many subsequent developments in concrete linings, expanded and bolted, with many variants of attachment. The current trend is towards flush-faced segments and for larger segments than for the earlier tunnels to take advantage of mechanization. Segments therefore require reinforcement, with possible need for protection against corrosion. Steel-fibre reinforcement has also been used for precast segments. Sealing of the joints between segments has prompted major developments of sealants. The most widely used form of seal has been the shaped neoprene gasket, originating from the Phoenix gasket in Germany in 1974, which girdles each segment and is sealed by pressure between segments, applied by bolting or by TBM rams. An alternative form uses a neoprene with hydrophilic additive which causes the seal to expand on contact with water.

Shotcrete

The earliest known major use of mortar linings for tunnelling was for the treatment of the arch of Alpine rail tunnels, preventing dripping of water onto the track,

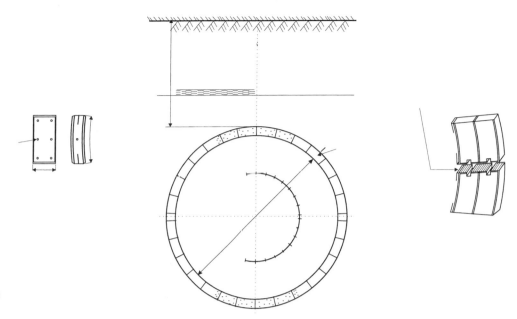

Figure 15.5   Heathrow Cargo Tunnel.

at the time of electrification, from 1917 onwards, the thickness, 75–120 mm, depending on the structural soundness of the original masonry lining, with additives promoting low permeability and high resistance against sulphates and other salts.[12]

The early history of shotcrete is described by Bussell in this volume. Rabcewicz[18] may well claim precedence in setting out in 1944 the virtues of a thin deformable concrete lining in achieving an economic stable rock/lining composite structure. This could only be achieved however with the advent of shotcrete which was used for the first time for the Lodano-Losogno tunnel in Switzerland for the Maggia hydro-power scheme (1951–55). Since that time there have been many developments in Informal Support of this nature, in many types of ground, including Sprayed Concrete Lined (SCL) tunnels in clay.[15] The term 'Informal Support' denotes the freedom to modify the elements of the system in relation to need and to geometrical variation of the lined cavity. The techniques include the New Austrian Tunnelling Method (NATM) but this term is avoided on account of the extraordinarily conflicting definitions of this form.

The principal objective of a shotcrete lining is to create a rapid form of support adequate to contribute to a stable ground/lining composite structure but sufficiently flexible to tolerate a calculated degree of deformation as ground movements, inevitable with the advance of the tunnel, occur. For the Snowy Mountains hydro-electric project, pneumatically applied mortar (with sand up to 4.7 mm size) with steel reinforcement was used from 1961.[19] The more general use of shotcrete as a primary form of ground support was developing in several European countries in the 1960s, usually with mesh and rock-bolts prior to an *in-situ* concrete lining. The concrete also served to provide a relatively smooth base to which to attach protective 'fleece' and a waterproof membrane outside the secondary concrete lining, as a protection against inflow particularly where icing might otherwise occur.

The development of mixing and placing equipment has kept pace with the market demand. For many years the debate has proceeded between the relative merits of the 'dry' and the 'wet' processes. In the latter, the concrete is mixed and pumped to the nozzle. Dry mix is impelled pneumatically in an earth-dry state to the nozzle, where the water is added. The general consensus[20] is now that wet-mix is preferred for reasons of:

- economy — capacity and rate of application;
- working environment — control of dust and reduction of 'rebound' (i.e. concrete which fails to adhere);
- quality — while average strength may be less than dry-mix, quality is less variable.

Following recent developments, shotcrete suitable for tunnelling normally contains accelerators and microsilica, having qualities of good adhesion, high early strength and ductility. It often contains steel-fibre reinforcement applied with the mix. Application over wide areas is undertaken by remote control or robotics to improve working conditions for operatives.

References

1. Popplewell, L., A Gazetteer of the Railway Contractors and Engineers of Scotland (1871–1914), Vol. 2, 1989. Published privately ISBN 0 906637 14 7.
2. Rickard, P., Tunnels on the Dore and Chinley Railway. Min. Proc. Instn Civ. Engrs, 1894, 116, 115–75.
3. Simms, F.W., In: D. Kinnear Clark ed., Practical Tunnelling, 4th edn. Crosby Lockwood and Son, 1896.
4. Ford, P.D., Discussion on Fitzgerald, 1894.
5. Fitzgerald, D., Lining a waterworks tunnel with concrete. Trans. ASCE, 1894, 31, 294–328.
6. Hewitt, B.N.M., Johannesson, S., Shield and Compressed Air Tunnelling. McGraw-Hill, 1922.
7. Lee, G.W., The cost of concrete tunnel lining and of tunnel excavation. Eng. News, 1903, 50, 531–32.
8. Anon., Construction work on the Pennsylvania Railroad between Harrisburg and Gallitzin. Eng. News, 1903, 50, 273–76.
9. Fitzgerald, J.H., Tunnelling Equipment — VI. Evolution of the concrete lining plant, Eng. News. Rec., 1931, 9, April, 616–19.
10. Kidd, B.C., The Orange-Fish tunnel in South Africa. Proc. Sth Wales Instn Engrs, 1977, 93, 27–33.
11. Hegnauer, H., Notes from the archives of Kaspar Winkler & Co from 1920–1930. Sika Promotion, 1984, 1 Zurich, 64–71.
12. Streuli, R., Tunneldichtung mit Sika an der Gotthardlinie, Sika-Nachrichten, 1943, 12, 1–8.
13. Jaeger, C., Present trends in the design of large pressure tunnels and shafts for underground hydro-electric power stations. Proc. Instn Civ. Engrs, 1955, 4(1), 116–200.
14. Anon., The McAlpine system of reinforced concrete tunnel lining. McAlpine, 1935.
15. Muir Wood, A.M., Tunnelling: Management by Design. E. & F.N. Spon, London, 2000.
16. Groves, G.L., Tunnel linings with special reference to a new form of reinforced concrete lining. J. Instn Civ. Engrs, 1943, 20, 29–41.
17. Scott, P.A., A 75-inch diameter water main in London: a new method of tunnelling in London clay. Proc. Instn Civ. Engrs, 1952, 1(1), 302–317.
18. Rabcewicz, L.v., The New Austrian Tunnelling Method. Water Power, 1964, 16, 453–57, 511–514; 17, 19–24.
19. Moye, D.G., Unstable rock and its treatment in underground works in the Snowy Mountains scheme. Proc. 8th Commonwealth Min. Metall. Congress, 1965, 6, 429–44.
20. Anon., Shotcrete in tunnelling. Status Report. International Tunnelling Association: Lyon, 1991.
21. Anon., Methods of work on the East Boston Tunnel extension of the Boston Subway. Eng. News, 1902, 47(4), 74–76.
22. Muir Wood, A.M. and Cribb, F.R. Design and construction of the cargo tunnell at Heathrow Airport. ICE Proc., 1971, 48, 11–34; 50, 187–201.
23. Bonnin, R., The new roof shield of the Metropolitan Railway Tunnel of Paris. Eng. News, 1905, 54, 324–25.
24. See Chapter 5.
25. Stevens, J.F., The Cascade Tunnel, Great Northern Railway. Eng. News, 1901, 45, 23–26.

16 Water-retaining structures in Britain before 1920

Michael Gould

General development

The construction of water-retaining structures, sewers, reservoirs, sewage tanks, and swimming baths, as well as bases for gasholders, followed a definite sequence: from brick, usually[1] with a clay puddle backing, through brick with mass concrete backing, then all mass concrete to reinforced concrete. However, development was not sequential, with some engineers still building brick structures when others were using concrete.

Experience revealed that brick, generally of local manufacture or sometimes produced on site, was frequently porous, and that ordinary lime mortar was both porous and susceptible to dissolution in water. From the 1850s, concrete began to be employed as backing on economic grounds, particularly with the spectacular strength being shown in tests of Portland cement concrete.[2]

The use of 'composite' walls, as they were described, became more general in the late 1860s, led, e.g. by designers such as Vivian Wyatt for the Chartered Gas Co. of London, but concern about the porosity of concrete meant that a puddle cradle continued to be employed. Puddle was also felt to have advantages in areas of poor ground to accommodate movement. Asphalt lining was used in porous new red sandstone at Liverpool gas works (1874) and as a seal between two layers of concrete 'wall', in Sheffield (c. 1899), to deal with movement from mining settlement.

As the puddle was an integral part of the construction, these tanks could not be tested for watertightness without the backfill, and the walls were designed for retaining only the soil.

Concrete was also employed for the base of tanks, although in areas of impermeable strata such as London clay, a base was sometimes dispensed with entirely.

Some brick gasholder tanks were strengthened by concentric rings of hoop iron, and this was also used on some concrete tanks. This practice was advocated by V. Wyatt who:

> 'considered that hoop iron was most essential in concrete walls. He found it had a tremendous affinity for cement; it did not rust; it never swelled, but took hold with the greatest avidity, and it was difficult to separate the two. They formed as it were one homogeneous mass from top to bottom',[3]

although he also puddled the floor and walls.

It was recognized that large lengths of mass concrete would crack and it was recommended that walls, e.g. should be subdivided by vertical divisions. However, the only joint treatment available initially was to fill the gap formed with strong cement.[4]

Problems with joints continued for some years. Lansdown water tower, Bath, built by Coignet in 1926, holds 100,000 gallons at 65 ft. Although still in use, this gave trouble from rusting at the joints and the tank was Gunite coated in 1932–34.[5] To offset such problems some tanks were lined, the low 40,000 gallon tower at Ham Green Hospital, Bristol, also by Coignet, was, e.g. rendered with Pudlo.[6]

It had been suggested that watertight concrete required cement sufficient to fill the interstices, although a stronger surface render was also advocated.[7] Later, other approaches were used.[8,9] One was the use of concrete additives, such as soft soap,[10,11] oil,[12] clay[13] or commercial preparations such as Ceresit.[14] An alternative was a waterproof surface layer, e.g. asphalt; or coating, e.g. sodium silicate.[15] Later, it was realized that many additives had no effect and that dense concrete would be watertight[16] if properly made with correctly graded aggregate.[17]

It was not until 1938 that the Institution of Civil Engineers, in consultation with the other Institutions, produced a code of practice for water-retaining structures.[18] This advocated the limitation of direct tensile stress and bay sizes, as well as the use of alternate bay casting, to prevent cracking. A minimum percentage of steel was proposed, although it was then permissible to put this all into the tension face.

The attitude of the Local Government Board (LGB), which sanctioned loans for public works, delayed the change from mass to reinforced concrete in the years after 1900. The LGB considered that reinforced concrete was less durable and loans were given on a 15-year repayment period, against 30 years for mass concrete.[19] For example, the high level reservoir at Exmouth had concrete walls lined with blue brick and a roof 6 in thick, in 9 ft 9 in bays between steel joists, reinforced by expanded metal.[20] This, the LGB suggested, was not reinforced concrete and a 30-year repayment period was allowed. In 1916, the reservoir at Wath upon Dearne was built with mass concrete/brick walls, with a 12 in clay cradle. Here, the loan on the reinforced roof was restricted to 10 years.[21] In a time of low interest rates, this represented a considerable increase in the rate burden.

Sewers

Apart from cement mortars used for brick sewers, concrete was first used in sewerage as a backing. Oxford sewerage in 1876, used brick sewers with a square concrete surround, flanged cast iron pipes (with concrete and iron filings joint filler) or stoneware with Stanford patent joints, the barrel being built up to a sliding fit.[22] Similarly, Croydon used earthenware, some with 3 in concrete surround; one ring of brick with a 1 in cement waterproofing collar, and surround; or two rings of brick with a cement collar between.[23] It was suggested that Portland cement was better than lime for resisting chemical action of sewage, but that an internal ring of brick was needed to protect the concrete backing.[24] Given the 'two-ring' brick construction used at Croydon with no backing, it is now unclear whether it was the one brick ring or the backing of the composite sewer which was considered as providing structural strength.

Concrete pipes had been used in USA since 1845, and some experiments were carried out in England by W. Buckwell in 1849. It was not until 1875[25] that Sharp, Jones & Co. started making 'rock concrete tubes', using broken stoneware aggregate. These were used at Bournemouth (where some cracked longitudinally 'in course of ordinary engineering precautions having been neglected'), Bromley and Frome.[26]

Elsewhere, in 1881, Peterborough Corporation laid 2 miles of 'concrete carrier'[27] to take settled sewage to the sewage farm.[28] Newhaven used fine cement concrete for cast *in-situ* sewers in 1887, in either 6 ft by 6 ft semi-circular sections with a 1 ft 6 in cunette, or 4 ft diameter pipes with a brick invert.[29]

In 1891, the Monier system for pipes reinforced by mesh was reported.[30] However, it was considered only suitable for large sewers due to difficulty in forming the bore.[31] This was partially overcome by the Bonna system, promoted by the Columbian Fireproofing Co. of Acton. A steel core-barrel was reinforced externally with longitudinal and a helix of plus-section bars before concreting. These two elements provided the necessary strength. Reinforced concrete was cast inside the barrel, but only as a protection. Collar joints were used (Figure 16.1).

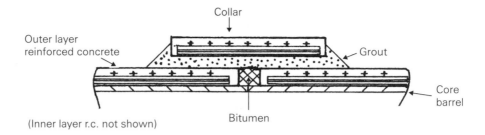

Figure 16.1   Bonna pipe joint.

Figure 16.2   Pipes at Swansea (C&CE Photo).

Bonna pipes, cast vertically (Figure 16.2) were used for water mains in Swansea,[32] for a sewer in Acton[33] (Figure 16.3) and for a sewage pumping main (117 ft head) in Norwich.[34] Thereafter, reinforced concrete was increasingly used for pipes; Professor Matthews reviewed progress in 1919.[35] Later, small diameter pipes became a reality with centrifugal spinning.

Culverts

Reinforced concrete proved suitable for culverts of various cross sections (Figure 16.4). Culverts at Birmingham and West Hartlepool were designed by the Expanded Metal Co.,[36] that at Birmingham was above ground and was surrounded by stonework. Kilton was a Trussed Steel Concrete Co. design,[37] while Ouseburn[38] and Belfast[39] were in Hennebique Ferro-concrete. The outfall culvert at Belfast (Figure 16.5) was on timber piles. Work was suspended by the Ministry of Munitions in 1917 and not completed until the 1930s.

Figure 16.3   Pipes at Acton.

Patent 'Aqueduct' blocks incorporating steel-ribbed granite-concrete pipes were used at Great Crosby.[40]

Reservoirs and tanks

Many early water supplies consisted of a well or river intake and a steam driven pump. This worked best with a constant flow against a constant head and a reservoir was often constructed as a pump-header; from these, the modern service reservoir developed. Such headers were open with sloping sides finishing at ground level. However, in 1852, the (London) Metropolis Water Act required that all filtered water reservoirs within 5 mi of St Paul's Cathedral had to be covered, and this provision gradually spread throughout the country. The slowness of this process is shown by the fact that some early Mouchel designs related to the covering of sloping sided reservoirs, e.g. Cowes in 1909,[41] while Camborne, built by the Expanded Metal Co. in 1910, was uncovered.[42]

Early reservoirs were covered by brick (usually barrel) vaulting or by brick jack-arching. Spans necessitated brick supporting pillars, although some jack-arching was carried on cast iron columns. Sloping sides were a hindrance when covering

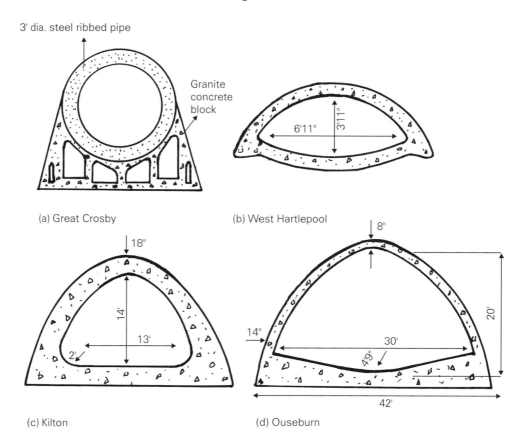

3' dia. steel ribbed pipe

Granite concrete block

(a) Great Crosby

6'11"

3'11"

(b) West Hartlepool

18"

14'

13'

2'

(c) Kilton

8"

20'

14"

30'

4'9"

42'

(d) Ouseburn

Figure 16.4   Reinforced concrete culvert sections.

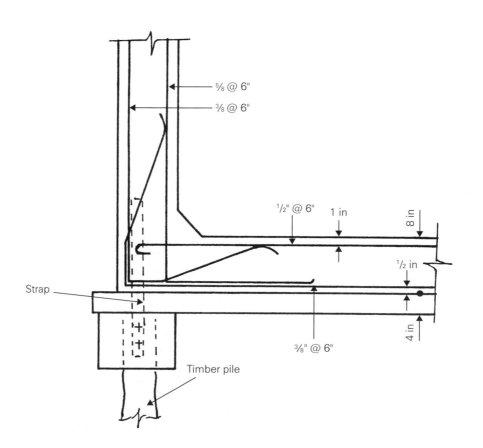

⅝ @ 6"

⅜ @ 6"

½" @ 6"

1 in

8 in

½ in

4 in

Strap

⅜" @ 6"

Timber pile

Figure 16.5   Belfast outfall culvert corner detail.

reservoirs and vertical sides were introduced, although a tank in Wolverhampton (1851) had an odd horse-shoe cross-section.[43] Vertical sides were of brick (occasionally stone) backed by clay puddle, brick backed by concrete, or, later, mass concrete, sometimes also backed by puddle.

Reinforced concrete was used from 1900; that much built was not publicised is a problem when discussing the history of water supply. In some reports the specialist designer for the reinforcement may not be credited, others partially so (Skegness reservoir, c. 1910, used 'Columbian' bars and is credited to Percy Griffith, perhaps employed by Industrial Constructions Ltd).[44] Two review papers on reservoirs were published in 1883[45] and in 1915,[46] and much of what follows is taken from them.

As mass concrete became acceptable, whether brick or concrete was used depended largely on material availability. Thus Sudbury, c. 1871, 55 ft × 55 ft, was of brick backed with concrete while Kettering, c. 1873, of similar size, was built of stone with clay puddle.[47] Kettering had a stepped floor (Figure 16.6).

A problem with barrel arching was the horizontal thrust at the wall; with jack-arching this is resisted by tie bars between the girders. In 1874 at New Cross, the wall of the 1.75 million gallon reservoir (brick backed by concrete with brick arches on brick piers) was seen to be rotating and an arch invert was added. Other tanks were built with floors arched between the lines of columns. Elsewhere, the outermost of the roof arches was carried round to meet the wall lower down, the thrust now being vertical. Woolwich Common, c. 1870, had the brick/concrete end-wall built as a series of vertical axis arches to brick piers at the apexes, although it is not recorded why this arrangement was used

Although most early reservoirs were rectilinear, circular examples are recorded. Greenwich Park, built as an open reservoir in 1845 with sloping sides and base diameter 154 ft, was lined with 12 in lime cement on 9 in puddle. This did not stand well and a brick lining was added. In 1871, it was covered by concentric barrel arches, the outer being carried to a point half way down the side slope.

As experience with concrete grew, its use became more general. Deptford reservoir, built in 1872 by covering a filter bed, used the bed materials to form concrete roof arches, tapering from 20 to 10 in in thickness, carried on longitudinal brick arches. At Oldham in 1910, a complete reservoir in this arrangement was built in reinforced concrete.[48] Addington (Croydon) was mass concrete, including the roof,[49] but the discussion suggests that this was not the first to be made thus. Fifteen vertical cracks in the walls had to be cut out and refilled with concrete. The construction in 1886[50] of the filters for Carlisle water-works in mass concrete passed with almost no comment (although as these tapered from 4 to 2.5 ft in a height of about 5 ft the design was very safe).

Some hybrid constructions were undertaken. Upleadon (Gloucester, c. 1900)[51] had roughly dressed masonry walls rendered internally (7 ft tapering to 2 ft 6 in over 18 ft), a floor of 1 ft 3 in of concrete and a concrete arch roof carried on steel girders, the arches tapering from 3 ft to 1 ft 6 in over the 12 ft 4 in span; the rise was 2.5 ft. The girders had brick piers with three courses of footing on the floor.

In Kent, a number of late 19th century circular concrete tanks were reinforced with three or four 1.5 in wide hoop-iron bands every foot (Figure 16.6). Initially these were horizontal (Knockholt, c. 1889, West Wickham and Eltham) but at Sundridge Park they were placed vertically. Southfleet (built after 1904) was similar but was the first of the design to omit a brick wall lining.

These tanks were not without problems. Eltham cracked at the wall/floor joint, the horizontal floor steel not being tied into the walls. Subsequently, a bitumen sheet was incorporated at this joint. Southfleet's concrete walls cracked, apparently due to the expansion of the steel roof joints[52] which were built rigid into the wall.

The construction of tanks for sewage paralleled those built for water supply, although the use of all brick construction seems to have continued for longer.

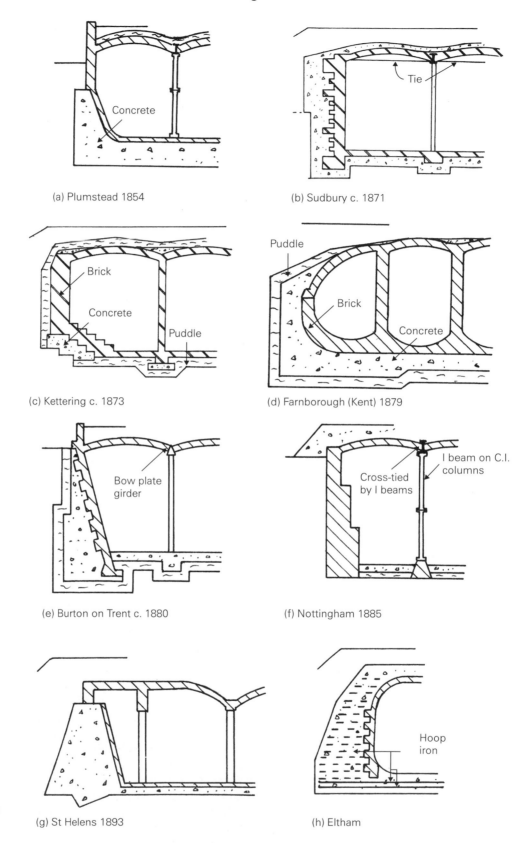

(a) Plumstead 1854

(b) Sudbury c. 1871

(c) Kettering c. 1873

(d) Farnborough (Kent) 1879

(e) Burton on Trent c. 1880

(f) Nottingham 1885

(g) St Helens 1893

(h) Eltham

Figure 16.6 Reservoir cross sections.

In 1884,[53] e.g. a 3.25 acre holding tank[54] was constructed at Portsmouth, built in brick with cement backing and 'a straight thick water-tight joint between' (i.e. of stronger cement).

Water towers

In 1900, Mouchel[55,56] commenced design of what is accepted as the first water tower in reinforced concrete in Britain. Completed c. 1903, this small structure was for green watering at Meyrick Park golf course, Bournemouth.[57] Thereafter the number of concrete water towers built rose steadily.

Early design followed one of three patterns. First were tanks on solid shafts, of several cross-sectional arrangements; this mirrored brick construction. Second were plain tanks, some on frail looking legs. Some, such as the Mouchel tank at Red Barn, Fareham (probably the oldest now surviving) (Figure 16.7), had no leg bracing but most had one or more ring bracing or, more commonly, wheel bracing. Bracing, normally horizontal, was occasionally vertical. The third design was of a tank on a central shaft and an outer ring of columns, as Mouchels 300,000 gallon tower at Newton-le-Willows (1905).

The tanks on these towers sometimes leaked, and many designers lined the tank as a precaution. Newton tank cracked horizontally between the hoop steel when filled. It was suggested that this was due to the sun's action on one side[58] but with Marsh,[59] a standard text of the time, suggesting that the wall spanned vertically between the hoop steel, it seems likely that the possibility of vertical bending in the wall was not catered for.

On small towers, the floor and roof simply spanned to the sides, although some towers had a domed floor. Mouchel often used a grillage of beams under the floor, clearly seen on the still standing tower at Trim, Ireland c. 1909.[60]

Expanded metal perhaps lent itself best to solid sided construction. Two ornamented circular examples stand in County Down (at Donaghadee and the Down Asylum, dating from 1912 and c. 1910).[61] This company also built square towers as at Drumcondra, Dublin and Holsworth, Devon (Figure 16.8).[62] Greenbank Gas Works, Blackburn, had a shallow tank on six asymmetrical spaced legs to accommodate a loading conveyor for the coke hoppers.[63]

Figure 16.7   Red Barn water tower (B. Otter Photo).

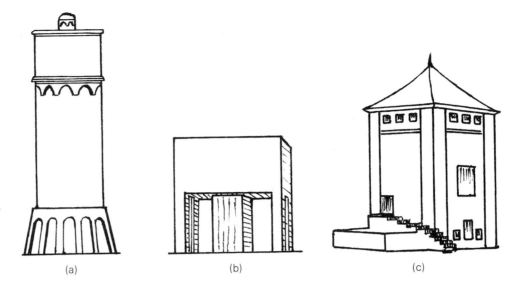

Figure 16.8 Expanded metal water towers. (a) Down asylum c. 1910; (b) Holsworthy; (c) Drumcondra, Dublin.

(a)   (b)   (c)

The patent Kahn bar was intended for beams and would appear to have little benefit for towers. However, a number were described as 'Kahn'. The earliest noted, at Severus Hill in York (1914,[64] Figure 16.9), built around a stand-pipe, was described as the largest built to that date (300,000 with a maximum height of 103 ft; the height of Newton was 83 ft).

The Indented Bar Co. also built towers. Cleethorpe (1907–10) had a steel tank set on a metal ring at a height of 110 ft to allow for expansion movements; the Ferbeck Chimney Construction Co. built the solid sided shaft.[65] It still stands, although the 250,000 gallon tank was beyond repair by 1962 and was replaced by a shorter concrete tank.[66] A smaller solid sided tower, also for the Great Grimsby Waterworks Co., was at Immingham. The company also built a number of multi-legged towers with a central core in the 1920s, culminating in Goole, being the largest in Europe in 1927 (750,000 gallons with an overall height of 145 ft).[67]

BRC designed that for the West Cheshire Water Co. at Whitby, c. 1915 (Figure 16.10). It was hexagonal with a diagonal length of 57 ft.[68] It was supported on thin square 42 ft long legs with one wheel bracing mid-height. The floor extended beyond the legs and was supported by down-stand beams. The pattern of attached beams and columns to the walls suggests that they were designed as slabs spanning horizontally and perhaps also vertically.

Some railway companies built water towers. The North Eastern Railway built at least two octagonal by Truscon[69] and three circular by Mouchel. Mouchel designed a rectangular tower for the North British Railway in 1908.[70] The London Brighton and South Coast Railway's Kahn tank at Lancing was on an arched circular beam over six 36 in by 9 in columns and curtain walls. The 6 in thick base was domed over the 25 ft span (with a rise of 4 ft), while the roof was only 4 in thick.[71]

Hospital towers were constructed at Warrington Union by BRC and at Aberdeen Poorhouse by the Indented Bar Co., both built by 1916.[72] The latter was square and held 110,000 gallons in two compartments — Mouchel & Partners also designed some square towers.

Some individuals commissioned private towers. The earliest designed by Mouchel, were at Roydon, Essex and Gopsall Hall, Leicester (both c. 1909).[73] A circular solid-sided example stands at Ridgmont, Beds., a private supply provided by the Duke of Bedford.[74] Milton Hill, near Oxford was built to the Kahn system in 1916[75] (Figure 16.11, although there is some doubt as to how much, other than

Sketch drawing prepared from old photograph.
Height 103 ft max., capacity 300,000 gallons. 1914, Kahn.

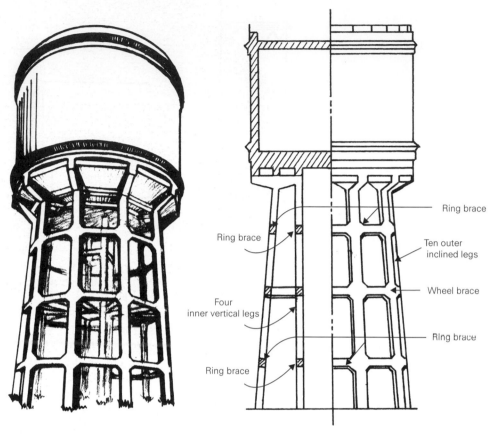

Ring brace

Ring brace

Ten outer
inclined legs

Four
inner vertical legs

Wheel brace

Ring brace

Ring brace

*Figure 16.9   Severus Hill
water tower, York.*

(Note: rear legs not shown)

the two water tanks, one above the other, was in concrete). A small drum on tall legs was provided by BRC at Brookmans Park[76] while a square tower over a well was built by the Indented Bar Co. at the Grange, Beaconsfield,[77] both before 1916.

In 1914, Coignet built an elegant tower for Sir Oswold Mosley at Rolleston on Dove; it stands but is unused. It was filled in 14 days and a 'slight dampness' at the floor/wall joint disappeared after 'a few days'.[78]

Swimming baths

In 1846, the construction of municipal baths and wash houses was permitted,[79] with set charges for bathing.[80]

By 1865,[81] the Act had been adopted by only eight metropolitan parishes (earliest St Marylebone, 7.11.1846) and 24 boroughs (earliest Birmingham, 7.10.1846). Birmingham provided three 'swimming baths' as the permitted 'open bathing places'.[82]

It was not until 1878[83] that the legislation was extended to 'covered swimming baths'. However, as this Act simply doubled charges set for classes 1, 2 and 3 and, as no definition of these classes was included, it appears that it only regularized the then current practice. The bath built by the Bury Improvement Commissioners in 1864, e.g. included first and second class swimming baths, each 52 ft by 20 ft.[84] At Burnley, two baths (72 ft by 37 ft, and 62 ft by 30 ft), were called 'plunge baths'.[85] Both sites appear to have been covered.

Figure 16.10   Whitby water tower (BRC Brochure).

In their next Report, the LGB noted that they, not the Treasury, now approved loans for baths. Surprisingly, they seem not to have issued any circular letter indicating what was considered acceptable. Even so, in 1879, loans were given to Aston Manor, Croydon and Nottingham Boroughs.

From the 1880s, although the construction of swimming baths became widespread, details were rarely reported, more account being given of the facilities. York baths, opened in 1880 and sized 70 ft by 25 ft, are simply reported[86] as having walls of white glazed brick and white glazed tiles on the floor — features common to most baths. However, construction techniques for swimming pools probably followed those for reservoirs. As late as 1907, the use of brickwork walls was being advocated in preference to concrete, although 'a good thickness of cement concrete' in the floor was considered acceptable.[87]

Elsewhere, all concrete construction was used. Carlisle baths,[88] 1883–84, had cement concrete walls 2 ft thick, with a floor 1 ft thick, and tiling. As discussed above, the watertightness of such construction was considered suspect. At Ealing,[89] hot bitumen was run between the glazing brick and the wall, while the open air bath at Tottenham[90] was rendered with Limmer asphalt. An asphalt lining (presumably

Figure 16.11   Milton Hill water tower.

(Kahn, Incorporated electricity generator, pump house and two water tanks in tower)

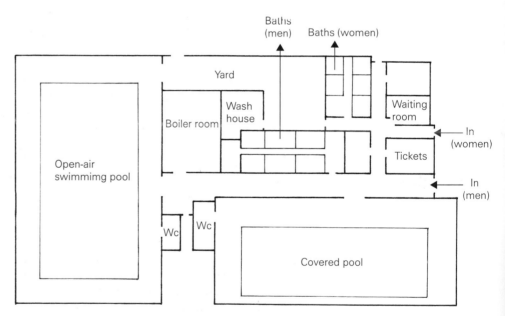

Figure 16.12   Sketch plan South Norwood baths, Croydon, c. 1888.

(Changing boxes arranged around each pool)

between the concrete and glazing) was used for Battersea,[91] said to be the largest covered bath when built (150 ft by 50 ft). Croydon baths,[92] 1889 (Figure 16.12) were constructed in concrete with 'no clay puddle', and tiles placed in neat cement.

As with tanks, the change to reinforced concrete was gradual. In 1907, two baths (100 ft by 30 ft, and 96 ft by 27 ft) were built for Chelsea by the Expanded Metal Co. The floor was supported on 24 in × 18 in ground beams at 10 ft centres, the wall and floor being 12 in thick, with an asphalt layer between the concrete and glazing. A 100,000 gallon storage tank was incorporated below the public foot-path (Figure 16.13).[93] Mesh sheets were used with rolled steel joists (RSJs) encased for beams and columns.

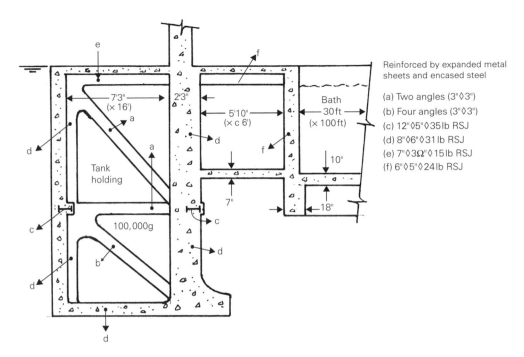

Figure 16.13   Cross section of Chelsea bath.

Gasholder bases

Thinner sections were used by Mouchel and Partners for Dundee baths.[94] Here, the floor rested on 8 in × 14 in or 6 in × 12 in beams spanning to piles. The floor was only 5 in thick and the walls 4–5 in. How watertight this pool was is not recorded. Mouchel's first ferro-concrete baths, dated to 1906,[95] were for Roedean school and for Goole UDC. Not all baths built were covered, Dundee, Northam and Pollockshaws, e.g. being open. Northam, Southampton,[96] replaced a mass concrete structure on the sea-shore. This had had insufficient weight and had cracked due to tidal variation on uplift pressures.

Early gasholders were of limited capacity and comprised of cast iron segments bolted together. They were quick to erect and some outlay could be recovered by selling them on to other, smaller concerns, or for scrap, when increasing demand necessitated their replacement by larger containers. The use of cast iron tanks was soon found to be uneconomic and excavated tanks lined with brick or stone walls with stone flag, brick or tile base became the norm. The tank was filled with water as a seal, and the gas stored in a cylindrical sheet iron vessel or bell, which floated in the water, rising and falling as the demand for gas fluctuated. The movement of the bell was controlled by side guides.

Although the excavation could be taken across the whole base, sometimes a trench was excavated around the perimeter to accommodate the wall, and the excavation completed with the wall in place. The centre could have been blasted out in rock, or excavated, leaving a rise or dome in the centre. This had the advantage of helping to support the bell when lowered but could, if carried too high, affect the storage and could be difficult to seal. This form of construction was by 1868 'not of a kind to be now recommended'[97] but it continued to be used, e.g. at Becton in 1890–91, see Figure 16.14.[98] Brick facing was generally used, increasingly on aesthetic grounds.

At Imperial Gas Works Bromley,[99] the whole of the bottom of the tank was covered by 'well prepared' blue lias lime concrete 12 in thick, of a 1 : 5 mix, although here the foundation was of blue clay. The Phoenix Gas Works, Kennington, used wall footings of Portland cement concrete, a 1 : 7 mix, placed in two courses

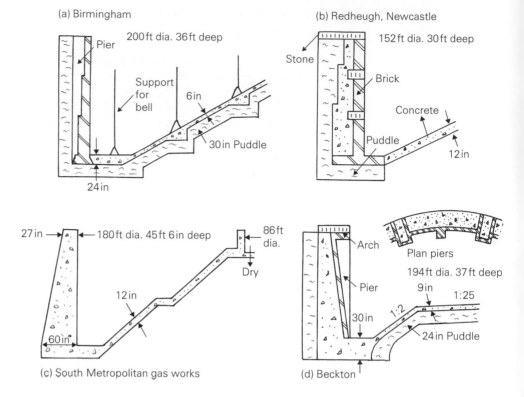

Figure 16.14   Sketches of gasholder tanks wall/floor sections.

each 2 ft thick, but relied entirely on the blue clay as a floor seal. Concrete piers, apparently unreinforced, ranging from 23 ft 6 in to 16 in and founded on the clay, carried the bell. Similarly, at Salford, c. 1880, concrete 12 in thick was used only in wall footings, the floor consisting of 2 ft of puddle. York stone was used for the coping, the rest stones (to support the bell when lowered) and as inserts in the brick to carry the guide rails. Birmingham, by contrast, had the brick wall founded on the puddle but concrete 6 in thick (2 ft under the piers to support the bell) over the floor puddle. The walls were reinforced with both hoop iron and bands of five courses of brick laid in Portland cement. At Bow a double brick skin was employed enclosing a concrete heart, the brick courses being carried across at intervals.

Finally, in 1871–72 George J. Livesey built a Portland concrete tank, 153 ft diameter, 35 ft deep, with a brick lining 9 in thick, bonded into the concrete, the wall and floor being puddled. The cost was stated as two-thirds that of an all brick tank.[100] This successful application encouraged other designers to make Portland cement concrete tanks. At Redheugh, Newcastle, puddle concrete walls with a single brick face and a concrete floor was used (although, oddly, the whole wall had a brick foundation). The whole wall thickness was reinforced with hoop iron. J. Douglas, engineer at Portsea, had information[101] from tests at Portsmouth Dockyard which indicated a 7-day strength of Portland cement concrete of $299.2\,lb/in^2$, 30-day strength of $417.3\,lb/in^2$, and 2-year strength of $675.8\,lb/in^2$ (earlier Medina cements had given $150\,lb/in^2$). Portsea tank was concequently built up using small moulds and made with West Medina Mills cement supplied by Charles Francis and Company. A 9 in brick facing, separate from the concrete, was employed purely for aesthetic reasons.

Livesey, in 1875–76, made a new tank for the South Metropolitan Company entirely of Portland cement concrete (1:7 mix) covered in a Portland cement render ½ in thick and with no puddle or brick. It was 180 ft in diameter, 45 ft 6 in deep, with the iron guides embedded in the concrete. It was founded in chalk at a depth of 42 ft, with water bearing sand above. The walls were 5 ft thick tapering to 2 ft 3 in. The tank was completed in 1875 and partially filled, showing no signs of leakage, but when, in December it was refilled with freezing canal water vertical cracks appeared, causing concerns about the viability of all concrete tanks, and their vulnerability to temperature shocks, although at the time it was realized the ground conditions may have contributed. Livesey himself felt he should have used hoop iron, although other engineers were sceptical of its value.

Hooped concrete walls were used at Waterford in Ireland where a tank had to be founded on a peat bog of 12–25 ft overlying silty sand. A 9 ft wide perimeter trench was excavated, a layer of planks was then laid. Onto this, a layer of 1 : 6 Portland cement mortar incorporating larger pieces of schist was spread. The first layer was 3 ft deep with an angled skewback at the centre to receive the dome base. Three hoops of 4 in by ½ in wrought iron were embedded in this layer to help resist thrust of the dome. The concrete wall was 4 ft thick at the base and 20 in at the top, with hoops at the guide supports. Over the winter, the walls were loaded with stone and there was differential settlement where the peat was deepest. The tank was then modified.

The wall for the gasholder at East Greenwich (c. 1890) stood 21 ft above ground and 13 ft into it. It was 3 ft 6 in thick at the top tapering to 4 ft 6 in just below ground, it then stepped to 5 ft continued to the base. The wall above ground was embanked, was not considered to be sufficiently strong in concrete and was reinforced with hoop iron bands, 5 in × ⅜ in riveted together at 2 ft centres. These were 'strutted outwards while filling the concrete around them'.[102] If this was not released until the concrete had set then it should have induced a modest degree of prestress.

The tank built by Wyatt at Beckton, concrete with a brick inner face, was unusual in that the taper was taken inwards as providing better soil support. Many tanks had outside piers giving added support, but at Beckton these were also inside, and they were finished with flat brick arches to carry the coping.

Unsurprisingly reinforced concrete was employed in the 1890s by British gas engineers for tanks, although the earliest example appears to have been the Monier system tank at Fredericksberg Works, Copenhagen. Some engineers were sceptical of the progressive nature of the design, however.[103]

Other structures

The ship test tank at the National Physical Laboratory, Teddington[104] (now British Maritime Technology Ltd.), see Figure 16.15, was completed in 1911. Towing Tank

Figure 16.15   NPL Tank.

No. 1[105] is 52 m long by 9.1 m wide, maximum water depth is 3.7 m. This is a large U-shaped channel incorporating a circulating water pipe on one side, the concrete varying in thickness from 600 mm to 1.2 m. Reference 104 implies that this was built in mass in lengths of 40 ft (12.2 m), the floor first, followed by the walls shortly after.

Concrete was also widely used in other structures subject to fluid pressure such as basements. By 1917, reinforced concrete gas mains were being tested.[106] Pipes of 14 and 24 in diameter were cast using the Coignet system and were impregnated with tar before installation.

Conclusion

By 1920, and in spite of the conservatism of the LGB, reinforced concrete construction was well established for use in water-retaining structures.

Appendix

Below is given the number of contracts for different classes of work undertaken by Mouchel and Partners up to 1919 (from Ref. 55):

- reservoirs 34 (of which 29 were public supplies):
  — coverings, 15,
  — lining, 1.
  (the reservoirs included the freshwater 'pipe reservoir', 500 ft long by 4.5 ft diameter, built thus due to difficulties over land acquisition);
- water towers 50 (of which only 17 were public supplies);
- sewage tanks 15;
- swimming baths 12 (6 public, 6 private);
- other tanks (a large miscellaneous group, probably mostly small units), 176, include tanks for water and rain water, chemicals, oil and tar, for filtering and settlement, and roof tanks.

References

For simplicity, the later title, the Proceedings of the Institution of Municipal Engineers, has been used throughout.

1. King, A., Treatise of coal gas. Reports gasholder bases in north west England built entirely of brick, 1870, 2, 62.
2. Douglas, J., Gasholder tanks in concrete. Br. Assn Gas Eng., 1874, p. 34.
3. King, A., Treatise of coal gas. King expresses some doubt about the wisdom of using hoop iron, 1870, 2, 32.
4. Newman, J., Notes on concrete. SPON, 1887, p. 76.
5. In RC in 1926 C&CE, 1927, 22. Repairs *ibid.*, 1932, 27, 655 and 1934, 29, 453. Also W&WE, 1926, 30, 15.
6. C&CE, 1916, 11, 47–49.
7. Newman, J., Notes on concrete, SPON, 1887. Pages 46 and 86 resp. The first probably had the same effect as the second, i.e., it produced a stronger concrete. Chatham Dockyard used a 4 : 1 (although later figures suggest this should have read 6 : 1) render on a 12 : 1 mass.
8. Gadd, W.L., Waterproofing concrete. C&CE, 1908, 3, 154–57.
9. Plumb, R.A., Waterproofing concrete. C&CE, 1910, 5, 497–501.
10. Memorandum. C&CE, 1911, 6, 320–1.
11. Grittner, A., Waterproof concrete. C&CE, 1912, 7, 810–13.
12. Page, LW., Waterproofing qualities of oil mixed concrete. C&CE, 1913, 8, 619–25.
13. Waterproofing concrete. C&CE, 1907, 2, 490.
14. Industrial notes. C&CE, 1910, 5, 451.
15. Memorandum. C&CE, 1910, 5, 933.
16. Memorandum. C&CE, 1912, 7, 865 & 932.
17. Waterproofing concrete. C&CE, 1911, 6, 7–9. A wide range of additives were tested.
18. Code of practice for the Design and Construction of Reinforced-Concrete Structures for the Storage of Liquids. Institution of Civil Engineers: London, 1938.

19. This subject appears throughout. C&CE, 1907–1908, 3–4.
20. Proc. Instn Munic. Engrs, 1909–10, 36, 136–51.
21. Proc. Instn Munic. Engrs, 1916–17, 43, 304.
22. Oxford main drainage. Proc. Instn Munic. Engrs, 1876–77, 3, 146–55.
23. Crump, W.S., Sewerage works for Croydon Rural Sanitary Authority. Proc. Instn Munic. Engrs, 1879–80, 6, 94–108.
24. Newman, J., Notes on concrete. SPON, 1887, p. 98.
25. C&CE, 1911, 6, 399.
26. Woods, C.F., Rock-concrete tubes. Proc. Instn Munic. Engrs, 1879–80, 6, 58–65.
27. It is possible that this was a concrete channel.
28. The Engineer, 1881, 51, 3.
29. Newhaven drainage. The Engineer, 1887, 64 & 92–96.
30. Given as an Appendix to Angell L. Mesh to cast iron beams. Proc. Instn Munic. Engrs, 1891–92, 18, 422.
31. Proc. Instn Munic. Engrs, 1904–1905, 31, 382.
32. Wyrill, R.H., Reinforced concrete watermains. C&CE, 1906, 1, 185–92.
33. Sewer at Acton. C&CE, 1907, 2, 248–53.
34. Proc. Instn Munic. Engrs, 1907–1908, 34, 185–87; C&CE, 1907, 2, 319.
35. Mathews, E.R., Use of reinforced concrete in sewer construction. C&CE, 1919, 14, 357–64.
36. Mathews, E.R., Reinforced concrete in aqueduct and culvert construction. C&CE, 1908, 3, 129–33.
37. RC culvert at Kilton. C&CE, 1910, 5, 207–11.
38. Proc. Instn Munic. Engrs, 1912–13, 39, 323–24.
39. RC outfall sewer construction for Belfast. C&CE, 1917, 12, 5–12.
40. RC sewers. C&CE, 1907, 2, 407–409; 489 (which illustrates a section supporting a 10 ton load).
41. Illustrated in reference 56, as is another in Portsmouth. Other reservoir covers are illustrated in various issues of C&CE.
42. Lamborne new waterworks. C&CE, 1910, 5, 516–20.
43. Given in discussion in reference 45.
44. Reservoir for Skegness waterworks. C&CE, 1912, 7, 436–38.
45. Morris, W., Covered service-reservoirs. Proc. ICE, 1882–83, Pt.III, 73, 1–62.
46. Morris, W., Circular covered service-reservoirs. Proc. ICE, 1914–15, Pt. 1, 199, 409–21.
47. Illustrated in discussion in reference 45.
48. Reservoir for Oldham Corporation. C&CE, 1912, 7, 432–35.
49. Walker, T., Some of the public works in Croydon. Proc. Instn Munic. Engrs, 1880–81, 17, 12–19.
50. Carlisle waterworks. The Engineer, 1886, 64, 515.
51. James, H.O., Newent waterworks of the Gloucester Corporation. Proc. ICE, 1901–1902, Pt. II, 148, 342–43.
52. The roof was apparently on the Barrett System with the wet cement supported on timber sheets between the bottom flanges of metal I beams.
53. Boulnois, H.P., Drainage of Portsmouth. Proc. Instn Munic. Engrs, 1884, 12, 128–31.
54. For discharge over one hour at high tide.
55. Mouchel-Hennebique. Ferro-concrete, list of works, 1897–1919. Produced by the company. The date given is sometimes the completion date and sometimes the enquiry date (for water towers the split is about 50:50).
56. Hennebique Ferro-Concrete Handbook. Produced by the company in c. 1921.
57. Referred to in C&CE, 1907, 1, 133 as new works, illustrated in Reinforced Concrete by Winn, 1907; C&CE, 1956, 50, 248 as first in reinforced concrete and in Concrete, September 1974, 47 as Historic Concrete No. 1. Although statutorily listed, it was demolished c. 1997.
58. Preliminary report on the committee on reinforced concrete. ICE, 1910, 176–78. Was first lined with bitumen protected by concrete in 1910. Although statutorily listed was demolished in 1979.
59. Marsh, C.F., Dunn, W., Reinforced Concrete, 3rd edn. Constable: London, 1906, 403–407 'for longitudinal steel assume wall acts as a slab spanning between the hoops and built in at the ends'.
60. Hennebique Ferro-Concrete, 1909, 1, 346.

61. Cox, R.C., Gould, M.H., Civil Engineering Heritage of Ireland. Thomas Telford: London, 1998; 144–45.
62. Expanded Metal Company Brochure No. 5, 1923.
63. Reinforced concrete coke hopper and water tank at Blackburn gasworks. C&CE, 1910, 3, 85. The coke hoppers braced the legs of the water tank.
64. In The York Waterworks Co. 1677–1968 published by the company. For repairs see Water Services December 1985, p. 519.
65. Water tower at Cleethorps. C&CE, 1908, 3, 77–80.
66. Historic concrete No. 48. Concrete, November 1987, 23.
67. Silcock, E.J., Goole water tower. W&WE, 1937, 41, 573–75.
68. Illustrated in C&CE, 1916, 11, 477.
69. Reinforced concrete water towers. The Engineer, 1911, 111, 86.
70. Reinforced concrete water reservoir at Edinburgh, NB. C&CE, 1908, 3, 86.
71. C&CE, 1916, 11, 545.
72. C&CE, 1916, 11, 477 & 482.
73. This date is probably that of the initial enquiry.
74. Mouchel Archives.
75. C&CE, 1916, 11, 544.
76. C&CE, 1916, 11, 478.
77. C&CE, 1916, 11, 482.
78. Rolleston on Dove water tower. C&CE, 1914, 9, 132–33. Also Historic concrete No. 71, Concrete, October 1980, 29.
79. An Act to Encourage the Establishment of Public Baths and Wash-houses 9&10 Vict c. 74, 1846.
80. 1847, 10&11 Vict c. 61. Charges set were one person over 8 years (including towel) cold 1d per bath warm 2d, several children (up to 4) (towel each) cold 2d per bath warm 4d, open bathing place where several persons bath in the same water — half-penny each. 9&10 Vict c. 87 extends the Act to Ireland but as this legislation was not altered as was the English Act, it seems probably that it was little used.
81. Return of Bathhouses. Parliamentary Papers, 1865, xlvii, 279.
82. Most likely for labouring men, and for men and women, the 1846 Act requiring separation of the sexes.
83. 1878, 41&42 Vict c. 14.
84. Proc. Instn Munic. Engrs, 1883–84, 10, 17. These were presumably used by each sex at different times.
85. *Ibid.*, 1889–90, 16, 114.
86. *Ibid.*, 1881–82, 8, 49.
87. *Ibid.*, 1907–1908, 34, 79.
88. Hepworth, J., The Carlisle public baths. *Ibid.*, 1887–88, 14, 274–78.
89. *Ibid.*, 1889–90, 16, 51.
90. Proc. ICE, 1905–1906, xxxii, 180.
91. *Ibid.*, 1905–1906, xxxii, 246.
92. Walker, T., Some of the public works of Croydon. Proc. Instn Munic. Engrs, 1890–1891, 17, 7.
93. Reinforced concrete swimming baths and storage tank at Chelsea. C&CE, 1907, 2, 235–40. In preparing the figure it has been assumed that the top of the tank is at the same level as the duct.
94. Swimming bath, Dundee. C&CE, 1910, 5, 595–96. This says walls were reinforced by columns 6 ft × 5 in; it is unclear if this means 6 ft counterforts or 6 in attached columns.
95. Ferro-concrete, list of works 1897–1919. This is probably the date of the initial enquiry.
96. Crowther, J.A., Northam open-air bath, constructed in ferro-concrete. Proc. ICE, 1912–13, xxxix, 534 ff.
97. Clegg, S., Practical Treatise on the Manufacture and Distribution of Coal Gas, 5th edn. Trubner & Co.: London, 1868: 231, footnote.
98. Trewby, G.C., Description of a four-lift gasholder at Beckton. Incor. Inst. of Gas Eng., Trans, 1892, 167–172 and plates. The 'four-lift' refers to the telescopic bell.
99. King, A., Treatise of coal gas, 1870, 2, 47–109.
100. Douglas, J., Gasholder tanks in concrete. Br. Assn Gas Eng., 1874, 28.
101. King, p. 31.

102. Livesey, F., A description of a six-lift gasholder and tank now being constructed at East Greenwich. Incor. Inst. Gas Eng. Trans., 1892, 156–166 and plates.
103. Marshall, F.D., The Monier System of cement construction as applied to gas works. Incor. Inst. Gas Eng. Trans., 1895, 145–65. The drawings are a little unclear as to how the tank, for tar and ammonia, was built, but the walls appear to have been flat arches to vertical I beams strutted off the floor.
104. Experimental tanks at the National Physical Laboratory, Teddington. C&CE, 1911, 6, 787–88.
105. Catalogue of facilities produced by the International Towing Conference.
106. Reinforced concrete gas mains. C&CE, 1917, 12, 612–16.
This illustrates women casting the pipes.

17   Historic concrete in dams

D.A. Bruggemann, K.J. Hollock, G.P. Sims

Introduction

This chapter deals with the historic use of concrete as applied to dams from the early days of masonry with thin mortar joints to modern roller compacted concrete and thin concrete arch dams.

The purpose of this chapter is to provide an outline of the development of concrete dams taking into account the progress made over the last century in the manufacture of cementitious materials, advances in concrete technology and the refinement of analytical techniques for the design of arch dams. The chapter is intended to provide a source of information for engineers who may be involved in the rehabilitation of and modification to old dams.

The chapter deals with the concrete in the body of dams rather than in the appurtenant structures such as draw-off works and spillways. These structures are covered in other chapters on structural use of concrete. Information on the concrete mix proportions is provided, and covers the mortar in early masonry dams up to modern roller compacted concrete. The references are intended to provide a basis for further, more detailed literature searches for specific issues related to rehabilitation or modification.

The development of cementitious materials, in relation to dam engineering, leading up to the introduction of pozzolans is described. This is followed by a discussion on the transition from masonry to concrete dams. The evolution of the various types of concrete dam (gravity, buttress and arch) is outlined, taking into account the driving factors, such as cost, increasing size of dams and the mechanisation of construction methods which led to roller compacted concrete (RCC). The chapter concludes with data on mix proportions from early masonry dams, gravity dams constructed in the 1950s and the 1970s, and arch dams constructed in the 1980s and 1990s.

Development of cementitious material

Rankine[1] has distinguished between fat lime, hydraulic lime and cement as shown in Table 17.1.

The distinction between these materials is useful in understanding the likely behaviour of an old dam, even though the boundaries between materials are sometimes hazy. The period of rapid growth in the use of cement began in the last years of the 19th century as a result of intense research and promotion. Thus dams in the 20th century have usually been constructed using cement, either in a masonry mortar, or as the basis of concrete.

Table 17.1   Cementitious materials[1]

Description	Silicate content	Hardens in air	Hardens under water
Fat lime	Little or none	Slowly	No
Hydraulic lime	10–30%	Yes	Slowly
Cement	40–60%	Yes	Quickly

Before that lime was the most common building material and some of the largest cathedrals in the world have been built with masonry cemented with it. Lime is made by burning limestone, in the process driving off carbon dioxide (CO_2), leaving a clinker of calcium oxide, quicklime. When the quicklime is slaked with water, it disintegrates into a fine grained powder or paste, depending on the volume of water added. The pure slaked lime formed in this way is also known as fat lime. It can be used for the construction of masonry but it hardens slowly in air and not at all when isolated from atmospheric CO_2. Masonry buildings made with fat lime have been demolished and the lime remote from the air surface has been found to be little harder than when it was placed.[2]

Roman masonry dams, of which there are several in Spain, may have used the stronger and more durable roman cement comprising a mixture of lime and a natural volcanic ash known as pozzuolana, from which the modern term pozzolan is derived. The roman cement, like some naturally made cements contain more silicate, often in the form of volcanic ash, or as clay minerals. When limestone and clay containing silicates are burned together, they form a material capable of setting in the absence of atmospheric CO_2. The practical benefit of this is that mortar made with the material would set hard even under water. Canal building at the end of the 18th century required structures that would last well under water and this led to the development of what came to be known as hydraulic limes.

The production of hydraulic lime by the addition of silicates has been practised in India for many centuries until fairly recently. This process comprised the addition of 'surkhi', a clay rich in silicates, to fat lime.[3] The setting process is speeded up in the air and continues, albeit slowly, in water. A great deal of time and effort was expended in producing the lime and 'surkhi' on dam sites.[4]

Hydraulic lime can be made into a satisfactory mortar that achieves a strength similar to that obtained by cement mortars, albeit after about a year; concrete made with it is less successful, presumably because of the length of time it takes for the concrete to achieve useful strength. Tests conducted on large samples of masonry show that the strength achieved by mortar made from hydraulic lime is not much different from that achieved by cement mortar or indeed of cement concrete.[5,6]

Portland cement was invented in 1825.[7] It was strongly promoted and its advantages over lime were gradually accepted. It was more expensive than hydraulic lime, needing grinding and quality control to give a reliable product. It can be made reliably into concrete. As the availability of skilled masons declined, and their cost rose, the possibility of replacing masonry with concrete became an economic necessity. In Europe, the use of lime in the construction of dams declined sharply in the second half of the 19th century and cement was the normal material from the beginning of the 20th century.

The first use of Portland cement for dam construction is attributed to Boyd's Corner Dam a 24 m high masonry dam, constructed in 1872, in Putnam County, New York and at Pérolles in Switzerland almost simultaneously.[8]

The early cements were produced in vertical kilns and were coarse ground by today's standards. The cement was slow setting and construction rates were low, thus the evolution of heat of hydration was low and little cracking occurred. From about 1905 to 1920 the coarse-ground Portland cement was giving way to fine-ground cement produced in rotary kilns. In the decade 1920–1930 there was a surge in the use of mass concrete, at a time when the technology of concrete was inadequately understood. The deterioration, after a short period of service, of many concrete dams in Europe and in the USA was causing concern. The 1920s were followed by rapid advances in concrete technology. Between the two World Wars, the techniques of concrete production developed from a rough-and-ready art producing a product of uncertain and often unsatisfactory quality, to the highly developed and rigidly-controlled processes used today to produce good quality concrete.[9]

In developing countries, however, cement remained too expensive and adequate manpower was available at an acceptable cost until the middle of the 20th century. Thus, the persistent use of locally produced hydraulic limes (lime-'surkhi') in India. Being locally made, the quality is often variable and a stronger more durable mortar is made with the addition of cement. Lime-surkhi mortar was specified for the Chikkahole Dam which was completed in masonry in 1966.[6]

During the 1930s, when large volumes of concrete started to be placed at fast rates, 'low heat' cements were developed in the USA to reduce the heat generated during hydration of the cement. This effect was achieved by limiting the combined proportion of the tricalcium silicate (C_3S) and tricalcium aluminate (C_3A) to less than 58%. This property is incorporated in cements classified as Type II and Type IV in US practice.[10]

Pozzolanic materials

As the 20th century progressed, efforts in dam engineering were made to reduce the amount of cement in concrete for dams. These efforts were made in the interests of economy and to reduce the temperature rise in concrete and led to the inclusion of pozzolans in concrete.

Pozzolans are siliceous substances, sometimes containing aluminium, that react with calcium hydroxide to form cementitious compounds. Pozzolans may be naturally occuring minerals such as clay or volcanic ash, or man-made materials such as fly ash or ground blast furnace slag. Pozzolans have several advantages over cement. They reduce the heat of hydration of concrete, enhance plasticity and workability, and can improve concrete quality and resistance to Alkali Aggregate Reaction (AAR).

As early as 1916, the United States Bureau of Reclamation (USBR) were concerned about the build-up of heat of hydration of cement in massive concrete dams. It was recognized that since the temperature rise in the interior of the mass was nearly adiabatic, a decrease in the quantity of cement would decrease the quantity of heat proportionately. On this basis, the USBR reduced the quantity of cement by half for Arrowrock and Elephant Butte dams. This reduction was done by the replacement of half of the cement with ground sand, which was a forerunner of the pozzolan replacements of the 1950s. The resulting concrete mass was as workable as all-cement concrete, but had lessened resistance to weathering.[11]

The USBR continued to use pozzolan in the concrete for their dams. Friant dam completed in 1942 included 20% pumicite pozzolan in the concrete. All the dams constructed by the USBR between 1952 and 1963 included pozzolan in the concrete. The pozzolanic materials included fly ash, calcinated clay, calcinated shale and pumicite.[10]

An example of the use of man-made pozzolans (flyash) in the UK is reported at the Lednoch Dam in Scotland, completed in 1958.[12] Two mixes were used in the construction of the dam and these are described later in the chapter.

Other advantages that have led to the increasing use of pozzolans to replace cement in the second half of the 20th century include a lower alkali content of the cementitious mass, hence a better resistance to Alkali Silica Reaction (ASR). The properties of pozzolans vary widely and it has been found necessary to undertake extensive tests on the pozzolanic mixes. It has been found necessary to include air-entraining agents. Pozzolanic concretes tend to gain strength more slowly than cement concrete. Some pozzolans cause large shrinkage and others reduce the durability of the concrete.

Transition from masonry to concrete

Concrete dams developed from masonry dams, which had been in operation for a substantial period. Schnitter[8] notes the existence of four masonry dams built in south-western Romania in the 1730s and 1740s and there are earlier examples dating back to the medieval period.

Masonry was popular for three reasons;[4] first, masonry was a building skill that has been readily available since the medieval period as demonstrated by churches constructed in that period. Second, masonry is a cheap and practical building technique when cement is scarce. Third, the spillway can be incorporated in the masonry structure and so there is no need for a separate structure. In addition to the above, the method has politically significant gains in terms of employment potential because construction is labour intensive. It is for this reason that masonry dams have been constructed in India until recent times.[4]

Transition took place from:

* masonry, where shaped blocks of stone were bedded on mortar and the joints packed with stone and mortar through;
* cyclopean masonry, where random pieces of stone were placed on and surrounded by concrete to;
* concrete construction, initially including large stone 'plums' or displacers placed into the concrete (cyclopean concrete).

A fine example of a masonry dam is Vyrnwy Dam, which was completed in 1891 and is shown during construction in Figure 17.1. Note the large cranes required to lift the blocks into place. The use of cyclopean masonry persisted in the UK into the 20th century as indicated by the construction of Angram Dam (completed 1912). The dam is illustrated in Figures 17.2 and 17.3, which show the dam during construction and after completion respectively. The first dam to be constructed in the UK from concrete with displacers was Blackbrook constructed in 1906.

Many masonry dams, such as Boyd's Corner dam, were constructed in the USA. The end of the masonry dam era in the USA was marked by the construction of

Figure 17.1 Vyrmwy dam under construction (Binnie Archive, ICE).

Figure 17.2 Angram Dam
under construction (courtesy
of Yorkshire Water plc).

Figure 17.3 Angram Dam
after completion (courtesy of
Yorkshire Water plc).

New Croton Dam (91 m high) completed in 1905, in New York state and Theodore Roosevelt Dam (85 m high) completed in 1911 and located on the Salt River in Arizona. It is interesting to note that about half of the cement required for these dams was Portland cement imported from England.[11]

The development of concrete dams

The development of concrete dams was driven by the following factors:

- the understanding of the influence of the water–cement ratio on the strength and durability of concrete;
- the need to control the temperature rise in the concrete as the size of dams increased and as the rates of construction increased;
- economy;
- improvements in design techniques.

These factors led to change in the types of dam broadly from massive gravity structures through buttress dams and cylindrical arch dams to double curvature arch dams and finally to gravity RCC dams. The evolution was gradual and not in strict sequence as much development was carried out in parallel.

There were some engineers who seemed to have understood the importance of the water–cement ratio before others. The engineer responsible for the Lower Crystal Springs dam, constructed in 1890 about 30 km south of San Francisco, paid special attention to the control of the water content in the concrete (some 250 kg/m^3 referred to bone dry aggregates).[8] Some 2 years after the completion of this dam, L. René Féret of the French 'Corps of Bridge and Highway Engineers' published results of tests carried out in 1892, showing the pre-eminent influence of the water–cement ratio on the quality of concrete.[13] The water–cement-ratio 'law' was rediscovered for the USA and the English speaking world in 1918 by Duff Abrams,[14] and after this date the need to control the water–cement ratio was generally accepted by the dam engineering fraternity.

The Hoover Dam shown in Figure 17.4 demonstrates the dramatic increase in the quantity of concrete in the dam and the rapid rates of placing concrete which were experienced in the early part of the 20th century. The dam was constructed on the Colorado River in the south-western USA between 1931 and 1936. It was designed by the USBR as a 221 m high massive curved gravity dam and contained about 2,500,000 m^3 of concrete.[10] At that time the dam was 60% higher and 2.5 times larger than any existing dam. Concreting at Hoover dam took barely two years with up to 8000 m^3 of concrete placed per day.[8,15] Figure 17.5 shows the fully automated batching plant used at Hoover dam and gives an indication of the magnitude of the concrete operation and would not look out of place on a modern construction site for a large concrete dam. The dramatic increase in the speed of construction was emphasized further at Grand Coulee Dam where concrete placing rates of twice that at Hoover were required. Grand Coulee Dam is a straight gravity dam containing about 7,400,000 m^3 of concrete and was built between 1933 and 1942 by the USBR.[8,15]

Several techniques were used at Hoover Dam to control the temperature of the concrete during the setting process. These techniques included the use of low heat cement and post cooling by circulating cold water through pipes embedded in the concrete. The post cooling technique was used for the first time

Figure 17.4   Hoover Dam completed 1936.

Figure 17.5 Fully automated concrete batching plant, Hoover Dam.

at Hoover, but low heat cement had been used for the first time in the USA at Morris Dam built between 1932 and 1934, near Pasadena, California and at Hoover dam.[15]

The practice of precooling concrete materials prior to mixing to reduce the maximum temperature of mass concrete began in the early 1940s and was extensively employed in the construction of large dams during the late 1940s and 1950s.[15]

Gravity dams reached their prime in the 1960s[8] but they were becoming too expensive in comparison with earth and rockfill dams, which had benefited from the development of earth moving technology during the Second World War and the subsequent construction of the interstate highway network in the USA.[16] In response to rising costs, the concept of Roller Compacted Concrete (RCC) was developed in 1970 in the USA by J.M. Raphael.[17]

The difference between RCC and conventional concrete dams is that earth moving equipment is used to construct RCC dams. Concrete is placed in layers of several hundred metres in length, and 300 mm thick, and consolidated by vibrating rollers. The construction period is much shorter, and consequently the labour and plant costs are reduced. The RCC construction process is much safer than conventional concrete dams because the layers are long and wide, and the height of each lift is minimized. There is little or no formwork needed. Machinery is readily available and so is effective and flexible. The first large RCC dam, Shimajigawa Dam, was constructed in Japan between 1978 and 1981.[18] By the end of 1998, 209 large (>15 m high) RCC dams had been constructed and were under construction.[19]

In order to reduce the quantity of concrete in gravity dams and thus reduce costs the buttress dam was developed from the massive gravity dam by introducing intermediate spaces. The buttress dam has several advantageous features arising from the introduction of the spaces such as reduced quantity of concrete, reduced uplift and heat dissipation. These features are discussed below.

The spaces allow water seepage through foundation and dam to discharge not only downstream, but also side- and upward into them, thus reducing the uplift pressures. For the classical triangular profile with a vertical upstream face, the total elimination of uplift pressures would theoretically have permitted a reduction of the dams mass by 40% without reducing its stability. However, the actual reduction proved considerably smaller, because the intermediate spaces had to be closed upstream with slabs, arches or a thickening of the buttress heads to make them contiguous. The saving in costs was even more modest due to the fact that buttress dams, per cubic metre of concrete, required more formwork of a more complicated shape and, often, a higher cement content in the less easily placed concrete. On the other hand, the intermediate spaces facilitated the dissipation of the heat produced during the hardening of the concrete, so that elaborate cooling measures were seldom needed during the construction of a buttress dam.[8]

The buttress dam proved popular in the early part of the 20th century in areas such as Australia and the western USA. These areas were undergoing similar pioneering developments far from manufacturing centres and thus favoured engineering solutions which minimized imports and transport costs.[20a]

The buttress dam type is designed in several forms:

- multiple arch,
- flat slab or Ambursen Dam,
- contiguous buttress.

A multiple-arch dam, Mir Alam, was constructed in 1804 in India by the Madras Engineering Corps for the water supply of Hyderabad. It was almost 100 years before the next multiple-arch dam was constructed. This dam was a 19 m high dam constructed across the Belubula River near Sydney between 1896 and 1897. At the same time, a similar design was being studied for a 31 m high structure near Ogden, north of Salt Lake City, Utah.[20a]

An interesting example of a multiple-arch dam because it was the first application of a double-curved dome-structure in dam engineering[7] is shown in Figure 17.6 which shows the 76-m high Coolidge Dam under construction in 1928, 140 km east of Phoenix, USA. The maximum arch span is 54.9 m and the thickness from

Figure 17.6   Coolidge Dam under construction, 1928.

1.2 to 6.1 m. In spite of its sophisticated formwork, the dome concept proved about 2% more economical than a conventional multiple-arch, 11% more economical than a conventional arch, and 27% cheaper than a gravity structure.[10]

At the beginning of the 20th century reinforced concrete was relatively new and spreading fast and a competitor to the multiple arch was the flat slab buttress dam developed and patented by Nils F. Ambursen.[8] This type of dam became quite popular and by the end of the 1920s, more than 200 around the world had been constructed, thus outnumbering multiple-arch dams by far. Among the flat slab dams, was the 41-m high La Prele (or Douglas) irrigation dam, built from 1908 to 1909 in eastern Wyoming.[8] The dam had to be provided with a new upstream slab in 1977 since the original one had disintegrated by about 20% of its thickness in the rough mountain climate.[8,21] Outside the USA, the flat slab buttress dam design was especially adopted after the Second World War for about 50 dams exceeding 15 m in height in Norway.[8,22]

Despite the temporary popularity of reinforced concrete dams, especially the Ambursen flat slab type, the rise in labour costs in relation to the prices of materials and the ensuing mechanisation, worked strongly against them. Moreover, several of them had deteriorated rapidly in severe climatic conditions to which they were sometimes exposed, or they otherwise developed cracks and leaks. By the late 1920s engineers were increasingly questioning the wisdom of using thin concrete members in dams at all.[8]

The problems with the flat slab reinforced concrete dams led to the development of the contiguous buttress dam. Examples of these are V. Carranza irrigation dam in north-eastern Mexico, the Burgomillodo dam, completed in 1929, north of Madrid and the Dixence dam in Switzerland, completed in 1935.[8]

An example of the contiguous buttress dam in the UK is Haweswater Dam completed in the 1940s and shown in Figure 17.7. A diamond headed configuration was adopted to concentrate the water load into each buttress and the buttresses were thickened at the downstream end to give a continuous downstream face for greater lateral stability and resistance against buckling under load.[23]

Figure 17.7   Haweswater Dam, completed 1941 (Binnie Archive, ICE).

The move towards minimum cost dams (always perceived as minimum-concrete dams) culminated in the design of arch dams. Arch dams may contain as little as 20% of the concrete of the equivalent gravity dam.[23] Arch dams can be defined as concrete (or masonry) structures, the base of which measured less than half their height and which thus had to rely on their curvature in plan for the transmission of part of the water load laterally into the valley flanks.[8]

Arch dams are designed, both in the single or double-curvature versions. The strength of the rock in the abutments is therefore extremely important and weaknesses have led to failures.[23] The early arch dams tended to be the single curvature type, e.g. Zola dam, 41-m high, completed in 1854. Zola dam was the first dam to be designed on the basis of a stress analysis. The stresses were calculated from the 'cylinder formula'. From the 1830s through to the 1930s the method of analysis of arch dams was refined from a series of independent arches to the 'trial load method'[24] which determined the load distribution in the horizontal and vertical directions by a trial and error method. Much of the research into methods of analysis of arch dams was carried out by the USBR as part of the investigations for the Hoover dam. After adaptation to electronic data processing and other refinements, the method remained the most frequently used. It has now tended to be replaced by the finite element method.[8]

G. Albert Pelletreau introduced the concept of the double curvature arch dam in a study in 1879.[8] He found that in order to minimize the volume of an arch dam, its radius of curvature should decrease from crest to base. The 'cylinder formula' also yields smaller arch thicknesses for a given allowable stress.[25] The idea was revived in the early 1900s in the USA: the variable radius design was applied on the 51-m high Salmon Creek dam in south-western Alaska, completed in 1914.[8]

The main break through for arch dams came in 1905 to 1925, mainly in the fast developing west of the United States. Important contributions also came from Switzerland, Italy and France, which took the lead in the field upon the decline of arch dam construction in the United States in the 1930s.[20b] By the end of that decade all-important elements had been assembled for the spectacular qualitative as well as quantitative development of arch dams after the Second World War and their proliferation around the globe. At present the age of the arch dam appears to be drawing to an end except for very high structures.[20b]

The move towards the minimum-cost dams culminated in the design of the thin arch dams in countries where suitable sites exist for such designs; in the Alps, Portugal, Spain, etc.[23] The design of theses structures has reached the extremes of sophistication and now involves complex computer analysis as a complement to (or substitute for) model tests in the laboratory. The degree of sophistication is such that many of these structures are expensive to monitor, operate and maintain. There are very few such structures in the UK; only Monar Dam is of any size and is shown under construction in Figure 17.8 and after completion in Figure 7.11.

A recent example of a double curvature arch dam is shown in Figure 17.9. This is the 122-m high Victoria dam constructed on the Mahaweli Ganga in Sri Lanka between 1980 and 1984. In order to control heat generation, low heat cement was used, mixing water was chilled to 5°C, some of the water was replaced with flaked ice and a post cooling system was installed. Concrete placement rates of up to 40,000 m³ per month were achieved.[26] It is interesting to note that the rate of placement achieved is comparable to rates of production which were achieved at Hoover dam some 50 years earlier and is a tribute to the organisational skills of the early dam construction contractors in the USA.

The application of post tensioned pre-stressed concrete has not had a wide application to the body of new dams. It has been used in special locations, e.g. the spillway, gate house, abutments and for remedial strengthening of old dams. An exception to this is Allt na Lairige dam, constructed between 1953 and 1956, which

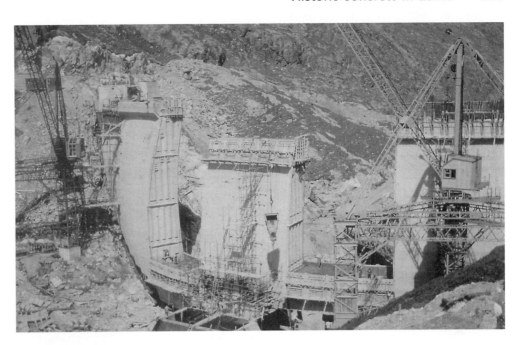

Figure 17.8   Monar Dam under contruction.

Figure 17.9   Victoria Dam, Sri Lanka, completed 1984.

was the first dam in western Europe to be constructed through out in prestressed concrete. The dam is shown under construction in Figure 17.10 and note the very thin section. An interesting fact about this dam is that the section of this dam is used as the logo for the International Commission on Large Dams (ICOLD).

Concrete mixes

In the early part of the century, concrete mixes were proportioned by volume. The mix often varied over the height of the dam and at the masonry dams Howden and Derwent (1912, 1916) the cement aggregate ratio varied as follows:[23]

* base section 1 : 6;
* middle section 1 : 5½;
* top section 1 : 5.

Figure 17.10   Allt na Lairige
Dam, a post-tensioned,
pre-stressed concrete dam
completed 1956.

Figure 17.11   Monar Dam
completed 1963.

The proportion of stone to concrete in these dams formed about 44% of the whole mass. The blocks were quarried Millstone grit, ranging from 0.5 to 6 t in mass, and rubble stone was used for the concrete aggregate. The strength of the concrete ranged from 9 to 24 MPa, with an average of 18 MPa at 1 year old, while the crushing strength of the stone was 43–53 MPa.[23]

Engineers realized that for gravity dams, only low-strength high-density concrete was required for the majority of the dam body. This realization in conjunction with considerations of cost of cement and its thermal properties, led to the specification of low cement-content concrete for the heart of the dam with 'normal' concrete on the exposed faces. The terms 'hearting' concrete and 'facing'

Table 17.2   Concrete mixes in two UK gravity dams

		Lednoch, 1958[12]		Cow Green, 1970[23]	
		Facing	Hearting	Facing	Hearting
Cement and fly ash (kg/m^3)		235	173	350	180
Fly ash proportion (%)		15	20	20	20
Aggregate (kg/m^3)	76 mm	500	529	–	640
	38 mm			830	426
	19 mm	500	529	560	426
	Coarse sand			406	575
	Fine sand	472	498	135	64
Water–(cement + fly ash) ratio		0.5	0.6	0.5	0.84
Water–cement ratio		0.6	0.75	0.63	1.05
Density (kg/m^3)		–	–	–	2520

concrete were coined, with an intermediate mix 'infill' concrete being used to infill pockets in the rock foundation.[23] The term 'infill' concrete was also used to infill between buttresses (in certain designs) or to infill around shafts and galleries when box outs had been left for their construction.[23]

Typical mixes for facing and hearting concrete are shown in Table 17.2.[12,23] It is interesting to note that the water–cement ratios are similar and that the proportions of cement and pozzolan are similar for the hearting concrete. The cementitious material content of the facing concrete is 350 kg/m^3 in 1970 compared with 235 kg/m^3 in 1958. The increase in cementitious material suggests that increased durability was sought at Cow Green.

Currently most gravity dams are constructed from roller compacted concrete (RCC). Dunstan 1999[19] notes that broadly there are five forms of RCC construction as shown below:

- High paste RCC in which the cementitious content (cement and pozzolan) is greater than 150 kg/m^3. An example of this type of dam is Upper Stillwater Dam in the USA, completed in 1987.[27]
- Medium paste RCC in which the cementitious content is between 100 and 149 kg/m^3. An example of this type of dam is Salto Caxias Dam in Brazil, completed in 1998.[28]
- Roller Compacted Dam (RCD) which is a technique unique to Japan. The cementitious content is generally 120–130 kg/m^3 and produces a relatively expensive, but high quality, structure suited to the high dynamic loading conditions in Japan.[19] An example of this type of dam is Shimajigawa Dam completed in 1980.[18]
- Lean RCC in which the cementitious content is less than 99 kg/m^3. An example of this type of dam is Willow Creek constructed in the USA in 1982.[27]
- Hardfill dams, which are essentially cement stabilized rockfill and are particularly suited to areas in which there is high seismic loading and relatively weak foundations.[18] The first large hardfill dam was Marathia Dam completed in Greece in 1993.[18]

It is interesting to note that the Portland cement content of RCC in the main countries of application (China, Japan, USA and Spain) is between 75 and 80 kg/m^3. The difference in cementitious content is created by the difference in pozzolan content.[19] These figures show that commonly a large proportion of the Portland cement is replaced with pozzolan generally with attendant cost savings and lower heat of hydration.

Table 17.3 summarizes the RCC types and gives the proportion of each type either under construction or completed. For purposes of comparison with

Table 17.3   Roller compacted concrete dam types[19]

Description	Proportion (%)
High paste RCC cementitious content >150 kg/m^3	46.5
Medium paste cementitious content 100–149 kg/m^3	21.5
RCD (Japan) cementitious content 120–130 kg/m^3	18
Lean RCC cementitious content <99 kg/m^3	12.5
Hardfill	1.5

Table 17.4   Concrete mix for two modern double curvature arch dams

	Victoria Dam, Sri Lanka, 1984[26]	Katse Dam, Lesotho, 1997[29]
Volume of concrete (m^3)	670,000	2,360,000
Cement and fly ash (kg/m^3)	230	220
Fly ash (%)	0	30
Aggregate maximum size (mm)	150	150
90-day Characteristic strength	–	24 MPa
28-day	23 MPa (specified), 30 MPa (achieved)	–

gravity dams, the concrete mixes for two modern double curvature arch dams are shown in Table 17.4. The total cementitious material content is similar in both cases except that Katse Dam in Lesotho incorporated fly ash. Arthur *et al.*[30] note that fly ash was readily available from nearby South Africa and it was a viable option to replace some of the cement with this pozzolan. Cole and Neal 1991[26] do not mention if cement replacement by fly ash was considered for Victoria Dam. The table also shows that the cementitious content is higher than for RCC dams and similar to the older gravity dams.

Conclusion

The chapter has examined the development of cementitious materials their application to dams including the use of pozzolans. The transition from masonry to mass concrete dams and the development of concrete dam types has been explored. Some general information on the concrete mixes used in masonry dams at the beginning of the last century and up to recent times has been provided.

Acknowledgements

The authors would like to express their thanks to Brown and Root Services for their support in the preparation of this paper. Thanks are due to Yorkshire Water PLC for providing a selection of historical photographs for the authors to select from.

References

1. Rankine, W.J.M., A Manual of Civil Engineering. Charles Griffen: London, 1865.
2. Scott, H.V.D. Account of the manufacture of a new cement. Corps of Royal Engineers, Papers , X, 1861, 136.
3. Hill, N., Holmes, S., Mather, D., Lime and Other Alternative Cements. I T Publications, 1992. ISBN 1 85339 178 6.
4. Doddiah, D., Selection, processing and specification of stones and mortar for the construction of Surki mortar masonry dams in India. In: Proc. 7th ICOLD, 1961: Q24, R32, 525–44.
5. Sims, G.P., Ageing of masonry dams. In: Proc. Instn Civ. Engrs, Wat. Marit. Energy, 1994: 106, 61–71.
6. Murthy, Y.K., Mane P.M., Pant B., Tensile failures in some stone masonry gravity dams in India. In: Proc. 13th ICOLD, 1979: Q49, R31, 461–77.

7. Eckel, E.C., Cements, Limes and Plasters, their Materials, Manufacture and Properties. John Wiley, New York, 1928.
8. Schnitter, N.J., A History of Dams, the Useful Pyramids. A.A. Balkema: Rotterdam, Brookfield, 1994.
9. Crossthwaite, C.D., Hunter J.K., The deterioration of concrete dams, 40 years experience in North Wales. In: Proc. 9th ICOLD, 1967: Q34, R13, 207–26.
10. Price, W.H., Higginson E.C., Bureau of reclamation practices in mass concrete. In: Symp. Mass Concrete, Am. Concr. Inst., 1963, Special Publication SP-6, Paper No. 4, 77–87.
11. Kollgaard, E.B., Chadwick, W.L., Development of Dam Engineering in the United States. Pergamon Press, New York, 1988.
12. The Engineer, 18 July 1958.
13. Féret, L.R., Sur la compacité des mortiers hydrauliqes, 1892.
14. Abrams, D.A., Design of Concrete Mixtures. Structural Materials Research Laboratory: Chicago, 1918.
15. Davis, R.E., Historical Account of Mass Concrete, Am. Concr. Inst., 1963, Special Publication SP-6, Paper No.1, 1–35.
16. Chadwick, L.W., In: Kollgaard, E.B., Chadwick, L.W. (eds), Influence of Some Related Technologies on the Technology of Dams, Development of Dam Engineering in the United States. Pergamon Press, New York, 1988.
17. Raphael, J.M., The optimum gravity dam, rapid construction of concrete dams. In: Proc. Eng. Res. Conf., Asilomar, 1970: 221–47.
18. Dunstan, M.R.H., The state-of-the-art of RCC dams. Int. J. Hydropower Dams, 1994, 1(2).
19. Dunstan, M.R.H., Recent developments in RCC dams, Int. J. Hydropower Dams, 1999, 6(1), 40–45.
20. (a) Schnitter, N.J., The evolution of buttress dams. In: Studies in the History of Civil Engineering, Vol. 4, Dams, Ashgate-Variorum, Aldershot, Brookfield USA, Singapore, Sydney, 1997; (b) Schnitter, N.J., The evolution of the arch dam. In: Studies in the History of Civil Engineering, Vol. 4, Dams, Ashgate-Variorum, Aldershot, Brookfield USA, Singapore, Sydney, 1997.
21. Tolle, J.M., Simard P.W., Brown L.A., Modern Engineering Saves Troubled Dam, Civil Engineering, New York, 1979, June, 78–80.
22. Berdal, A.B., Slab and buttress type dams. In: Concrete Dams in Norway, Norconsult AS, Oslo, 1968: 5–8.
23. Kennard, M.F., Owens, C.L., Reader, R.A., Engineering guide to the safety of concrete and masonry dam structures in the UK, Report 148, CIRIA, 1996.
24. USBR, Trial Load Method of Analysing Arch Dams, United States Bureau of Reclamation: Denver, 1938.
25. Pelletreau, A., Barrages cintrés en forme de voûte. Annales ponts et chaussées, 1879, 1er sém., 198–218.
26. Cole, K.L., Neal, P.C., The Victoria Project, Sri Lanka: Construction of Victoria Dam. In: Proc. Instn Civ. Engrs, Part 1, Vol. 90, Paper 9528, 1991.
27. ICOLD, Roller Compacted Concrete for Gravity Dams — State of the Art. Bulletin No. 75, 1989.
28. Kamel, K.F.S., Babá, L.J.N., Machado, B.P., Salto Caxias: Brazil's largest RCC Dam. Intern. J. Hydropower Dams, 1999, 6(1), 56–60.
29. Develay, D., Hagen, R.J., Bestagno, R., Lesotho highlands water project — design and construction of Katse dam. Suppl. Civil Eng., 1997, Vol. 120, Special Issue 1.
30. Arthur, L.J., Wagner, C.M., Hein, B., Lesotho highlands water project — design of the 'Muela' hydropower station. In: Proc. Instn. Civil Engrs, Suppl. Civil Eng., 1997, Vol. 120, Special Issue 1.

18　Concrete roads

Graham West

Synopsis

Considered as historic artefacts, concrete roads lack much of the visual appeal of other concrete structures such as dams and bridges. Nevertheless, concrete roads are a part of our civil engineering heritage and have an interesting history which is worth studying in its own right. This chapter deals with the history of concrete roads from earliest times to the present day, but before dealing with their history a basic description of the structure and composition of concrete roads, including some definitions and explanations of the technical terms used, will be given so that the reader may understand the descriptions that follow.

Structure and composition

The basic structure and main components of a concrete road are shown diagrammatically in Figure 18.1. The slab and the sub-base taken together are referred to as the *pavement*. The soil on site beneath the road is referred to as the *subgrade*.

The main structural function of the pavement is to spread the load applied by the traffic over a sufficiently wide area of the subgrade such that the bearing capacity of the soil is not exceeded. The principle element of a concrete road, the concrete *slab*, is not generally cast directly on the subgrade, but on a thin layer of granular material (e.g. gravel, sand) called the *sub-base* which is placed on the soil first. (In the road industry, concrete pavements are sometimes called *rigid pavements*, in contrast to bituminous pavements which are called *flexible pavements* — this terminology reflects their different material properties.) The load-carrying capacity of the concrete road structure depends mainly on the structural rigidity of the slab and so long as a concrete slab remains in sound condition it will perform its structural function satisfactorily. Since the late 1980s, if the bearing capacity of the subgrade soil is very low (California bearing ratio of 5 per cent or less) a capping layer is spread on the subgrade before the sub-base is laid. The *capping layer* is a material of lower quality than the sub-base but of higher quality than the subgrade. As well as assisting in load-spreading, the capping layer provides a working platform for construction of the road. One section of concrete slab is separated from another by regularly spaced, vertical *joints* and the traffic load is transferred across the joints by horizontal steel connectors called *dowel-bars*.

Concrete is strong in compression but weak in tension. Because of this plain concrete slabs tend to be prone to cracking, and to prevent this from happening the concrete slab is often reinforced with steel. Traditionally, therefore, there have been two basic methods of constructing concrete road pavements, and these are

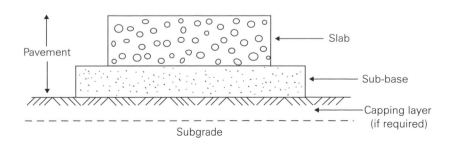

Figure 18.1　Concrete road structure (diagrammatic).

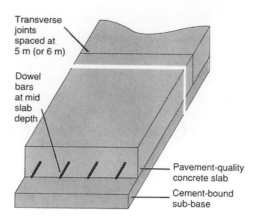

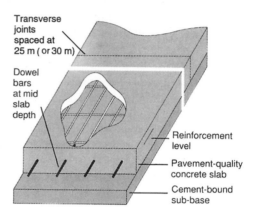

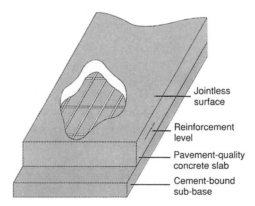

Figure 18.2   Jointed unreinforced (top) and jointed reinforced (centre) concrete road pavements. Unjointed continuously reinforced concrete road pavement (bottom).

the *jointed unreinforced concrete road pavement* and the *jointed reinforced concrete road pavement*: they are shown in Figure 18.2, and make up the majority of concrete roads in the United Kingdom. Joints tended to be troublesome, so that in addition to these two methods there is another type of construction that came into use in the 1980s, the *unjointed continuously reinforced concrete road pavement*, which is also shown in Figure 18.2; as its name indicates, this slab has no joints.[1] (It should be added that roads made from bituminous surfacing materials sometimes have a lean concrete roadbase but this use of concrete will not concern us in this chapter.)

A typical road-slab concrete would have had mix proportions of 1 : 2 : 4 (1 part cement : 2 parts fine aggregate : 4 parts coarse aggregate), a water : cement ratio of 0.5 : 1 and a 28-day compressive strength of about 35 MN/m^2. The thickness of the slab of a concrete road pavement depended on whether it was reinforced or not and on the traffic loading, but for an unreinforced slab it varied from 230 mm for light traffic to 330 mm for heavy traffic. All types of concrete road pavements can be laid using either fixed-form or slipform pavers. Traditionally, roads made of unreinforced or reinforced concrete were designed to have a life of 40 years.

Precursors of concrete roads

Concrete has been known since about 1350 BC when it was used to construct the foundations for the city of Akhet-Aten in ancient Egypt.[2] The Romans were great users of concrete, employing it in building and civil engineering to great effect almost everywhere in their empire where they could.[3] A cross-section through the pavement of a Roman road is shown in Figure 18.3, with dimensions given for the various layers as measured in the archaeological excavation. The Romans' names for the layers of the pavement are also given.

The section of road is the Fosse Way near Radstock in Somerset, and shows that for this important road the Romans had used concrete for both the *rudus* and *nucleus*. The cement used to make the concrete was probably lime-based and it is likely that it was manufactured locally. The total thickness of pavement was a massive 863 mm, of which the concrete comprised 635 mm. The Fosse Way was a Roman main road and the pavement structure shows that it was constructed to their highest standard.

The use of concrete almost died out with the decline of the Roman Empire, and apart from some isolated instances, it was not until the middle of the 18th century, with the construction of Smeaton's Eddystone Lighthouse, that a revival in the use of concrete came about. With introduction of improved cements in the 19th century, concrete as a construction material began its steadily rise in importance which has continued to the present day.

Modern history of concrete roads

The first concrete road in Great Britain were constructed in Inverness and Edinburgh, Scotland, in 1865–66; the concrete was laid fairly dry and compacted by roller, and performed satisfactorily until 1875 when the concrete was probably damaged by frost. In the early years of the 20th century experimental lengths of concrete pavement were constructed by County and Port of London Authority engineers. The first concrete road in the United States was constructed in 1893 at Bellefontaine, Ohio. This was followed in 1909 by a concrete highway built at Wayne County, Michigan; and another in 1912 at Saltney in Chester City, Ohio. However, in the United States by 1920, concrete pavements were being laid at the

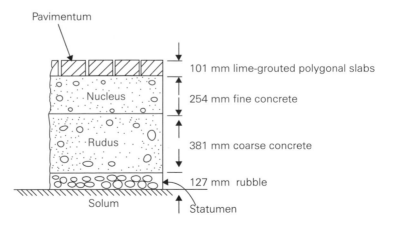

Figure 18.3 Section through a Roman road, the Fosse Way at Radstock, Somerset.

rate of 2000 miles per year, and the Experimental Branch of the Ministry of Transport thought that this form of construction would be suitable for use in southern England where sand and gravel aggregates were plentifully available. As a result, a number of by-passes and other roads were constructed in concrete, often incorporating experimental features aimed at developing a general specification for concrete pavements.[4] Some of these pre-war concrete roads were:

- Kingston Bypass, Kingston-upon-Thames, Surrey, A3 (Figure 18.4);
- Great West Road, west of London, A4;
- Caterham Bypass, Caterham, Surrey, A22;
- Great Chertsey Road, near Chiswick, Middlesex, A316;
- Colnbrook Bypass, Middlesex, A4.

Many of these bypasses were still satisfactorily carrying traffic after over 40 years service, although they have for various reasons been overlaid with bituminous surfacing, and it is not now readily apparent that these were once concrete roads.

Notably, the Lancashire County Council constructed many concrete roads in the 1930s, such as the Maghull Diversion, built in 1936, and pictured in Figure 18.5. Many of the early concrete road slabs were laid and finished predominantly by

Figure 18.4   Kingston Bypass, Surrey, 1934.

Figure 18.5   The Maghull Diversion, Lancashire, 1936.

Figure 18.6 Finishing with smoothing board on the Rickmansworth to Denham Road, Buckinghamshire, pre-1935.

Figure 18.7 The Dingler compaction and paving train at work on the Mickleham Bypass, Surrey, in the early 1930s.

hand methods (Figure 18.6), but on some roads a forerunner of the modern paving train was making its appearance (Figure 18.7).

The service experience of these early concrete roads, however, cannot have been wholly satisfactory because when the motorway network was designed and constructed in the 1950–60s; the Ministry of Transport did not allow concrete-surfaced roads to be built. However, during this period the Road Research Laboratory carried out a number of full-scale concrete road experiments which resulted in renewed interest in concrete roads.[5] The concrete industry made successful representations against the Ministry's policy, and after 1969 contractors were allowed to use either concrete or bituminous construction for the Department of Transport's trunk roads and motorways. The proportions, and corresponding lengths, of concrete roads in the United Kingdom, as estimated in 1986, are given in Table 18.1.

In July 1992, following concern about the high level of noise generated by vehicle tyres running on concrete road surfaces, the Department of Transport once again imposed a ban on new concrete-surfaced construction for heavily trafficked

Table 18.1   Concrete roads in the United Kingdom

Category of road	Proportion (%)	Length (km)
All roads	5	17 500
All trunk roads and motorways	20	3000
Trunk roads and motorways constructed since 1969	22	530
M25 motorway	46	90

motorways and trunk roads, although it still allowed concrete roadbase to be used with an asphalt surfacing.[6] In 1996 the Department of Transport announced that an exposed-aggregate concrete surface, which was quieter than the conventional brushed concrete surface, would be an allowable option for new road construction.

The situation in the United Kingdom, where the minority of trunk roads and motorways roads are of concrete, can be contrasted with that in the United States where 53 per cent of interstate highway pavements are of concrete.

Examples of historic concrete roads

Five examples of historic concrete roads are now given: the first is from Italy and the remainder are from England.

Milan to Lake Maggiore autostrada

The first motorway (*autostrada*) in Europe was planned in 1921 by an Italian engineer, Piero Puricelli. His idea was to connect the city of Milan in northern Italy with the neighbouring towns of Como, Varese and Lake Maggiore with a road tailored specifically to the needs of the motor car. The Italian government approved the project, construction work began in June 1923 and the whole project was complete in September 1925; the total length being 84 km. The motorway consisted of a single two-way carriageway varying in width from 8.5 to 10.5 m, separation of opposing traffic being by a continuous dividing line of white paint — looking back it is rather amusing to think that this was considered adequate. After studies of the various kinds of pavement used for motor roads in the United States, the Italian motorway was constructed with a concrete pavement, special continuous mixers and pavers laying the whole width of the carriageway in one pass. The mix proportions were such that each cubic metre of concrete contained $0.75\,\mathrm{m}^3$ crushed stone aggregate, $0.5\,\mathrm{m}^3$ of sand and 350 kg of cement.[7]

Alconbury Hill

In 1957 the Road Research Laboratory conducted a full-scale pavement design experiment on a section of dualling on the Great North Road, A1, at Alconbury Hill, Huntingdonshire. Part of the experiment involved the construction of 1.6 km of concrete pavement. The main variables studied were: thickness of slab (127–279 mm); thickness (76–229 mm) and type (natural sandy gravel and lean concrete) of sub-base; amount of reinforcement; type of aggregate (crushed rock and natural gravel); and strength of concrete (44 and $64\,\mathrm{MN/m}^2$ compressive strength at 28 days). A central batching and mixing plant was erected on site for preparing the concrete, which was delivered by side-tipping lorries to a train or rail-mounted concrete-laying machinery spanning the 7.9 m width of the carriageway. The reinforcement in the reinforced concrete slabs consisted of oblong mesh placed at a depth of 64 m below the surface. The subgrade was a heavy clay having a California bearing ratio of 5–9 per cent at the time of construction. Figure 18.8 shows the concrete slab under construction, and Figure 18.9 the

Figure 18.8   Laying the concrete slab from a rail-mounted paving train at Alconbury Hill, Huntingdonshire, A1, 1957.

Figure 18.9   Near side: the just-completed new concrete carriageway at Alconbury Hill. Far side: the old Great North Road still carrying two-way traffic, 1957.

completed carriageway just before opening to traffic. The main conclusions after six years of traffic (4000 commercial vehicles per day) were that only reinforced slabs of 203 mm thickness and over, laid on a granular sub-base at least 76 mm thick were likely to give a satisfactory long-term performance. Also, it was found that satisfactory pavement-quality concrete could be made using either crushed rock or natural gravel aggregate.[8]

St Albans Bypass

The St Albans Bypass in Hertfordshire, 27 km long, was part of the M1 motorway, Britain's first full-length motorway, and built in 1958–59. Although the rest of the M1 (92 km) was of flexible construction, the St Albans Bypass was of rigid construction so that the characteristics and performance of the two types of pavement under traffic could be compared. The pavement consisted of a 279 mm thick reinforced concrete slab overlying a 178 mm thick hoggin sub-base, giving a total pavement thickness of 457 mm. Figure 18.10 shows a train of concreting plant in operation laying the slab on the St Albans Bypass, and Figure 18.11 gives a general view of the completed road.

Padiham Bypass

The 1 km long section of concrete pavement on the Padiham Bypass, A6068, Lancashire, was constructed in 1963 to provide the Lancashire County Council with experience in preparation for the forthcoming construction in concrete of the Broughton to Hampson Green section of the M6 motorway, using the same paving machine and source of aggregates as were to be used for the motorway. The total slab thickness was 254 mm, made up of a 178 mm thick layer made with a siliceous limestone coarse aggregate, and a 76 mm thick upper layer made with

Figure 18.10 Train of concreting plant in operation on the St Albans Bypass, Hertfordshire, M1, 1958–59.

Figure 18.11   General view of the completed St Albans Bypass, 1958–59.

siltstone coarse aggregate. The fine aggregate used in both layers was a sea-dredged natural sand. The reinforcement in the slab consisted of a mesh laid between the two layers, 9.5 mm diameter cranked tie bars 0.76 m long, and dowel-bars at the contraction joints. The joint spacing was 12 m. The slab was laid in two 3.65 m wide runs, using an SGME box spreader, a compactor and Jaeger finisher, and was placed on a 152 mm thick sub-base of wetmix crusher-run limestone. The two carriageway slabs were held together by the tie bars running across the longitudinal joint. The completed road, therefore, consisted of a two-lane single carriageway, 7.3 m wide. Map-cracking was noticed on the road surface in 1979, but by 1989 had become serious enough, particularly near the transverse joints, to merit an investigation, which was carried out by the Transport Research Laboratory.[9] Cores were taken and thin-sections of the concrete were examined under the polarizing microscope (Figure 18.12). It was found that the slab was affected by alkali–silica reaction, both the coarse aggregates being reactive. The reactive constituent was found to be microcrystalline quartz which was present in both coarse aggregates. The source of alkali was identified as the deicing salt applied to the road surface in the winter.[10]

M25 motorway

The London orbital motorway, M25, constructed between 1976 and 1986, and 195 km long, was the last full-length motorway to be built in Great Britain. The Department of Transport decreed that for the M25 they would allow contractors to submit tenders for either flexible or rigid construction for the road pavement. The outcome was that 54 per cent of the motorway has a flexible pavement and 46 per cent has a rigid pavement — this result is due to the very close tenders which in turn must reflect the very close costs of the two forms of construction a the time. On the Epping section, in the north-east quadrant of the M25, the road pavement was an example of rigid construction. There, the continuously-reinforced concrete slab had a thickness of 250 mm with the reinforcement placed at

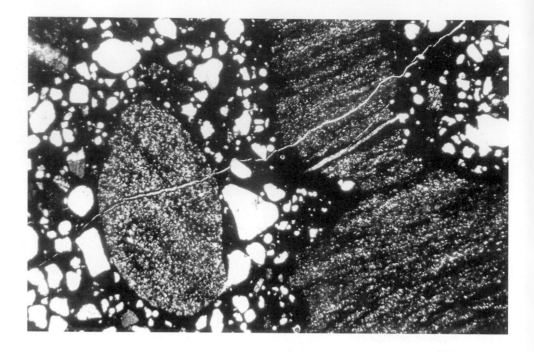

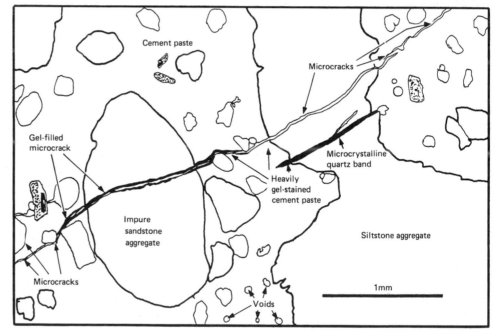

Figure 18.12   Photomicrograph of thin-section of concrete from the Padiham Bypass, Lancashire, A6068. A microcrack can be seen passing across the field of view. The reactive constituent is microcrystalline quartz in the siltstone aggregate.

mid-depth, the concrete was made using flint gravel aggregate, and the cement content was 320 kg/m^3. Figure 18.13 shows the Westerham to Sundridge section of the M25 where, in 1979, the unreinforced concrete slab was laid with a conventional rail-mounted paving train; and Figure 18.14 shows a fixed-form paving train being used on the Wisley to Leatherhead section of the M25 in 1984. Some of the concrete on the M25 was later given a bituminous surface dressing so that its concrete nature is not obviously apparent now.

The future

At the risk of oversimplification, it can be said that experience with increasingly heavy traffic has shown that flexible pavements need to be stiffer to resist deformation, but that rigid pavements are, perhaps, too stiff and liable to crack. One solution to

Figure 18.13   Westerham to Sundridge section of M25, 1979. The unreinforced concrete slab was laid with a conventional rail-mounted paving train.

Figure 18.14   Fixed-form paving train on the Wisley to Leatherhead section of M25, 1984.

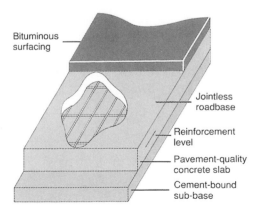

Figure 18.15   Composite concrete and bituminous pavement.

these problems would be to combine the best elements of rigid and flexible pavements in a single composite pavement: namely a concrete roadbase with hot-rolled asphalt surfacing, as shown in Figure 18.15, and this may well be the road pavement of the future. In which case, concrete roads will still be with us in the new millennium, but they will be hidden beneath the surface.

Acknowledgement

The British Cement Association is thanked for providing Figures 18.2, 18.4–18.11 and 18.13–18.15.

References

1. British Cement Association, Concrete Pavements for Highways, 2nd edn. Publication 46.030. British Cement Association: Crowthorne, 1994.
2. Hammond, N., Excavation shows pharaohs knew the secret of concrete. The Times, 11, October 1993.
3. Stanley, C.C., Highlights in the History of Concrete. Publication 97.408. Cement and Concrete Association: Slough. 1980.
4. Croney, D., The Design and Performance of Road Pavements. HM Stationery Office: London, 1977.
5. Road Research Laboratory, Concrete Roads: Design and Construction. HM Stationery Office: London, 1955.
6. Department of Transport, Minister announces steps to tackle road noise problem. Press Notice No 204. Department of Transport: London, 1992.
7. Mangarano, A., Pellizzi, G., Italy's first motorways. AIPCR-PIARC 1909–1969. In: Permanent International Association of Road Congresses, Paris, 1970, 129–40.
8. Croney, D., Loe, J.A., Full-scale pavement design experiment on A1 at Alconbury Hill, Huntingdonshire. Paper No 6848. In: Proceedings of the Institution of Civil Engineers, 1965, 30, February, 225–70.
9. Sibbick, R.G., West, G., Examination of concrete from the Padiham Bypass, Lancashire. Research Report 304. Transport Research Laboratory: Crowthorne, 1992.
10. West, G., Alkali-Aggregate Reaction in Concrete Roads and Bridges. Thomas Telford: London, 1996.

19 Military

John Weiler

Introduction

Military uses of concrete in construction were shaped by circumstances that differed in at least three important respects from those prevailing in civil projects. Most obvious was the overriding concern for national defence and the achievement of war aims. Then, there was the regimen of military discipline under which the armed services worked coupled with the government bureaucracy that controlled military contracts awarded to the private sector. Finally, by its very nature, military construction engaged in a continuous game of technological leap-frog with the fire-power of weaponry and its means of delivery by land, sea and air.

The story of concrete in military works extends from early 19th century experiments to the Second World War and its aftermath. Royal Engineers figured prominently in developments, both as individuals and as a corps. Most often they worked in collaboration with civil engineers and others engaged in the business of construction. During the 20th century, the private sector took increasingly greater responsibility in the design and construction process. Although the clear emphasis in this narrative is on British achievements, frequent reference is made to significant developments in other countries to provide historical perspective.

Early experiments with concrete

In the early 19th century, Charles William Pasley and other Royal Engineers joined a controversy surrounding the introduction of concrete and artificial stone in England, particularly concerning the work of William Ranger. Pasley was director of the Royal Engineer Establishment at Chatham and a prodigious experimenter and recognized authority on cementitious materials as well as developer of an artificial Roman cement that bore his name. What gave particular impetus to the Royal Engineers' involvement in the issue was a directive from the Board of Ordnance to ascertain the fitness of Ranger's patent concrete for casemate arches.

On orders from the Inspector-General of Fortifications, Lieutenant Colonel George Harding oversaw the construction of a model vault by Ranger at Woolwich Marshes in 1835 (Figure 19.1). Concrete specifications were seven parts gravel and sand mixed with one part Dorking lime and one part and a half of boiling water (a peculiarity of Ranger's formula). The model vault had a span of 17 ft and a rise of 9 ft and contained 5947 ft^3 of concrete. It stood up well under bombardment notwithstanding the fact that the arch had already cracked because of inadequate foundations on the marshy soil and even though the core of the structure was still soft (Ranger's concrete was quick setting but slow hardening). Harding reported that he was prepared to recommend Ranger's concrete be adopted in the arches of small magazines and casemates by virtue of its strength and economy, but he was doubtful about using it for the core of brick piers because of the difference of compression and expansion in the two materials; he also concluded that a 4 ft thick arch of concrete would be bombproof.[1] There is no evidence that Harding's recommendations were accepted or implemented.

Pasley argued that in the face of artillery fire brick casemates laid in cement mortar and coated with pure cement would be stronger than concrete and less prone to dampness.[2] In 1836 he began experiments to determine how any given lime was fit for making concrete on Ranger's principle, and his curiosity induced him

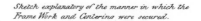

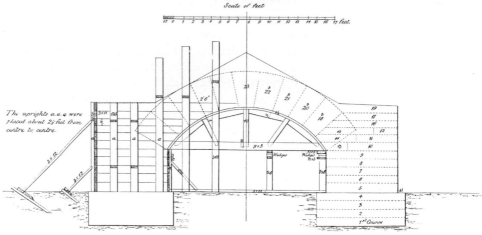

Figure 19.1   Method of constructing Harding's concrete model casemate, 1835. (Source: Ref. 1, Plate 1.)

to investigate cement concrete. Strength tests on concrete using various materials and methods led him to the following conclusions: concrete, especially Ranger's, was inferior to all natural building stones and even well-made bricks; concrete should be confined to building foundations, backing of wharf walls and the formation of retaining walls; and cement should not be used for concrete because the adhesion of mortar was reduced by a greater amount than was lime mortar by the inclusion of sand.[3] It is interesting to note that in 1836 General Joseph Gilbert Totten of the US Army Corps of Engineers undertook experiments on limes, cements and concrete that included artillery resistence trials on model structures in which these materials were used.[4] Pasley's fellow officer William Denison, superintending engineer at Woolwich naval dockyard, also investigated concrete. In 1837 Denison experimented with making concrete blocks using slightly hydraulic lime to see if the blocks would break down under the action of sea water. Later he sent these blocks to Pasley who performed crushing tests. Denison also observed the performance of Ranger's concrete both in blocks and in mass that had been used in the construction of wharf walls at Woolwich. He concluded that exposed concrete, especially Ranger's, was unsuitable in marine works.[5]

Pasley and Denison were not alone in their views on concrete. Robert Smirke restricted the material to building foundations.[6] Another authority who concurred was Dennis H. Mahan, professor of military and civil engineering at West Point; his 1837 textbook was a standard reference for the US Army Corps of Engineers.[7] Royal Engineers only began to change their opinions about concrete in the 1860s. Henry Scott, instructor at the Royal Engineer Establishment and the inventor of selenitic cement (1854), refuted Pasley's earlier teachings by demonstrating through experiment that higher proportions of sand weakened both lime and cement mortars equally; in 1862 he offered a design for a cellular revetment in concrete.[8]

The forts of the 1859 Royal Commission

At the end of the 1850s, Britain faced an arms race with France in fast moving, increasingly manoeuvrable, ironclad warships equipped with ever more accurate and powerful guns. The weak link in Britain's defence was inadequate coastal fortifications and obsolete methods of constructing batteries. The Palmerston government's response was to appoint a Royal Commission. W.F.D. Jervois, Assistant Inspector-General of Fortifications, was secretary to the commission and he largely wrote its report of 1860. Recommended were the fortification of the Royal Navy dockyards and Woolwich Arsenal and the harbours of Portland, Dover and Cork, and the construction of coastal batteries to resist sea attack and a ring-fortress system to defend against land assault. The Commission forts were designed by the Royal Engineers under the overall direction of Jervois.[9]

Newhaven Fort, begun in 1865, included the first use of mass concrete above foundation level in permanent fortifications in Britain (Figure 19.2). It was designed and constructed under Lieutenant John Charles Ardagh. Early in 1864, Gerald Graham, Commanding Royal Engineer at Brighton, experimented with a variety of limes and cements to determine which material would be best for mass concrete in Newhaven's revetments, and Scott's selenitic cement was selected.[10] The revetments contained some 20,000 m^3 of concrete; specifications called for one part Scott's cement to one part course sand, shingle or flint.[11] This accomplishment had actually been anticipated the year before by Royal Engineers stationed at Halifax, Nova Scotia. Under Lieutenant Colonel F.C. Hassard, they used Portland cement concrete in the construction of an escarp wall at Fort Ogilvie, one of the outlying defences of Halifax harbour.[12]

Except for Newhaven Fort, the Commission Forts used concrete in conventional ways, for a seal to brick casemate arches, for floors in magazines, in foundations, for bombproofing works from plunging fire and other minor applications.[13] The most extensive use was in Portland cement concrete blocks for the foundations of the great ironclad coast forts of Plymouth Breakwater, Portland Breakwater, and the Spithead Forts, Portsmouth (Spit Bank, Horse Sand, St Helens and No Man's Land Fort). Foundations for the Spithead Forts (Figure 19.3) were massive rings of stone work in the sea bed executed by engineering contractors and filled in with concrete blocks by the Royal Engineers afterwards. Horse Sand Fort, for example, contained 15,000 t of concrete blocks weighing 3–7 t each.[14] Iron shields on the Spithead Forts had Portland cement concrete backing between the shields' armour plate layers; in some shields, iron concrete was used.[15] Portland cement concrete was also used to fill the wrought iron pier casings which supported the armour on the Spithead Forts.[16] The first large scale use of Portland cement concrete in fortification superstructures in the British Isles was in the defences of Cork Harbour, beginning about 1873; this included sea walls, casemated batteries, powder magazines and dwelling houses.[17] The Chatham ring of land forts, begun in 1876 but not completed until 1899, were built primarily of Portland cement mass concrete but casemates and tunnels still had brick walls with concrete arches.[18] More or less contemporaneously Portland cement concrete was being

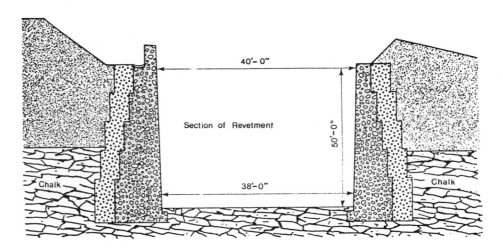

Figure 19.2 Plan and section of Newhaven Fort, 1865. (Source: Ref. 11, p. 14.)

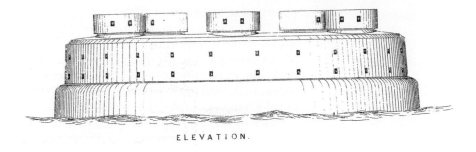

ELEVATION.

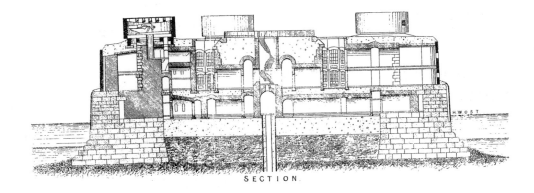

SECTION.

Figure 19.3   Elevation and cross section for one of the proposals for the Spithead Sea Forts. (Source: J. Roy. United Serv. Inst., 1869, XII, Plate XXXV.)

employed in the Chatham Dockyard extension works,[18a] the dock walls being entirely of Portland cement concrete, of varying mix proportions. Viewed in context, the Royal Engineers were late and conservative in their adoption of structural concrete. From 1859 the US Army Corps of Engineers used natural hydraulic cement concrete for superstructure construction, in the strategically important forts Richmond and Tomkins at New York harbour.[19]

The First World War

New technological developments in weaponry, particularly the explosive shell, led to the replacement of brick and masonry in fortifications, first by mass concrete and soon after by reinforced concrete. The French forts at Verdun were the most advanced with thick concrete of superior quality and reinforced.[20] The Germans were the first to use reinforced concrete in fieldwork emplacements and shelters.[21] Indeed, the use of reinforced concrete by Germany on the Western Front was probably the first use of the material on such a vast scale.[22] In addition to this, as part of a series of submarine shelters on the Belgian coast, Germany built at Bruges a forerunner to the famous submarine pens of the Second World War.[23]

British concrete shelters were well below the standards set by the Germans. Nonetheless, Britain's adoption of prefabricated defences and the air-space theory deserve comment. Corrugated iron was used for permanent formwork on concrete emplacements. The Ryes plate, bolted together in a part-hexagon shape, was the most sophisticated development.[24] The Moir Pill Box (Figure 19.4), the ultimate British prefabricated design, was developed at the end of the First World War but did not see service. It was constructed from interlocking concrete blocks and steel parts.[25] Application of the air-space theory was in machine gun posts and other shelters. In this approach, a reinforced concrete slab detonated the shell and the air space or soft absorptive material such as earth in a concrete sandwich construction, cushioned the blast.[26]

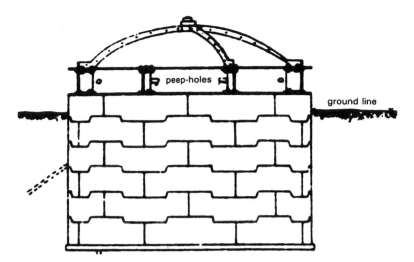

*Figure 19.4 'Moir' Pill Box.
(Source: Ref. 24, p. 80.)*

The inter-war period

Although there was no significant British military construction in the inter-war period, notable developments elsewhere deserve brief mention. Begun in 1930, the Maginot Line built by France adopted the air-space theory used by the British in the First World War. The final thickness of reinforced concrete was 3.5 m but exposed concrete was limited to combat blocks; 1.5 million cubic meter and 150,000 t of steel were used in this unique feat of engineering.[27] The West Wall or Siegfried Line constructed by Germany along its western border in 1938–39 drew on the diverse experiences of the First World War concrete defences and the construction organization developed by Dr Fritz Todt for the autobahn system built in the 1930s. Within 18 months, 6 million tons of concrete were used to construct 22,000 individual works.[28] Finally, with the growing threat from Japan in the Pacific, the Americans reinforced their defences in the Philippines through fortifications and large, tunnelled underground installations. Construction began in the early months of 1941. Amongst these works, the US Army Corps of Engineers built Fort Drum on El Fraile Island in Manila Bay. Shaped like a battleship, it had reinforced concrete with thicknesses of between 5 and 9 m.[29]

The Second World War

The Second World War sparked a number of innovations in military concrete construction. As early as October 1939, the Royal Engineers began building reinforced concrete pill boxes for the British Expeditionary Force that featured standard steel formwork and reinforcing bars bent and bundled into 'pill box packs' for delivery to sites. The designs and methods were utilized for standard hexagonal pill boxes in British coastal defences in 1940 and 1941.[30]

War also called for an immediate and drastic change in the construction of Britain's military barracks. Concrete became the almost necessary solution with the shortage of timber and steel. During the winter of 1939–40, investigations were made by the Directorate of Fortifications and Works and it adopted as the standard design for small single huts of short span, pre-cast concrete units for roof trusses and purlins.[31] Patented pre-cast unit designs were soon approved by the War Office.[32] By 1942, the British Concrete Federation had developed a largely prefabricated standard design for a concrete hut (Figure 19.5). The design featured a lightweight raft foundation into which were fitted reinforced concrete posts, slotted for the insertion of wall panels, and a roof of reinforced concrete beams and breeze concrete slabs.[33] Also in 1942, the Ministry of Works and Planning developed the MOWP Standard Hut which provided a basic prefabricated, reinforced concrete frame structure in which several types of cladding could be used (Figure 19.6).[34] Finally, Major Waller designed the C'tesiphon hut that used

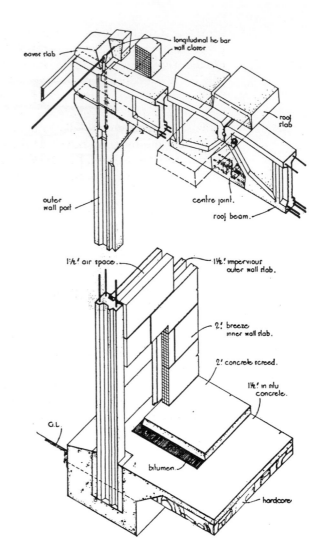

Figure 19.5   British Concrete Federation prefabricated barrack hut design. (Source: Ref. 33, p. 236.)

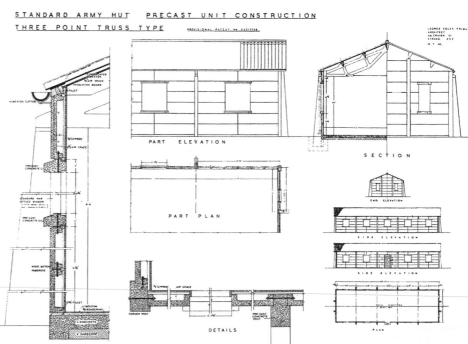

Figure 19.6   Assembly of an MOWP prefabricated standard hut. (Source: Ref. 34, p. 194.)

for the first time in hutting an all compressive thin concrete shell structure; cement grout render of 50 mm was applied on fabric that completely avoided the use of steel reinforcement.[35] A system similar to this hut was developed during the war by Waller and Kurt Billig for the construction of ends to blister hangers of 90 ft span.[36]

A great leap forward with air power created a corresponding need to protect civilian populations as well as military installations against the devastating effects of areal bombing. The Blitz and the public rush in London to the Underground for protection eventually shook the British government out of its lethargy concerning the development of effective air raid shelters. Both reinforced concrete and reinforced concrete block work were employed in shelter construction. The Borough of St Pancras used an innovative constructional process using a system of steel shuttering originally designed for building reinforced concrete culverts on a travelling basis; sections of rectangular concrete tube were laid in the streets to form air raid shelters.[37] Some compartmental cellular shelters were built at Finsbury despite the government's official rejection of the design. These were constructed in two stages. The first was a reinforced concrete slab shelter of minimum thickness placed about 1.5 m below ground level and covered over with earth; when the opportunity afforded itself, the initial shelter was re-excavated and a second stage of reinforced concrete protection added.[38] In contrast to British efforts, the Germans constructed (from 1940) extensive, heavily protected reinforced concrete bombproof bunkers for the civilian population. Original roof slabs had 1.5 m of concrete but this was later increased to 2.5 m.[39]

The design and construction of Britain's sea forts in 1941–42 stands as a celebrated wartime engineering achievement (Figure 19.7). Four naval forts were erected in the North Sea, opposite the Thames estuary (Figure 19.8); six army forts were constructed, three in the Thames estuary and three in the Mersey estuary. A prefabricated, amphibious type of fort solved the problems of time, labour, materials and the risk of enemy attack during construction. G.A. Maunsell, a civil engineer, designed the sea forts and the projects were executed under his supervision

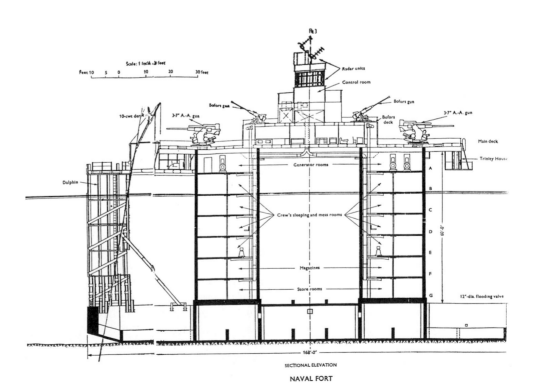

Figure 19.7   British Naval Fort under tow to action station. (Source: Ref. 40, pp. 132–62.)

Figure 19.8   British Army Fort being lowered to sea bed in Thames estuary. (Source: Ref. 40, pp. 132–62.)

together with the Civil-Engineer-in-Chief of the Admiralty; reinforced concrete design was worked out and supervised by G. Smith, a structural engineer.[40] The naval forts comprised a reinforced concrete pontoon from which rose twin cylindrical reinforced concrete towers surmounted by a steel superstructure. These forts were floated to and sunk at their respective action stations. An innovation in construction practice was the pre-assembly of reinforcing steel in mats (slabs) and cages (towers) that were partly wired and partly spot welded to form rigid units; plain round bars were used and hook ends and bends avoided.[41] The army forts comprised seven separate towers interconnected by walkways. Each tower had a hollow reinforced concrete base consisting of four members in the form of an Oxford picture frame. At the base's four intersections rose four hollow reinforced concrete cylindrical legs (each leg pre-cast in three separate sections), surmounted by a concrete cap in which main joists were embedded to support a triple-decked steel house. Because the foundation design was not self-buoyant, army forts needed special plant to float the bases out and deposit them on the sea bottom.

Another British accomplishment was with the amphibious 'Phoenix' reinforced concrete caissons that formed the major part of the 'Mulberry' Harbour breakwaters used to support the D-day invasion (Figure 19.9). The 'Phoenix' units were built at eight sites in Britain, and laid down in 12 specially excavated basins along the Thames. Responsibility for design rested on a War Office team headed by Captain W.J. Hodge that included leading designers drawn from outside.[42] A fast-track construction system based on standard designs was essential. Six types were originally constructed with two others later for the supplementary units. Construction began in December 1943 and 213 units were completed in 150 days; the first caisson arrived at Arromanche in Normandy on D-day plus one. The caissons were of plain symmetrical cellular reinforced concrete construction; reinforcing was almost entirely of straight bars to save time and labour; internal walls had large openings to lighten the structures; caissons were divided into 22 compartments and solid concrete blocks were formed at the four corners to take towing or mooring strains; and torsional diagonal tie bars were provided across the tops of each open compartment (substituted by concrete roofs in supplementary units).[43] A gale which broke on June 19 and lasted for three days damaged or destroyed several of the original units; supplementary units were strengthened accordingly.[44]

Notwithstanding what Britain accomplished with concrete in the Second World War, it is important to see this in perspective. Germany lead the way in both the scale and innovative uses of the material. Submarine pens were particularly

Figure 19.9 'Phoenix' units for Mulberry Harbour breakwaters.

outstanding. Construction started with the Deutsche-Werft pens at Hamburg, and involved extensive use of precast, prestressed concrete trusses manufactured by Wayss and Freytag. After June 1940, pens using the same system were built on the Atlantic coast in France and other captured territories. At Brest these featured closely spaced trusses with up to 5.5 m of concrete in a roof slab reinforced with steel bars, and fortified with a 1.5 m 'burster' slab over part of the roof.[45]

Epilogue

During and after the Second World War, British experts observed wartime bomb damage and carried out trials on the effects of bombs and other explosive devices on structures, including those constructed of mass concrete and reinforced concrete. The Research and Experiments Department of the Ministry of Home Security handled most of the work, and trials ranged from simple to sophisticated structures. British researchers shared findings with their American allies. Amongst the more interesting experiments were Anglo-American bombing trials on the roofs of German submarine pens.[46] American observations in the Pacific theatre included those on unusual failures of reinforced concrete caused by atomic bombs at Hiroshima and Nagasaki. Atomic bombs were detonated at a considerable height above the buildings and blast effects were distributed fairly evenly as opposed to the localized patterns seen with conventional explosives.[47] All the same, hardened fortifications had become redundant in mobile war; the focus shifted to hardened protection for civilian populations in the event of nuclear war and to nuclear missile silos.[48] The contest between concrete and weaponry was virtually over.

References

1. Harding, G.J., Description of a concrete bomb-proof erected at Woolwich with detailed experiments as to the effect produced on it by artillery fire. Papers on subjects connected with the duties of the Corps of Royal Engineers, 1837, 1, 33–38.
2. Pasley, C.W., Observations on Limes, Calcareous Cements, Mortars, Stuccos and Concrete etc. London, Weale, 1838: 22.

3. Pasley, C.W., Observations on Limes, Calcareous Cements, Mortars, Stuccos and Concrete etc. London, Weale, 1838: 85–86, 144–45.

4. Gillmore, Q.A., Practical Treatise on Limes, Hydraulic Cements and Mortars. New York, 1874: 80, 184–85, 252–53.

5. Denison, W., Notes on concrete. Papers on Subjects Connected with the Duties of the Corps of Royal Engineers, 1838, 2, 266.

6. Pasley, C.W., Observations on Limes, Calcareous Cements, Mortars, Stuccos and Concrete etc. London, Weale, 1838: vii.

7. Mahan, D.H. and Barlow P. (eds). An Elementary Course of Civil Engineering. Edinburgh, MacMarton, 1845.

8. Scott, H.Y.D., Observations on limes and cements, their properties and employments. Papers on Subjects Connected with the Duties of the Corps of Royal Engineers etc. New Series, 1862, XI, 15–94.

9. Powter, A., Conservation of concrete in fortifications and gun batteries. Unpublished thesis, Diploma in Conservation Studies, University of York, 1979, 40–46.

10. Graham, G., Experiments on limes and cements. Papers on Subjects Connected with the Duties of the Corps of Royal Engineers etc. New Series, 1865, XIV, 155–61.

11. Hamilton-Baillie, J.R.E., Nineteenth century concrete and the Royal Engineers. Concrete, 1980, 14, 15–16.

12. Vincent, E., Substance and Practice: Building Technology and the Royal Engineers in Canada. Canada Communications Group — Publishing, Supply and Services Canada: Ottawa, 1993, 73–75.

13. Powter, A., Conservation of concrete in fortifications and gun batteries. Unpublished thesis, Diploma in Conservation Studies, University of York, 1979, 40.

14. The Engineer, 1864, 18, 163.

15. The present condition of our armoured defences. The Engineer, 1881, 51, 326, 334.

16. The present condition of our armoured defences. The Engineer, 1881, 51, 326, 334.

17. Maj, M., Notes on Portland cement concrete. Papers on Subjects Connected with the Duties of the Royal Engineers etc. New Series, 1874, XXII, 149–53.

18. Powter, A., Conservation of concrete in fortifications and gun batteries. Unpublished thesis, Diploma in Conservation Studies, University of York, 1979, 46.

18a. Bernays, E.A., Portland cement concrete and some of its applications. Min Proc. Ice, 62, 1880, 87–95.

19. Gillmore, Q.A., Practical Treatise on Limes, Hydraulic Cements and Mortars. New York, 1874: 246.

20. Mallory, K., Ottar, A., The Architecture of War. Pantheon Books: New York, 1973: 29.

21. Mallory, K., Ottar, A., The Architecture of War. Pantheon Books: New York, 1973: 45.

22. Mallory, K., Ottar, A., The Architecture of War. Pantheon Books: New York, 1973: 51.

23. Mallory, K., Ottar, A., The Architecture of War. Pantheon Books: New York, 1973: 69.

24. Taylor, G., The general design and use of hardened defences in twentieth century warfare. Royal Engineers Journal, 101, 1987, 87.

25. Taylor, G., The general design and use of hardened defences in twentieth century warfare. Royal Engineers Journal, 101, 1987, 87.

26. Mallory, K., Ottar, A., The Architecture of War. Pantheon Books: New York, 1973: 59.

27. Taylor, G., The general design and use of hardened defences in twentieth century warfare. Royal Engineers Journal, 101, 1987, 80–83.

28. Taylor, G., The general design and use of hardened defences in twentieth century warfare. Royal Engineers Journal, 101, 1987, 87–88.

29. Dodd, K.C., The Corps of Engineers: the war against Japan. United States Army in World War II, Technical Services. Office of the Chief of Military History, United States Army, Washington, DC, 1966.

30. Taylor, G., The general design and use of hardened defences in twentieth century warfare. Royal Engineers Journal, 101, 1987, 90.

31. Singer, C.M., Notes on alternative materials and methods of construction for war hutting. Roy. Engrs J. 1940, 54, 180–87.

32. Precast unit construction for hutting. The Builder, 1941, 160, 90–91.

33. Clear span hostels of prefabricated type. The Builder, 13 March 1942, 236.

34. The M.O.W.P standard hut. J. Roy. Inst. Br. Archit. September 1942: 193–94.

35. Mallory, K., Ottar, A., The Architecture of War. Pantheon Books: New York, 1973: 197.

36. Billig, K., Concrete shell roofs with flexible moulds. J. Inst. Civil Engrs, 1945–1946, 25, 228–31.
37. Mallory, K., Ottar, A., The Architecture of War. Pantheon Books: New York, 1973: 233.
38. Mallory, K., Ottar, A., The Architecture of War. Pantheon Books: New York, 1973: 225.
39. Mallory, K., Ottar, A., The Architecture of War. Pantheon Books: New York, 1973: 223–43.
40. Posford, J.A., The construction of Britain's sea forts. The Civil Engineer in War, Vol. 2. Institution of Civil Engineers: London, 1948: 133, 161.
41. Posford, J.A., The construction of Britain's sea forts. The Civil Engineer in War, Vol. 2. Institution of Civil Engineers: London, 1948: 137, 161.
42. Wood, C.R.J., Phoenix. The Civil Engineer in War, Vol. 2, Institution of Civil Engineers: London, 1948: 342.
43. Wood, C.R.J., Phoenix. The Civil Engineer in War, Vol. 2, Institution of Civil Engineers: London, 1948: 342–46.
44. Jellett, J.II., The lay-out, assembly and behaviour of the breakwaters at Arromanche Harbour (Mulberry 'b'). The Civil Engineer in War, Vol. 2. Institution of Civil Engineers: London, 1948: 300–307.
45. Mallory, K., Ottar A., The Architecture of War. Pantheon Books: New York, 1973: 252–55.
46. Walley, F., The effect of explosions on structures. Proc. Inst. Civil Engrs, Struct. Build., 1994, 104, 325–34.
47. Thomas, W.N., The effects of impulsive forces on materials and structural members. J. Inst. Civil Engrs, The Civil Engineer in War, 3, 1948, 72.
48. Taylor, G., The general design and use of hardened defences in twentieth century warfare. The Civil Engineer in War, 99.

Appendix I: sources of further information on historic concrete

Concrete as a material has been in use for thousands of years yet the term does not appear to have come into use in England until the late 1820s, having been rediscovered by English architects and engineers a decade or so earlier. Since that time, as with so much technical literature, published information on the use of concrete has grown exponentially, making it very difficult for the inexperienced researcher to identify and locate useful sources of information. This book is intended to fill this gap, but inevitably many readers will need to research further. In terms of further reading some indication is provided by the references cited in individual chapters, and some authors, notably Bussell, have provided further guidance. Having identified likely references one has still to find them, and what follows is intended to provide guidance on the most important collections, as well as a general reading list.

It is important to distinguish the purpose of this reading list from that of the references at the end of each of the 19 chapters in this book which are seen as windows opening on to further information on specific points in the text, whereas the general reading list is intended not only to broaden the outlook of engineers dealing with specific tasks but to appeal to historians or to the general reader who may be unaware of the great impact of concrete on almost every aspect of life today. No list can possibly be complete but it is hoped that this reading list will at least act as a starting point.

With dual readership in mind the list has been split into three sections. The first, based on a broad readership, gives some of the key historical books or articles on the adoption of concrete as a material and on the introduction of both reinforced and prestressed concrete. The second, intended mainly for engineers involved in appraisals of existing construction, provides a short list of sources of information on the historical investigation and assessment of actual concrete structures. Most of these sources have extended bibliographies. The final section covers published guidance on practical methods of appraising, repairing and maintaining existing concrete structures. The contents of these three sections are summarized below:

A General reviews of the growth of understanding of structural concrete
1 Broad historical reviews
2 Books and papers on cements, mass concrete and fireproof flooring up to about 1900 (before the reinforced concrete period)
3 Books and papers on pioneer reinforced concrete 1900–20
4 Books and papers on reinforced concrete 1920–45
5 Books and papers on early prestressed concrete 1945–65
6 Periodical publications: 1830s to the present day
B Practical guidance on the identification of concrete structures
1 Sources of information on buildings and structures
2 General directories/guides to archives, etc.
3 Organizations holding archives on individual structures as well as general records (possibly including drawings)
4 Organizations which may be worth consulting for general technical and historical data on concrete structures
5 Websites and bulletin boards

6   Trade literature and published synopses of details of proprietary concrete systems

C   **Publications giving guidance on methods of appraisal, repair and maintenance of existing concrete structures**

A: General reviews of the growth of understanding of structural concrete

A1 Broad historical reviews

Texts on cement and concrete in the first half of the 19th century were descriptive rather than analytical, but nonetheless the better examples such as Pasley's provide a wealth of information on contemporary knowledge and practice. By the 1870s increasing data were available on the strength and chemistry of concrete. Similarly the earliest textbooks on reinforced concrete had much useful descriptive material in them — Christophe in France, and Marsh in England being obvious examples. These do contain some general analytical material, but the reinforced concrete textbooks most would recognize today generally appear after the First World War, consolidating a lot of early research on the continent and in the USA, as well as design experience. For those who are able to tackle some German the *Handbuch der Eisenbetonbau* — a multi-volume encyclopaedia, begun by Emperger is a fascinating insight into continental practice — generally years ahead of Britain. This comprehensive approach was reflected in the Concrete Publications series which began publication in the 1920s (qv).

American Concrete Institute. A selection of historic American papers on concrete 1876–1926. Detroit: ACI, 1976 (*ACI Special Publication 52*, includes papers by J.O. Draffin, T. Hyatt, W.E. Ward, A.N. Talbot, A.R. Lord, C.A.P. Turner, E.L. Ransome, A. Saurbrey and D.A. Abrams)

Hamilton, S.B. A note on the history of reinforced concrete in buildings, *National Building Studies, Special Report, 24*, 1956.

Idorn, G.M. *Concrete Progress from Antiquity to the Third Millennium*. Telford, 1997.

Mainstone, R.J. *Developments in Structural Form*, 2nd edn. Architectural Press, 1998.

Revolution or evolution? Historic concrete: vital information for all involved in carrying out appraisals and alterations. Catalogue for the exhibition at the Institution of Civil Engineers. ICE, 1996.

Stanley, C. and Bond, G. *Concrete Through the Ages*. BCA, 1999 (New edition of: Stanley, C.C. *Highlights in the History of Concrete*. Cement and Concrete Association, 1979.)

A2 Books and papers on cements, mass concrete and fireproof flooring up to about 1900 (before the reinforced concrete period)

Baker, I.O. *Treatise on Masonry Construction*. New York, various editions. c. 1890–1910.

Bristow, I.C. Exterior renders designed to imitate stone. *Trans. ASCHB*, 1997, 22, pp. 13–30.

Grant, J. Experiments of the Strength of Cement, Chiefly in Reference to the Portland Cement Used in the Southern Main Drainage Works. London: Spon, 1875.

Hamilton, S.B. A short history of the structural fire protection of buildings. *National Building Studies Special Report 27*, London: HMSO, 1958.

Lawford, G.M. Fireproof floors. *Trans. Soc. Engrs*, 1889, pp. 43–70.

Pasley, C.W. *Observations on Limes, Calcareous Cements, Mortars, Stuccos, and Concrete*. London: Weale, 1838.

Pasley, C.W. *Observations on Limes, Calcareous Cements,…* Part 1 only, 2nd edn, 1847.

Pasley, C.W. *Outline of a Course of Practical Architecture…* Royal Engineers Establishment, repr. 1862 of 1828 lithographed notes. Reprint in facsimile by Donhead, 2001 (see Hurst).

Potter, T. *Concrete: Its Uses in Buildings*. Various editions. 1870–1907.

Sutcliffe, G.L. *Concrete, its Nature and Uses*. London: Crosby Lockwood, 1893.

Vicat, L.J. *A Practical and Scientific Treatise on Calcareous Mortars and Cements, Artificial and Natural…* translated by J.T. Smith. London: J. Weale, 1837 (also later French edn, 1865). Reprint in facsimile by Donhead, 1997.

Webster, J.J. Fire proof construction. *Instn Civil Engrs, Min. Proc.*, 1891, CV, pp. 249–88.

White, G.F. Observations on artificial, hydraulic or Portland cement; with an account of the testing of the brick beam erected at the Great Exhibition, Hyde Park, *Instn Civil Engrs, Min. Proc.*, 1852, 11, pp. 478–502.

A3 Books and papers on pioneer reinforced concrete 1900–20

Berger, C. and Guillerme, V. *La Construction en Ciment Armé*. Paris: Dunod, 1902.

Christophe, P. *Le Béton Armé et ses Applications*, 2e edn. Paris: Béranger, 1902 (originally published in *Annales des Travaux Publics de Belgique*).

Coignet, E. and Tédesco, N. de. Du calcul des ouvrages en ciment avec ossature métallique. *La Societé des Ingènieurs Civils de France*, Paris, 1894.

Considère, A. *Experimental Researches on Reinforced Concrete*. Trans. L.S. Moisseiff. New York: McGraw-Hill, 1903.

Emperger, F. *Handbuch für Eisenbetonbau*. Berlin: Ernst, various editions. c. 1908–39.

Marsh, C.F. and Dunn, W. *Reinforced Concrete*, 3rd edn. London: Constable, 1906 (excellent description of systems and early structures).

Marsh, C.F. and Dunn, W. *Manual of Reinforced Concrete and Concrete Block Construction*. London: Constable, 1908 and various other editions.

Morsch, E. *Concrete-Steel Construction*. Trans. from 3rd (1908) edn revised and enlarged by E.P. Goodrich. New York: Engineering News, 1909.

Tédesco, N. de and Forestier, V. *Manuel Théorique et Pratique du Constructeur en Ciment Armé*. Paris: Béranger, 1909.

Twelvetrees, W.N. *Concrete-Steel Buildings*. London: Whittaker & Co., 1907.

A4 Books and papers on reinforced concrete 1920–45

A very large number of publications on reinforced concrete were issued in this period amongst which the series published by Concrete Publications is outstanding. A list of these prepared by Dr L.G. Booth with dates of revisions is given below. This was first published in *Proc. Instn Civ. Engrs Structs & Bldgs*, 1998, 128, pp. 398–99.

Concrete Publications advertised the 'Concrete Series' as 'practical and useful books on various branches of concrete and reinforced concrete design and construction, estimating, cost-keeping, pre-cast concrete, etc.' It appears that the Series began in 1918 with *Concrete Cottages, Bungalows, and Garages* by Albert Lakeman, a

Lecturer at Woolwich Polytechnic. During the next 50 years, until it ceased publication, the company published some 85 titles (see table), most of which went into several editions or reprints.

For the first decade, the books tended to concentrate on the practical applications of the material, particularly pre-cast work. The results of research and the current design of reinforced concrete beams were covered by Faber (1924) and the design and construction of formwork by Wynn (1926). During the 1930s the emphasis moved to design manuals (the first of the 'Charley' Reynolds in 1932) and the design of special structures: e.g. Gray (1931) on reservoirs and Terrington (1937) on arches and domes. The first of the commentaries on the codes (Scott and Glanville in 1934) also appeared. A notable immediate post-war addition to the list was a translation of Magnel's pioneering work on *Prestressed Concrete* (1948). Ultimate load methods were introduced by Baker (1949), who was professor of concrete technology at Imperial College, London.

Structural analysis was never strongly represented: the well-established publishing houses provided stiff competition. There were, however, minor contributions from Shepley (1942), Manning (1952) and Smolira (1955), and a major (unappreciated) contribution from Gartner (1942). Gartner's book introduced the so-called δ_{ik} (influence coefficient) method, but the full potential of the method was only fulfilled when Jenkins introduced matrices to the structural analysis of shells and skeletal frames and when Henderson began to spread the gospel through the Concrete Technology diploma course (DIC) in the late 1940s at Imperial College.

Some later well-known names appeared as authors early in their careers: (Sir) Ove Arup (1939), (Professor) A.L.L. Baker (1937), (Sir) Charles Davis (1924), (Dr) Oscar Faber (1924) and (Sir) William Glanville (1934).

The writer's search for a complete set of titles has been made over many years. It is hoped that the first edition titles and dates are complete and correct (a mathematics book — an unusual venture — has been omitted). Some of the dates of later editions are missing despite a thorough trawl of the book reviews in the publisher's monthly magazine *Concrete and Constructional Engineering (C&CE)* and an examination of the catalogues of various libraries. In the table opposite, titles in the ICE Library are marked with an asterisk (*) and those in the British Library by a dagger (†). Reprints of books have not been included nor have all reprints of articles. The writer has not seen copies of all the titles in the table.

Care must be taken in defining the dates of the editions of a book. The title page is never dated but since about 1950 its reverse has been dated. If the reverse is dated, it is used to date the book: if this date is not given then the date of the preface (or foreword), is quoted and it is given in square brackets. These dates may, or may not, be the same. A code and its commentary may have the same or different dates. A book may be a reprint of articles that had appeared in previous issues of *C&CE:* they would usually have different dates. The nomenclature for the different editions is also confusing. A history of printing is sometimes given on the reverse of the title page: it is possible for a book to be described as 2nd edn, 2nd edn revised, 2nd edn revised and enlarged, or even as reprinted.

It would be interesting to learn if anybody has a complete set. It may be that the ICE Library with about 60 first editions has the most. Additions to the list would be welcome.

The conventional alphabetical list has been reorganized as a tool for the refurbishing engineer by presenting it in chronological order under 11 headings. The method of use is to work backwards in time to the first edition devoted to the topic of interest and then to work forwards through the later editions to the date in question and so find the current practice. (For example, a chimney in 1965: working backwards yields a first edition of Taylor and Turner in 1940: working forward from 1940 yields a second edition dated 1960: both editions are in the ICE Library (*) and the British Library (†).)

Titles published by Concrete Publications Ltd in the Concrete Series. Chronological classified list of authors, titles and dates of editions

Bridges
Anon., Road Bridges in Great Britain, 1939*
Legat, A.W., Dunn, G., Fairhurst, W.A., Design and Construction of Reinforced Concrete Bridges, [1948*†]

Cement
Davis, A.C., A Hundred Years of Portland Cement: 1824–1924, 1924*
Kuhl, H., Cement Chemistry in Theory and Practice, 1931*
Davis, A.C., Portland Cement, 1934*; 2nd edn, 1943*
Platzmann, C.R., International Dictionary of Cement. English–French–German–Spanish, [1935*†]
Watson, W., Cradock, Q.L., Cement Chemists' and Works Managers' Handbook, [1940*†]; 2nd edn, 1952*†
Davis, Sir Charles, Portland Cement, 1948

Commentaries on Codes of Practice
Scott, W.L, Glanville, W.H., Explanatory Handbook on the Code of Practice for Reinforced Concrete as Recommended by the Reinforced Concrete Structures Committee of the Building Research Board [of the DSIR], 1934*; 2nd (rev.) edn, 1939*; 2nd (further rev.) edn, 1948*
Scott, W.L., Glanville, W.H. Thomas, F.G., Explanatory Handbook on the BS Code of Practice for Reinforced Concrete Recommended by the Codes of Practice Committee Formed Under the Aegis of the Ministry of Works, 1950*† [commentary on CP 114: 1948 code]; 1957*†; 1961*† [commentaries on CP 114: 1957 code]; 1965*† [commentary on CP 114: 1965 code]
Walley, F., Bate, S.C.C., A Guide to the B.S. Code of Practice for Prestressed Concrete: No. 115: 1959, 1961*†

Concrete making and finishes, and pre-cast products
Wynn, A.E., Andrews, E.S. Modern Methods of Concrete Making, 1926*; 2nd edn, 1928; 3rd edn, 1939*
Childe, H.L., Manufacture and Uses of Concrete Products and Cast Stone, [1927*]; 5th edn, 1930; 6th edn, 1939; title changed in 1940 to Concrete Products and Cast Stone*, 8th edn, 1949*; 9th edn, 1961
Childe, H.L., Pre-cast Concrete Factory Operation and Management, [1929*†]
Anon., Moulds For Cast Stone and Pre-cast Concrete. 1930. 1937*
Wynn, A.E., Estimating and Cost Keeping for Concrete Structures, [1930*]; 2nd edn, 1944*
Wynn, A.E., Making Pre-cast Concrete for Profit: A System of Cost-Keeping and Determining Profits, 1930*
Anon., Concrete Surface Finishes, Renderings and Terrazzo, 1935. Gray W.S., H.L. Childe, 2nd edn, 1948*
Baumgarten, R.H., Childe, H.L., Manufacture of Concrete Roofing Tiles, [1936*†]; 2nd edn, 1947
Burren, F., Gregory, G.R., Moulds for Cast Stone and Concrete Products, [1936*]
Walsh, H.N., How to Make Good Concrete, [1939*†]; 2nd edn, 1955*
Fielder, F.H., Estimating and Costing Pre-cast Concrete Products and Cast Stone, 1943*; [2nd edn, 1963*†]
Gray, W.S., Childe, H.L., Concrete Surface Finishes, Renderings and Terrazzo, 2nd edn, 1948*
Pennington, A.M., Concrete Fences, 1950*†
Childe, H.L., Concrete Finishes and Decoration, 1964*
Nesbit, J.K., Structural Lightweight-Aggregate Concrete, 1966*

Design and special structures
Faber, O., Reinforced Concrete Beams in Bending and Shear, [1924*†]
Lakeman, A., Elementary Guide to Reinforced Concrete, 1925†; 6th edn, 1930*; 12th edn, 1950†
Gray, W.S., Reinforced Concrete Reservoirs and Tanks, [1931*†]; 2nd edn, 1942*; 3rd edn, 1954*†; Revised by G.P. Manning 4th edn, 1960*†
Reynolds Chas. E., Reinforced Concrete Designers' Handbook, [1932*†]; 2nd edn, 1939*; 3rd edn, 1946; 4th edn, 1948*; 5th edn, 1957*; 6th edn, 1964†
Adams, H.C., The Elements of Reinforced Concrete Design, [1933*†]; 3rd edn, 1947*; 1950
Gray, W.S., Reinforced Concrete Water Towers, Bunkers, Silos and Gantries, [1933*†]; 2nd edn, 1944*; 1953* (with G.P. Manning); 4th edn, 1964*, see Gray, Manning
Terrington, J.S., Design of Domes, 1937*
Terrington, J.S., Design of Arch Roofs, 1937*
Reynolds, C.E., Practical Examples of Reinforced Concrete Design, [1938*†]
Arup, O.N., Design, Cost, Construction and Relative Safety of Trench, Surface, Bombproof and Other Air-Raid Shelters, [1939*†]
Terrington, J.S., Design of Pyramid Roofs, 1939*
Lee, D.H., Design and Construction of Air-Raid Shelters, [1940*†]
Taylor, C.P., Turner, L., Reinforced Concrete Chimneys, [1940*†]; 2nd edn, 1960*†
Pennington, A.M., Concrete Farm Silos, Granaries and Tanks, [1942†*]
Fairhurst, W.A., Arch Design Simplified, [1945*†]; 1954
Baker, A.L.L., Reinforced Concrete, 1949*†
Terrington, J.S., Design of Arch Ribs for Reinforced Concrete Roofs, 1950*
Ashdown, A.J., Design of Prismatic Structures, 1951†; 2nd edn, 1958*†
Reynolds, C.E., Examples of the Design of Reinforced Concrete Buildings, 1952*†
Eriksen, B., Theory and Practice of Structural Design Applied to Reinforced Concrete, 1953*†

Continued from previous page

Manning, G.P., Reinforced Concrete Arch Design, 1st edn, 1933*; 2nd edn, 1954*

Pennington, A.M., Concrete Farm Structures, 1954[†]

Baker, A.L.L., Ultimate-Load Theory Applied to the Design of Reinforced and Prestressed Concrete Frames, 1956*[†]

Childe, H.L., Editorial Notes from Concrete and Constructional Engineering, 1958*

Bennett, J.D., Design of Eccentrically Loaded Columns by the Load-Factor Method, 1958*

Bennett, J.D., Reinforced Concrete Members Subjected to Bending and Direct Force: Design by the Load-Factor Method, 1962*

Reynolds, C.E., Basic Reinforced Concrete Design, 2 vols, 1962*[†]

Gray, W.S., Manning, G.P., Watertowers, Bunkers, Silos and Other Elevated Structures, 1964*

Terrington, J.S., Turner, F.H., Design of Non-planar Concrete Roofs, 1964*[†]

Pannell, F.N., Design Charts for Members Subjected to Biaxial Bending and Thrust, 1966*[†]

Baker, A.L.L., The Inelastic Space Frame, 1967*

Manning, G.P., Concrete Reservoirs and Tanks, 1967*

Formwork and construction

Wynn, A.E., Design and Construction of Formwork for Concrete Structures, [1926*[†]]; 2nd edn, 1930*; 3rd edn, 1939*[†]; revised by G.P. Manning, 5th edn, 1965[†]*

Turner, L., Lakeman, A., Concrete Construction Made Easy, [1929*]; 3rd edn, 1942[†]

Reynolds, C.E., Concrete Construction, [1938*[†]]; 2nd edn, 1945*; 3rd edn, 1967*

Childe, H.L., Introduction to Concrete Work, 1943*[†]; 3rd edn, 1951[†]

Hunter, L.E., Construction with Moving Forms, 1951*[†]

Disney, L.A., Steel Reinforcement, 1954*[†]

Gill, H.R., Concrete Formwork Designer's Handbook, 1960*

Foundations

Baker, A.L.L., Raft Foundations: The Soil-line Method of Design, 1937; 2nd edn, 1942*; 3rd edn, 1957*[†]

Wentworth-Sheilds, F.E., Gray, W.S., Reinforced Concrete Piling, [1938*[†]]; with H.W. Evans, Reinforced Concrete Piling and Piled Structures, 2nd edn, 1960*[†]

Lee, D., Sheet Piling, Cofferdams, and Caissons, 1945*

Manning, G.P., Design and Construction of Foundations, 1961*[†]

Szechy, K., Foundation Failures, 1961*[†]

Irish, K., Walker, W.P., Foundations for Reciprocating Machines, 1969*

Prestressed concrete

Magnel, G., Prestressed Concrete, 1948*[†]; 2nd edn, 1950*[†]; 3rd edn, 1954*[†]

Abeles, P.W., Turner, F.H., Prestressed Concrete Designer's Handbook, 1962*[†]

Abeles, P.W., An Introduction to Prestressed Concrete, 2 vols, 1964*[†]

Roads

Anon., Concrete Roads and Their Construction, 1920*; 2nd edn, 1923*

Smith, R.A.B., Design and Construction of Concrete Roads, [1934*[†]]; with T.R. Grigson, 2nd edn, [1946*[†]]

Small buildings

Lakeman, A. (ed.), Concrete Cottages, Small Garages, and Farm Buildings, 1918[†]

Lakeman, A., Concrete Cottages, Bungalows, and Garages, 1918; 2nd edn, 1924*[†]; 3rd edn, 1932*[†]; title changed to Concrete Houses and Small Garages, 4th edn, 1949*[†]

Structural analysis

Eriksen, B., Influence Lines for Thrust and Bending Moments in the Fixed Arch, 1939*

Shepley, E., Continuous Beam Structures, [1942*[†]]; 2nd edn, 1962*

Gartner, R., Statically Indeterminate Structures, [1944*]; 2nd edn, 1947*; 3rd edn, [1957*[†]]

Manning, G.P., Displacement Method of Frame Analysis, 1952*[†]

Smolira, M., Analysis of Structures, 1955*[†]

Rygol, J., Nomograms for the Analysis of Frames, 1957*[†]

Yearbook

Concrete Yearbook, 1924 onwards

---

* Titles held by Institution of Civil Engineers Library.
[†] Titles held by British Library.

The history of the publishers themselves, with their office initially off Aldwych and from the early 1920s in Dartmouth Street off Victoria Street, the consultant's home, should also be recorded. It is hard to think of any other house that had such an influence on structural engineering: particularly important were the commentaries on the codes and Reynolds' books. The contribution of H.L. Childe, who was managing editor from 1922 to 1959, as well as being a regular author, was equally noteworthy.

It is very likely that every structural engineer who practised between 1925 and 1975 owned at least one of the dark-blue-backed series and used it heavily: fortunately they belonged to an era when things were built to last and many of them — books and readers — still survive.

Other relevant publications on reinforced concrete issued in this period and since can usually be found without difficulty in technical libraries. No attempt has been made to cover them here.

A5 Books and papers on early prestressed concrete 1945–65

The following publications together with references in Dr Francis Walley's chapter give a good idea of the state of knowledge on prestressed concrete as first introduced in Britain.

Freyssinet, Eugene. Prestressed concrete: principles and applications. *Instn Civ. Engrs, J.*, 1949–50, 33(4), pp. 331–80.

Guyon, Y. Prestressed concrete. *Contractors Record*, 1955.

Prestressed concrete: proceedings of the conference held at the Institution, February, 1949. ICE, 1949.

Magnel, Gustave. Applications of prestressed concrete in Belgium. *Instn Civ. Engrs, J.*, 32(6), 1948–49, pp. 161–74.

Walley, Francis. *The Progress of Prestressed Concrete in the United Kingdom.* Lecture, Institution of Civil Engineers, 3 July 1962, C&CA, 1962.

As with reinforced concrete, there should be no difficulty in locating current books and papers on prestressed concrete in technical libraries.

A6 Periodical publications: 1830s to the present day

With the introduction of reinforced concrete in the 1890s the general engineering literature of the time was soon challenged by the specialist engineering press — both from specialist suppliers such as Hennebique supplying information to their licensees and commercial publishers. This was a world-wide phenomenon and in the UK the lead was taken by *Concrete and Constructional Engineering*, the *Transactions of the Concrete Institute* (founded 1908, name changed to Institution of Structural Engineers 1922) and *Ferro-concrete*, the latter being a promotional magazine for L.G. Mouchel and Partners and the Hennebique system. Whilst one cannot ignore the more general magazines, such as *Engineer, Engineering, The Builder* or the proceedings of the learned societies such as the Institution of Civil Engineers (founded 1818), the Society of Engineers (founded 1854), the Royal Institute of British Architects (founded 1834), the Institution of Water Engineers (founded 1895), etc., they were not specialist publications and their coverage of the rapid developments in reinforced concrete in the first two decades of the 20th century was inadequate. In recognition of these developments another general weekly, the *Builder's Journal* issued special 'Concrete and Steel' supplements. Important at the time they were soon overtaken by *Concrete and Constructional Engineering* which remained the chief trade periodical for 60 years. After the Second World War it was joined by the more attractive architectural approach of *Concrete Quarterly* but ceased publication in 1966. *Concrete*, the journal of the Concrete Society commenced publication in 1967.

Major early titles thus are:

- *Concrete and Constructional Engineering* (1906–66);
- *Transactions of the Concrete Institute* (1909–22);
- *Ferro-concrete* (1909–39);
- *Concrete* (1967 onwards);
- *Concrete Quarterly* (1947 onwards);
- *Civil Engineering and Public Works Review* (1906–89).

Despite their often inadequate coverage of concrete, the more general architectural journals published at the end of the 19th and during the 20th century often include useful building studies and an excellent introduction to them is included in the publication Victorian periodicals and Victorian society edited by J. Don Vann and Rosemary T. VanArsdel, Aldershot, Scolar, 1994 and Toronto, University of Toronto Press, 1995. The paper on architecture by Ruth Richardson and Robert Thorne includes a bibliography, locations and a select annotated list. The titles which follow are fully described in that list and are arranged in order of publication:

- *The Architectural Magazine* (1834–39);
- [Royal] Institute of British Architects' Publications: *Transactions* 1835–92; *Proceedings* 1878–93; merged and recast as *Journal of the RIBA* (1894 onwards);
- *Civil Engineer and Architect's Journal* (1837–68/9);
- *The Surveyor, Engineer and Architect* (1840–43) entitled *Architect, Engineer and Surveyor* 1843;
- *The Builder* (1842/3–1966) entitled *Building* thereafter;
- *The Architect and Building Operative* (1847–68) became *Architect and Building Gazette* in 1850, was then absorbed by *Civil Engineer and Architects' Journal*;
- *Building News* (1854–1926): originally *Freehold Land Times and Building News* (1854–56) continued as *Building News and Architectural Review* (1860–62); then *Building News and Engineering Journal* (1863–1962) absorbed by *Architect and Building News*, 1926;
- *The Engineer* (1856 onwards);
- *The Builders' Weekly Reporter* (1856 onwards) absorbed *Building and Engineering Times* in 1886 and renamed *Builders' Reporter and Engineering Times*;
- *Engineering* (1866 onwards);
- *The Architect* (1869 onwards) entitled *The Architect and Contract Reporter* (1893–1918) then *Architect and Building News* (1919–70; new series 1971–80);
- *The British Architect* (1874 onwards) absorbed by *The Builder* in 1919;
- *Illustrated Carpenter and Builder* (1877 onwards);
- *Architectural Association Journal* (1887 onwards) entitled *Notes* (1887–1905);
- *The Builder's Journal and Architectural Record* (1895–1905); became in succession: *The Builders Journal and Architectural Engineer* (1906–1909); *The Architects and Builders Journal* (1910–19); and then *The Architects Journal* (1919 onwards);
- *The Architectural Review* (1896 onwards).

B: Practical guidance on the identification of concrete structures

B1 Sources of information on buildings and structures

Listed below are some of the sources which can be used when trying to track down the records of a building or structure:

- Local Authorities:
  — Building Control Department;
  — Architects Department;
  — Planning Department;
- Members of the original design team or their successors: architects; engineers; surveyors; quantity surveyors; party wall surveyors; contractors; specialist sub-

contractors. (For many buildings and structures, the designers' etc. names may be found in contemporary technical periodicals.)

- Owners: current; previous; original, e.g. Railtrack for railway structures;
- Occupiers: current; previous; original;
- Owners' and occupiers' professional advisers: solicitors; managing/estate agents; architects; engineers; surveyors, etc.;
- Maintenance manuals and Health and Safety files;
- Public utilities and statutory undertakers;
- Public libraries:
  — Reference libraries;
  — Local studies libraries;
- Local museums;
- Local history societies;
- Regional record offices such as the London Metropolitan Archives;
- National record offices such as the Public Record Office; English Heritage National Monuments Record;
- Libraries with national coverage such as the Institution of Civil Engineers, the Institution of Structural Engineers, the Royal Institute of British Architects;
- Museums with national coverage such as the Science Museum and its library; the Victoria and Albert Museum (housing the RIBA Drawings Collection);
- Museums with subject coverage such as The Museum of Docklands.

B2 General directories/guides to archives, etc. (most published works will be available in large public libraries)

The Aslib Directory of Information Sources in the United Kingdom, edited by Keith W. Reynard and Jeremy M.E. Reynard, 11th edn. Aslib, 2000 (biennial).

A guide to libraries, record offices and organizations with contacts, information on coverage, special collections and publications.

The Directory of Museums and Special Collections in the United Kingdom, edited by Peter Dale, 2nd edn. Aslib, 1996.

Similar in format to the Aslib Directory.

Directory of British Associations & Associations in Ireland, 14th edn. CBD Research Ltd, 1998.

With brief details on each organization: contact details, information resources, publications.

The Business Archives Council
101 Whitechapel High Street London E1 020 7247 0024.

Does not hold archives or carry out research. Publishes a journal which includes location lists of archives deposited in public collections.

National Register of Archives
Quality House, Quality Court, Chancery Lane, London WC2A 1HP; Tel: 020 7242 1198; website: www.hmc.gov.uk

A source of information on archives held outside public collections maintained by the Royal Commission on Historical Manuscripts.

Visitors can consult the indexes in person, or searches can be carried out via the website. Telephone enquiries are not undertaken, although written enquiries are.

Record Repositories in Great Britain, Royal Commission on Historical Manuscripts. TSO (updated at intervals).

Guide to publicly available record collections in local and national record offices, universities and research centres.

The Building Museums Guide, Interbuild Publications for the Building Museum Project, 1993

Now very dated but still useful. Brief details of each museum are given.

Museums around the UK on the Web can be found at www.mda.org.uk/vlmp/

B3 Organizations holding archives on individual structures as well as general records (possibly including drawings)

Cadw
Welsh Historic Monuments
Brunel House, 2 Fitzalan Road, Cardiff, Wales CF2 1UY; Tel: 01222 465111.

Publications and documentation on histotic buildings in Wales.

Concrete Archive
See Institution of Civil Engineers Library and Archives.

Construction Industry Resource Centre Archive (CIRCA)
Curator: John Keenan, 3 Cranleigh Gardens, Stoke Bishop, Bristol BS9 1HD; Tel: 0117 968 7850.

Aims to collect drawings and documentation including text books and trade literature which might otherwise be lost.

English Heritage: National Monuments Record Centre
Main offices: Kemble Drive, Swindon, Wilts SN2 2GZ; Tel: 01793 414600; London search room: 55 Blandford Street, London W1H 3AF; Tel: 020 7208 8200; website: www.english-heritage.org.uk

Drawings, photographs, publications and documentation on historic buildings of all ages.

Historic Scotland
Longmore House, Salisbury Place, Edinburgh EH9 1SH; Tel: 0131 668 8668; website: www.historic-scotland.gov.uk

Publications and documentation on historic buildings in Scotland.

Institution of Civil Engineers Library and Archives
Great George Street, London SW1P 3AA; Tel: 020 7222 7722; website: www.ice.org.uk

The ICE is the oldest established library in the construction industry, and it has material on the use of cements and concrete reflecting this. These traditional strengths have been enhanced by the development of the 'concrete archive' with the support of the Concrete Society. In addition to published material, drawings, specifications, calculations in the original and microfilm, have been placed in the archive, together with samples of concrete and reinforcement.

London Metropolitan Archives (formerly the Greater London Record Office)
40 Northampton Road, London EC1R 0HB; Tel: 020 7332 3820; Fax: 020 7833 9136; website: www.cityoflondon.gov.uk/archives/lma

No entry formalities, but there is usually a wait for documents.

Public Record Office
Ruskin Avenue, Kew, Surrey, TW9 4DU; Tel: 020 8876 3444;
website: www.pro.gov.uk

Holds the official records for all government owned buildings in the United Kingdom and much material deposited by government departments dealing with buildings. Indexes can be searched on-line. A reader's ticket is required and can be obtained on site. There is usually a wait for documents.

Royal Institute of British Architects
66 Portland Place, London W1B 1AD; Tel: 0906 302 0400 (premium rate public information line); website: www.architecture.com

The early users of concrete were often architects rather than engineers, and whilst modern textbooks on reinforced concrete tend to focus on its engineering analysis, it is a basic building material so, inevitably, the RIBA Library has an important collection covering developments from earliest times to the present day. There is a charge for use of the Library but the library catalogue (covering accessions since 1984) is available online.

The Drawings Collection of the RIBA is being rehoused and will be available in the Victoria & Albert Museum from 2004.

Science Museum
Exhibition Road, London SW7; Tel: 020 7942 4000; website: www.nmsi.ac.uk

The Museum has a collection of artefacts connected with construction and its Library has a good collection of technical journals, publications and trade literature.

Transport Research Laboratory
Old Wokingham Road, Crowthorne, RG45 6AU; Tel: 01344 773131;
website: www.trl.co.uk

Originally established in 1933 as the Road Research Laboratory, the Government research body producing specifications and guidance on roads and pavements, bridges, etc.

B4 Organizations which may be worth consulting for general technical and historical data on concrete structures

Amberley Museum (formerly known as the Chalk Pits Museum)
Amberley, Arundel, West Sussex BN18 9LT; Tel: 01798 831370;
website: www.amberleymuseum.co.uk

Houses a display on the use of concrete, including exhibits and artefacts on loan from the Science Museum.

British Cement Association
Century House, Telford Avenue, Crowthorne RG45 6YS; Tel: 01344 762676;
website: www.bca.org.uk

The BCA, successor to the Cement and Concrete Association which for 50 years was the major source of information and research on cement and concrete, houses an important specialist collection, not only of its own published literature, but on cement and concrete generally, together with an impressive archive of audio-visual material. Although it charges for its services it is an essential and helpful resource.

British Library
96 Euston Road, London NW1; Tel: 020 7412 7676; website: www.bl.uk

Coverage is national and includes books, maps and patents. A reader's ticket is required and can be obtained on site. There is usually a wait for documents, but this can be avoided if they are requested before visiting.

British Standards Institution
389 Chiswick High Road, London W4 4AL; Tel: 020 8996 9001; website: www.bsi-global.com

For copies of current and earlier codes and standards. Their online catalogue shows their status.

Building Research Establishment
Garston, Watford WD2 7JR; Tel: 01923 664000; website: www.bre.co.uk

The BRE was for approximately 80 years the Government's centre of excellence for building research, a position which enabled it to collect information on concrete on a world-wide basis. Their own publications are a mine of useful information and pioneering research on concrete in the early years. For material eluding you elsewhere it is still worth trying.

Construction History Society
c/o The Chartered Institute of Building, Englemere, King's Ride, Ascot, Berks SL5 8BJ; Tel: 01344 23355; website: www.construct.rdg.ac.uk/chs/

A grouping of people interested in the history of construction.

Publishes a Journal which includes abstracts of periodical literature. Newsletter includes a digest of major accessions to Repositories relating to building and construction from the Royal Commission on Historical Manuscripts annual listing. Newsletter may be used to request information of members.

English Heritage (Headquarters)
23 Savile Row, London W1X 1AB; Tel: 020 7973 3000; website: www.english-heritage.org.uk

The head office of English Heritage. Source for regional office addresses, publications.

Institution of Structural Engineers Library
11 Upper Belgrave Street, London SW1X 8BH; Tel: 020 7235 4535; website: www.istructe.org.uk

With its origins as the Concrete Institute the IStructE is particularly strong in the development of structural concrete from the late 19th century onwards. Whilst material may also be held at ICE, IStructE often has material which has disappeared from ICE over the years, particularly standards.

The Lime Centre
Long Barn, Morestead, Winchester, Hants SO21 1LZ; Tel: 01962 713636;
website: www.thelimecentre.co.uk

Newcomen Society
Science Museum, Exhibition Road, London SW7 2DD; Tel: 020 7371 4445;
www.newcomen.demon.co.uk
A society for the study of engineering and technology.

Twentieth Century Society
70 Cowcross Street, London EC1M 6BP; Tel: 020 7250 3857;
website: www.c20society.demon.co.uk

Victorian Society
1 Priory Gardens, Bedford Park, London W4 1TT; Tel: 020 8994 1019;
website: www.victorian-society.org.uk

B5 Websites and bulletin boards

Website addresses have been included for many of the organizations listed. However, it must be stated that many organizations change their servers and their addresses and therefore the accuracy of the addresses given cannot be assured for the future. Being able to carry out searches at a distance is extremely convenient and many more organizations will doubtless soon have their own sites with links from them to other related sites.

The online catalogues of major universities and national libraries such as the British Library and the Library of Congress are easily accessible and are excellent guides to published information.

When the major sources have been exhausted, a medium which should not be ignored is bulletin boards on the internet. Many organizations and groups which have a website also have bulletin boards on which enquiries can be posted. These may form part of a service to members on a professional institution's website or may be open to a wider participation.

Links from professional institutions websites can also be a more efficient way of searching the internet than simply using the major search engines. However, amongst the major search engines, the Google site, which scans and rates other sites, is an excellent place to start: www.google.com

B6 Trade literature and published synopses of details of proprietary concrete systems

Useful as textbooks and periodicals are, they rarely provide details on the many commercial reinforced and prestressed concrete systems employed in buildings and other structures from the middle of the 19th century onwards. Details of these must be sought in trade literature issued by the patentees or their licensees. Unfortunately, the survival of such material is unpredictable, although better for the more recent technology of prestressed concrete; firms tend to discard old trade literature as new revised literature arrives, whilst most libraries have never collected promotional literature, unless it has a high technical content. Well-organized firms with a long history of involvement in reinforced concrete such as L.G. Mouchel and Partners have excellent collections, but for every Mouchel there are dozens of firms who have ceased trading or reorganized so frequently as to destroy the integrity of their records. In the past the Science Museum has collected trade literature, but this collection contains relatively little on reinforced concrete. In these circumstances one can pursue information contained in original patents, obtainable in the British Library or the various local depositories, or seek business records in libraries and record offices local to the firms concerned.

Some information on systems can be obtained from *Specification* (published annually since 1898); compilations such as *Diamant's on Industrialised Buildings of the 1960s*, the BCSA publications on *Prefabricated Floors for Use in Steel Framed Buildings* (1964, 1965, 1977), and various other 19th and early 20th century texts on concrete and fireproof construction. For the inter-war period Jaggard and Drury's *Architectural Building Construction*, 3 vols, CUP, 1916–47, is a good starting point. Unfortunately these rarely provide 'safe load tables' or the like and if they do, they should be applied with caution as the criteria used for testing are unlikely to be known.

There is good coverage of prestressing systems incorporating hac and used in the 1960s in Addendum 2 of Building Regulations Advisory Committee Report by Sub-committee P — High Alumina Cement Concrete, Dept of the Environment, 1976.

The Institution of Civil Engineers Library (ICE) is a good first point of contact for finding trade brochures, followed by the Institution of Structural Engineers (IStructE), The British Library (BL) and the Science Museum Library.

Trade literature can be found in these major libraries as follows:

Reinforced concrete systems
- Mouchel-Hennebique: *Betón Armé* (ICE some issues; BL) *Ferro-concrete* (ICE incomplete; BL);
- Monier: 19th century brochures (ICE; BL);
- BRC: Early brochures and journals (ICE);
- Considere: Early brochure (ICE);
- Coignet systems: Early brochure (IStructE);
- Expanded metal: No known source for 19th century; various booklets (ICE; IStructE);
- Indented Bar Company: Brochures (IStructE, complete set; ICE some);
- Kahncrete/Truscon (TrussedConcrete Steel): Kahncrete periodical (BL, 1 issue ICE);
- Kleine (ICE archives);
- Ritchie system (ICE archives);
- Truscon leaflets (ICE, incomplete; IStructE).

Prestressed concrete: the early systems are described by Francis Walley
- PSC-Freyssinet: Brochures (ICE; IStructE); House journal (ICE, incomplete, IStructE);
- Gifford-Udall: Book (ICE; IStructE);
- Hoyer;
- Laingspan: Brochures (ICE; IStructE);
- Lee-McCall/McCalloy: Brochures (ICE; IStructE);
- Rylands: Brochures (ICE; IStructE).

Piling systems
- See also 'reinforced concrete systems' above;
- Frankipile: Books (ICE; IStructE);
- Lee-McCall: (ICE).

This select short list is intended to provide sources for engineers and others involved with existing concrete structures (principally buildings).

C: Publications giving guidance on the investigation, appraisal, repair and maintenance of existing concrete structures by Michael Bussell

C1 General: investigation and appraisal

The Institution of Structural Engineers. *Appraisal of Existing Structures*. ISE, 2nd edn. 1996.

Building Research Establishment. Digest 366. *Structural Appraisal of Existing Buildings for Change of Use*. BRE, 1991.

Beckmann, P. *Structural Aspects of Building Conservation*. McGraw-Hill, 1995.

Health & Safety Executive. Guidance Note G58. *Evaluation and Inspection of Buildings and Structures*. HMSO, 1990.

Institution of Structural Engineers. *Guide to Surveys and Inspections of Buildings and Similar Structures*. ISE, 1991.

C2 General: repair and maintenance

Currie, R.J., Robery, P.C. Report BR254. *Repair and Maintenance of Reinforced Concrete*. BRE, 1994.

Fédération Internationale de la Précontrainte. *Inspection and Maintenance of Reinforced and Prestressed Concrete Structures*. FIP, 1986.

C3 Information on original construction

The Building Research Establishment has published numerous reports and pamphlets on proprietary, prefabricated and large-panel housing systems; these describe the construction, defects and inspection of systems used in the UK since the Second World War.

C4 Investigation, testing and repair of concrete

General

Bungey, J.H. Technical Note 143. *Testing Concrete in Structures: A Guide to Equipment for Testing Concrete in Structures*. CIRIA, 1992.

Strength

British Standards Institution. *BS 6089: Guide to Assessment of Concrete Strength in Existing Structures*. BSI, 1981.

Concrete Society. Technical Report No 11. *Concrete Core Testing for Strength*. Concrete Society, 1987.

Watkins, R.A.M., McNicholl, D.P. Statistics applied to the analysis of test data from low-strength concrete cores. *Struct. Engr.*, 1990, 68(16), pp. 327–32.

Cracking

Concrete Society. Technical Report No 22. Non-structural Cracks in Concrete. Concrete Society, 1992.

Corrosion

Building Research Establishment. Digest 444. Corrosion of Steel in Concrete: Part 1: Durability of Reinforced Concrete Structures; Part 2: Investigation and Assessment; Part 3: Protection and Remediation. BRE, 2000.

Building Research Establishment. Digest 434. *Corrosion of Reinforcement in Concrete: Electrochemical Monitoring*. BRE, 1998.

Building Research Establishment. Digest 389. *Concrete Cracking and Corrosion of Reinforcement*. BRE, 1993.

Lea, F.M. Building Research Technical Paper 7. *Slag, Coke, Breeze and Clinker as Aggregates*. HMSO, 1929.

Brady, F.L. Building Research Special Report 15. *The Corrosion of Steel by Breeze and Clinker Concretes*. HMSO, 1930.

Concrete Society. Technical Report No 26. *Repair of Concrete Damaged by Reinforcement Corrosion*. Concrete Society, 1984.

Pullar-Strecker, P. *Corrosion Damaged Concrete: Assessment and Repair*. CIRIA/Butterworth, 1987.

Carbonation

Building Research Establishment. Digest 405. *Carbonation of Concrete and its Effect on Durability*. BRE, 1995.

Parrott, L.J. Report BR114. *A Review of Carbonation in Reinforced Concrete*. Building Research Establishment/Cement and Concrete Association, 1987.

Roberts, M.H. Information Paper 6/81. *Carbonation of Concrete Made with Dense Natural Aggregates*. BRE, 1981.

Chlorides

Roberts, M.H. Information Paper 21/86. *Determination of the Chloride and Cement Content of Hardened Concrete*. BRE, 1986.

High alumina cement

Building Research Establishment. Digest 392. *Assessment of Existing High Alumina Cement Concrete Construction in the UK*. BRE, 1994.

Bate, S.C.C. Report BR235. *High Alumina Cement Concrete in Existing Building Superstructures*. HMSO, 1984, repr 1992.

Dunster, A.M. and others. Report BR386. *Durability of Ageing High Alumina Cement Concrete. A Literature Review and Summary of BRE Research Findings*. CRC, 2000.

Alkali silica reaction

Institution of Structural Engineers. *Structural Effects of Alkali–Silica Reaction: Technical Guidance on the Appraisal of Existing Structures*. ISE, 1992.

Hobbs, D.W. *Alkali Silica Reaction in Concrete*. Thomas Telford, 1988.

Mundic

Royal Institution of Chartered Surveyors. *The Mundic Problem: A Guidance Note: Recommended Sampling, Examination and Classification Procedure for Suspect Concrete Building Material in Cornwall and Devon*. RICS Books, 1994.

Load testing

Building Research Establishment. Information Paper 2/95. *Guidance for Engineers Conducting Static Load Tests on Building Structures*. BRE, 1995.

Building Research Establishment. Digest 402. *Static Load Testing: Concrete Floor and Roof Structures within Buildings*. BRE, 1995.

Fire damage

Concrete Society. Technical Report No 33. *Assessment and Repair of Fire-Damaged Concrete Structures*. Concrete Society, 1990.

Surface treatment

Construction Industry Research And Information Association. Technical Note 130. *Protection of Reinforced Concrete by Surface Treatments*. CIRIA, 1987.

Bridges

Mallett, G.P. *Repair of Concrete Bridges: A State of the Art Review on the Repair, Maintenance and Protection of Concrete Bridges*. Telford, 1994.

Appendix II: 19th century proprietary floor systems

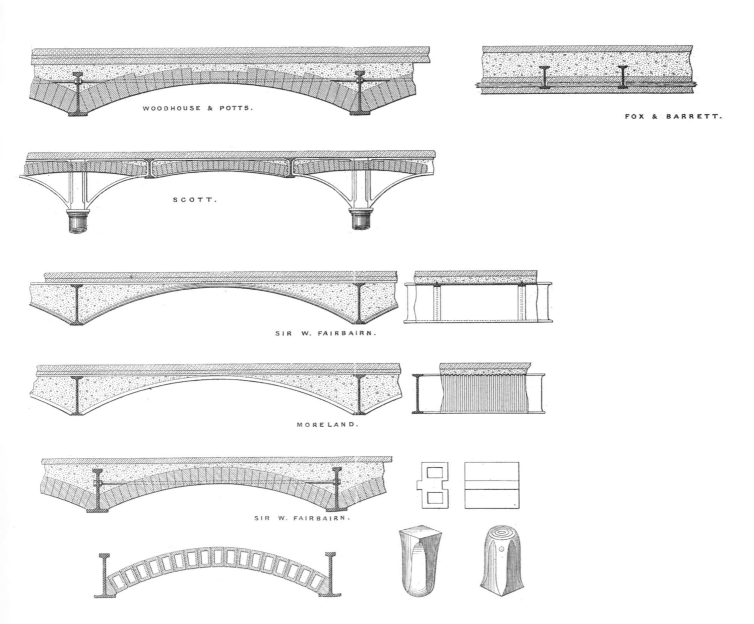

WOODHOUSE & POTTS.

FOX & BARRETT.

SCOTT.

SIR W. FAIRBAIRN.

MORELAND.

SIR W. FAIRBAIRN.

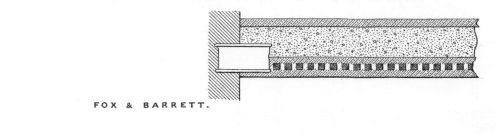

FOX & BARRETT.

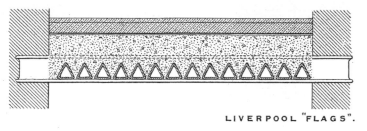

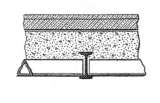

LIVERPOOL "FLAGS".

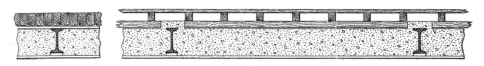

DAWNAY.

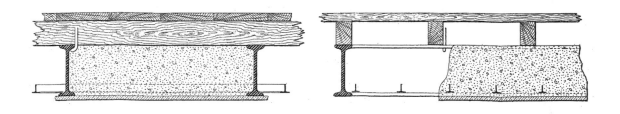

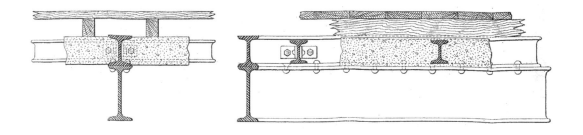

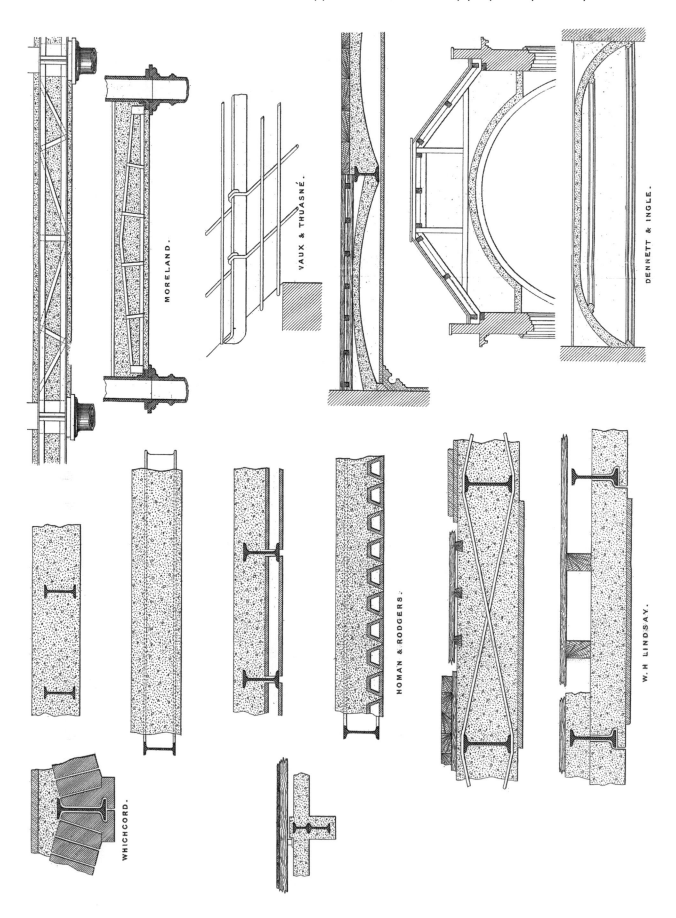

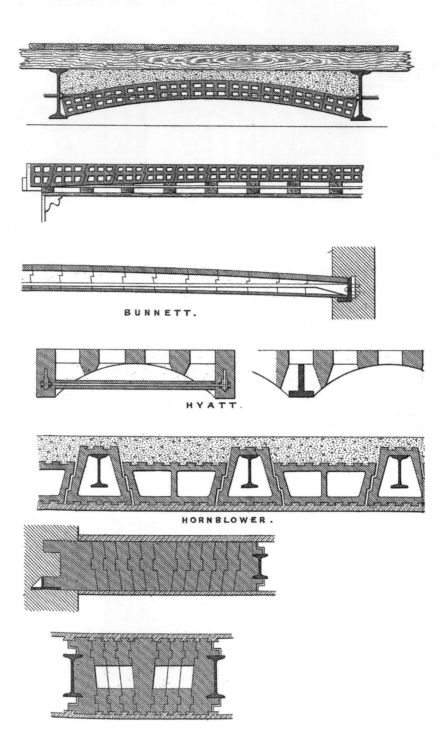

BUNNETT.

HYATT.

HORNBLOWER.

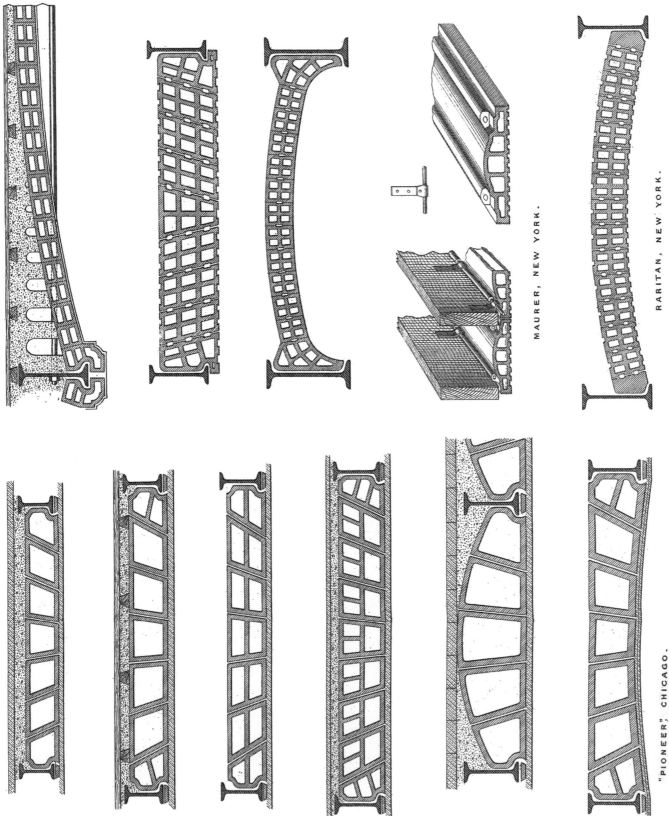

MAURER, NEW YORK.

RARITAN, NEW YORK.

"PIONEER," CHICAGO.

Appendix III: proprietary reinforcing and flooring systems

Introduction

This appendix describes and illustrates many of the proprietary reinforcing and flooring systems used in Britain during the greater part of the 20th century, excluding more recent developments. It aims primarily to help those dealing with concrete structures to identify such reinforcement and flooring systems. Identification is a necessary preliminary to seeking further information on the particular system, using specific chapter references and the more comprehensive advice on information sources listed in Chapters 3–5 and in Appendix E.

In addition, it is hoped that the general reader will find interest in the sheer variety of such systems, as compared with the relatively limited range of reinforcement and flooring types in common use today.

The systems listed here include most of those known to have been in general use, but certainly not all of the many systems that were patented, advertised, and promoted. Quite a few of these, particularly from the earlier part of the century, in practice were used only rarely, and some not at all.

As explained in Chapter 4, the early concrete specialists had their own reinforcing systems, and often provided much more, culminating in what today can be described as a complete 'design and build' service, as exemplified by the Hennebique system. Reinforcement for such systems came in many shapes and sizes in the early 20th century, before patents expired and the round or square bar became the norm for reinforcement, specified by designers and used by contractors without the constraints and costs imposed by patent protection. Such changes in reinforcement types are described more fully in Chapter 5.

Early 20th century flooring systems evolved from those of the second half of the 19th century, which saw the development, patenting, and use of many so-called 'fireproof' floor systems. The commonest of these are described in Chapter 3, which also lists some specific references to aid their identification.

Such floors commonly used beams or joists of cast or wrought iron (later steel) at relatively close spacing, between which spanned infill blocks, timber joists or lathing, or flat or shallow-vaulted (but essentially unreinforced) concrete. The blocks, of fired clay, or clinker or breeze concrete, were often hollow to save both weight and cost. They often had 'tongues' at either end; these butted together to protect the beam flanges from fire. Alternatively, the timber joists and the beams were fire-protected by plastering the floor soffit. Concrete might be laid as 'topping' over the beams and blocks to form a level floor.

These forms of flooring construction continued into the 20th century. However, as the use of reinforced concrete increased, the infill blocks or timbers were replaced by longer-span proprietary precast concrete floor units, themselves often used in conjunction with infill blocks spanning between the units, and usually with an in situ concrete topping. In situ concrete slabs were a widely-used alternative. These were typically solid for shorter spans and ribbed for longer spans, the latter often employing hollow blocks or 'pots' to infill the spaces between the ribs. The supporting beams and columns could be of steel, or equally of reinforced concrete.

Many building contractors produced their own flooring units and – together with specialist precast concrete manufacturing firms – offered 'off-the-shelf' standard units for floor, roof, and stair construction as well as other elements (including architectural cladding).

The remainder of this appendix comprises three illustrated lists. The first two, of early reinforcing systems and early-20th century flooring systems, are arranged alphabetically. To aid identification, each particular system is illustrated, described, and/or cross-referred to similar systems. The third list, of mid-20th century flooring systems, is arranged within generic types of floor construction, in view of the proliferation of such systems; to describe and illustrate each is beyond the available space in this volume. For each type, a representative proprietary floor is illustrated, and the larger or more widespread manufacturers are listed.

Early reinforcement systems

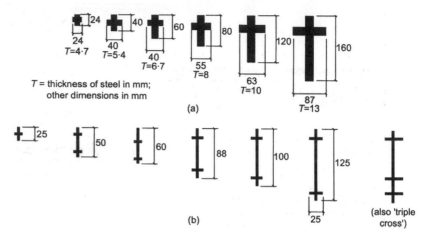

Figure III.1. Bonna (also known as Columbian or Feresco): (a) sizes of 'Latin Cross' sections used in building frames; (b) 'Double Cross' or 'Cross of Lorraine' and also 'triple cross' bars used in floors (see also Figure III.29)

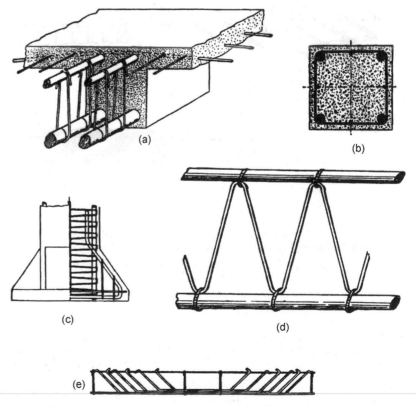

Figure III.2. Coignet: (a) spiral links in columns; (b) stirrups sometimes formed of spiral rod wired to top and bottom bars; (c) bent-up bars used in small diameters as shear reinforcement; (d) beam and slab; (e) column. Columbian: see Bonna, Figure III.1

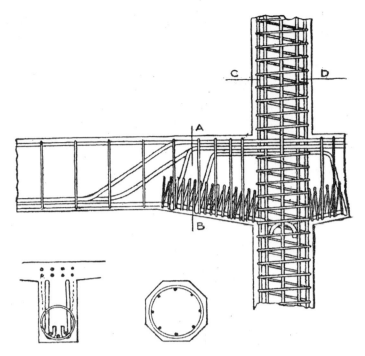

Figure III.3. Considère: showing spiral binding to main bars

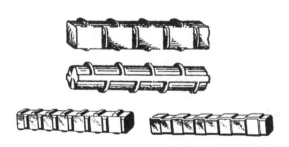

Figure III.4. Types of Corr or Corrugated bars (Indented and Johnson similar or identical); some were early deformed bars of high carbon steel with a yield strength of 65,000–70,000 lb/in² (448–482 N/mm²) (see also Unit, Figure III.24)

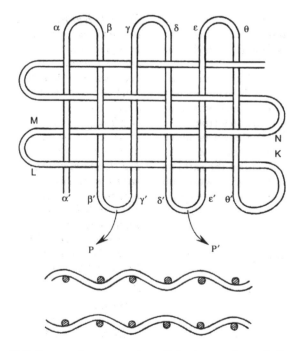

Figure III.5. Cottancin: woven interlinked mat of small diameter rods; system closer to ferro-cement than 'true' reinforced concrete, with thin slabs reinforced by ribs often acting as folded plates

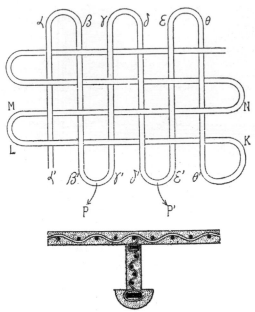

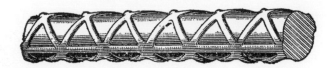

Figure III.7. Expanded metal (Expamet): John F. Golding patented 'slashed metal screening' in 1883 in the USA. Feresco: See Bonna, Figure III.1

Figure III.6. Diamond: rolled ribbed bar with constant cross-sectional area

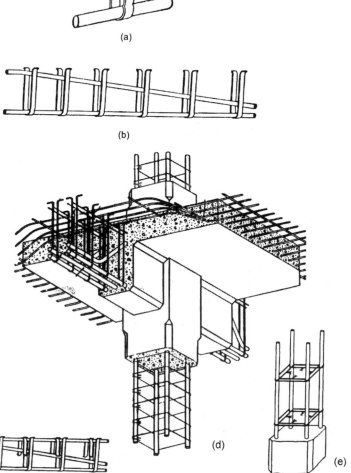

(a)

(b)

(c)

(d)

(e)

Figure III.8. Hennebique: very widely used system, with round main bars with 'fishtail' ends, thin slabs and walls, distinctive deep narrow beams, often square columns, and chamfers on beam and column arrises: (a) stirrup round tension bar; (b) tension bars and stirrup; (c) bottom bars cranked up to form top steel over supports in continuous spans; (d) stirrups and links originally flat strips, later round bars; (e) columns linked by 'sausage' bars around main bars in some cases

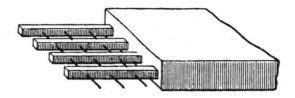

Figure III.9. Hyatt: early American reinforcement, flat bars pierced with round bars, used in slabs and later in beams (also known as Jackson bars)

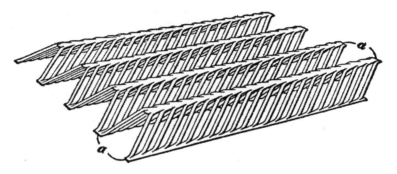

Figure III.10. Hy-rib: also known as herring-bone metal lath; folded expanded metal that could act as both permanent formwork and reinforcement Improved Construction: plain round bars with spiral shear reinforcement (not widely used). Indented: see Corr, Figure III.4. Jackson: see Hyatt, Figure III.9. Johnson: see Corr, Figure III.4

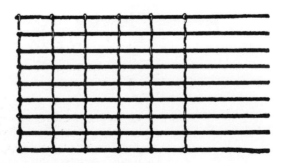

Figure III.11. Johnson's lattice: triangular or rectangular woven steel fabric

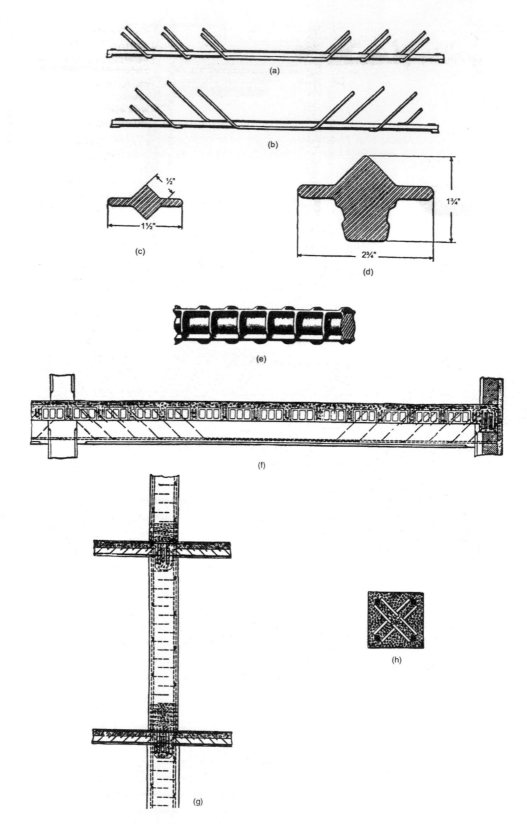

Figure III.12. Kahn (also Truscon): unique 'trussed bars' with fins rolled on opposite corners of plain square bar that could be bent up to improve anchorage: (a) paired fins; (b) staggered fins; (c) typical bar profile; (d) larger bar section; (e) a later and more conventional profile was the Kahn cup bar, with yield strengths 33,000–37,000 lb/in^2 (227–255 N/mm^2); (f) fins bent at 45° as beam stirrups; (g) and (h) fins bent at 90° as column links

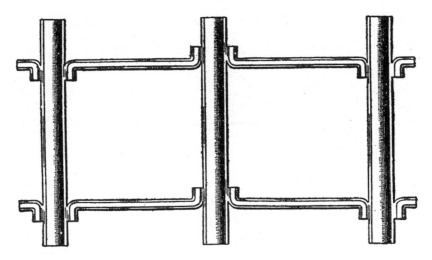

Figure III.13. Kahn expanded metal: a sheared expanded rolled profile with the main bars normal to the secondary steel (unlike expanded metal, cf. Figure III.7)

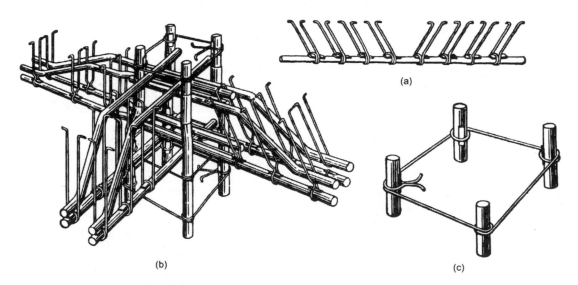

Figure III.14. Keedon: stirrups and links mechanically 'keyed on' to main bars by wedges inside fully enclosing loops (a) stirrups inclined to improve shear resistance in beams; (b) some bottom bars bent up in continuous spans; (c) column links

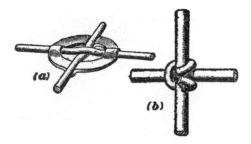

Figure III.15. 'Lock-woven' or 'lock-tied' mesh: (a) wires secured by mechanically-fixed wires or (b) by staples, in lieu of welding to provide an anchored mesh system

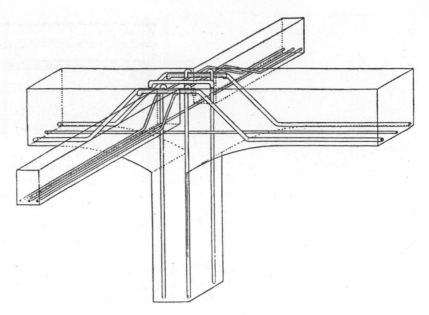

Figure III.16. Monier: plain round bars, in beams bent up over continuous supports; more commonly used in Germany, where known as Wayss und Freytag system

Figure III.17. Monolith: a grooved bar into which angled stirrups could be fitted, being held in place by swaging

(a)

(b)

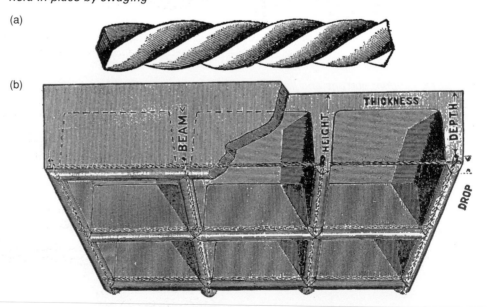

Figure III.18. Ransome: (a) the archetypal square-twisted cold-worked bar (also hot-rolled), with a yield strength of 53,000 lb/in^2 (379 N/mm^2) or better, much employed in the USA; (b) early use of coffered slab construction for longer slab spans

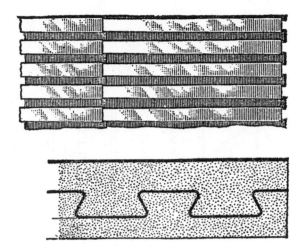

Figure III.19. Ridley-Cammell: corrugated 'dovetail' steel sheeting [sic.], with angle bars as main steel in beams, and vertical sheeting panels for shear reinforcement riveted or bolted to main steel; sheeting also used in columns and walls; served also as permanent formwork where required

Figure III.20. Spiral: similar to Ransome, but with corner 'beads', with a yield strength 65,000–80,000 lb/in² (448–551 N/mm²)

Figure III.21. Square twisted Lug: a further deformed (if not tortured) version of the Ransome square-twisted bar with additional lugs to improve bond

(a)   (b)

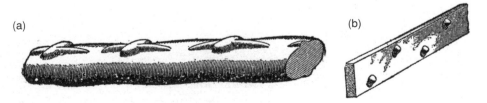

Figure III.22. Thacher: deformed bar; (a) constant cross-sectional area without sharp reentrant corners; (b) an early form was of flat strip with indentations

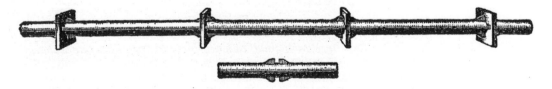

Figure III.23. Truscon: a deformed bar with notches to accommodate washers to provide mechanical anchorage; also used in the Kahn reinforcement system (see Figure III.12)

(a)

(b)

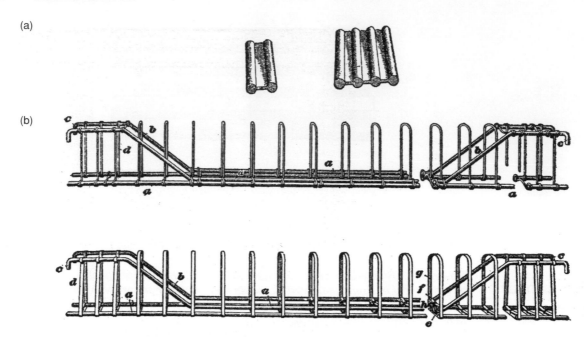

Figure III.24. Unit: (a) rolled bars (two or four) within linking web; (b) prefabricated reinforcement system using first Corr bars (see Figure III.4), later plain round bars

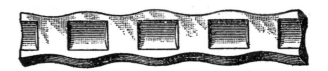

Figure III.25. Universal: an indented bar (not widely used)

(a)

(b)

(c)

(d)

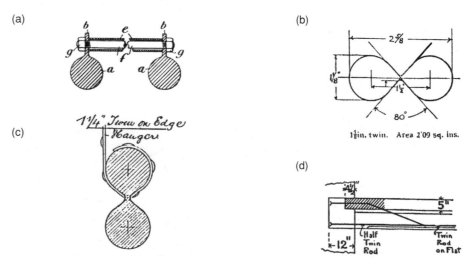

Figure III.26. Wells: (a) reinforcement profile originally a single round bar profile with a flat continuous fin, which could be perforated to accommodate transverse bars (spacers in beams, links in columns, etc); (b) later developed into twin round bar section joined in an hour-glass profile; (c) perforated to receive indented or perforated hanger bars; (d) or slit longitudinally allowing one bar to be cranked up over supports as anti-crack steel; the system itself later used part-precasting of beam soffits as permanent formwork, with roughened top surface for keying to in situ topping

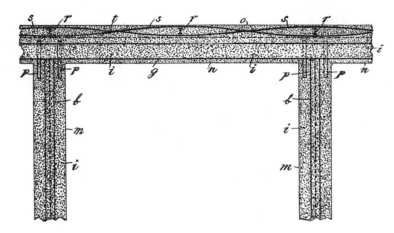

Figure III.27 Williams: filler joist floors reinforced with draped transverse steel rods

Early 20th century flooring systems

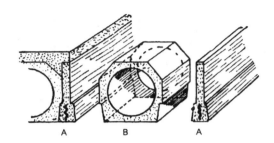

Figure III.28. Armoured Tubular: (a) precast concrete ribs with corrugated strip reinforcement; (b) fired clay, or coke or breeze concrete hollow infill blocks; also known as Herbst floor. Caxton: see Figure III.35

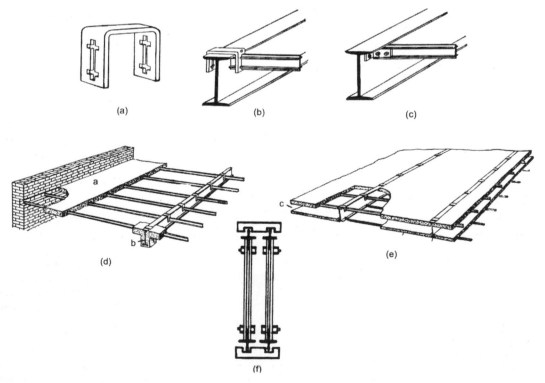

Figure III.29. Columbian (known also as Feresco, using Bonna bars, see Figure III.1): distinctive cross-profiled rolled bars used as filler joists: (a) and (b) supported on hangers or (c) by bolting to beams; (d) used to form single and (e) double floors, the latter giving excellent sound-proofing and an early 'services void'; (f) bars in beams were spaced and framework could be supported by cross braces

Figures III.30, III.32, III.34, and III.35 are reproduced from Jaggard, W. R. and Drury, F. E., Architectural Building Construction, Volume 3, 1937 (2nd edn) by courtesy of the publishers, Cambridge University Press.

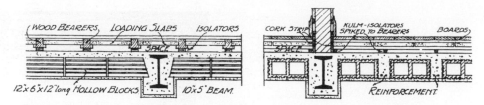

Figure III.30. Cullum: variant of Kleine floor (see Figure III.35) with sound-proof raised timber floor)

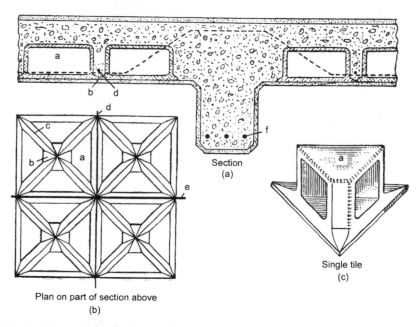

Figure III.31. Dentile: triangular clay mitre tiles interlocked to form voids within one-way or two-way spanning floor: (a) section; (b) plan on part of section; (c) single tile

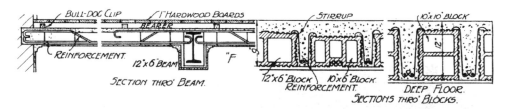

Figure III.32. Diespeker: variant of Kleine floor (see Figure III.35)

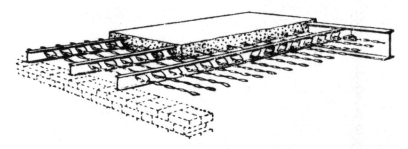

Figure III.33. Fawcett (Monolithcrete): unusual floor with early perforated-web steel beams with enlarged bottom flange and transverse reinforcement from twisted flat bars (a form of filler joist construction, successor to the Fawcett hollow clay block and iron/steel joist system, see Chapter 3)

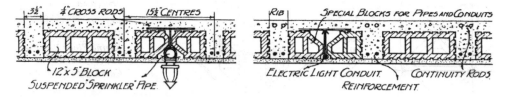

Figure III.34. Helicon: hollow tile floor similar to Kleine (see Figure III.35). Herbst: see Figure III.28

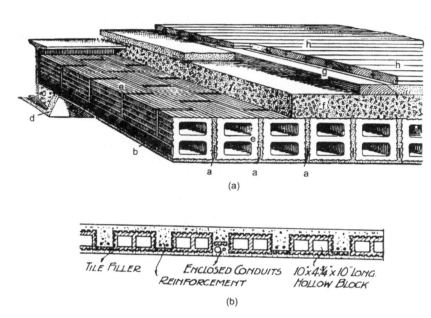

Figure III.35. Kleine: early hollow clay tile floor, first used in Britain by Higgs and Hill c.1905; (a) reinforcement was flat steel on edge in cement mortar joints between blocks with coke-breeze concrete topping; (b) later, clay blocks and slip tiles were used with in situ concrete ribs and topping, usually with steel beams, often concrete-cased; Caxton floor similar

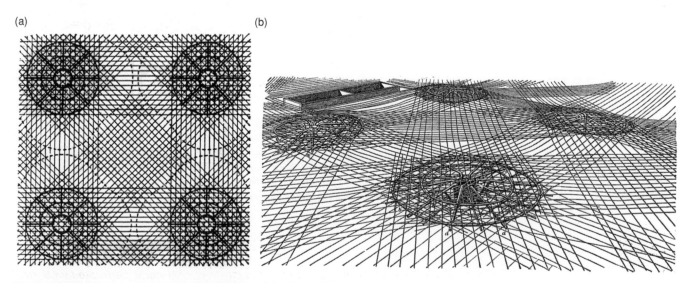

Figure III.36. Mushroom: originally developed by C.A.P. Turner, US pioneer of flat slab construction c.1906, notable both for this and for four rather than two layers of slab steel: (a) plan with broken lines suggesting 'mushroom heads'; (b) perspective indicating draping of slab reinforcement

Figure III.37. Siegwart: among the first hollow precast concrete units, originally reinforced with corrugated sheeting, later with round bars

Figure III.38. Stolte floor: German system using reinforced hollow concrete blocks with flat bar embedded reinforcement

Typical proprietary concrete floors used in the mid-20th century

Figures III.39–III.48 are reproduced from Prefabricated Floors for Use in Steel Framed Buildings (2nd edn, 1965) by courtesy of the publishers, the British Constructional Steelwork Association Limited, and Figure III.49 is reproduced from Mitchell's Structure and Fabric: Part 2 (3rd en, 1983) by courtesy of Longman Scientific & Technical.

Various generic types of floors were in use during the middle part of the 20th century. Typical floors are shown for each type, and makers are identified. Some of the types and their makers remain active today.

Prestressing was often used in precast ribs, beams, and planks.

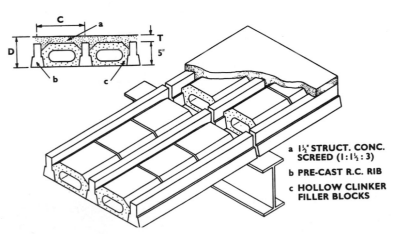

Figure III.39. Precast ribs or joists with/without infill blocks (usually of fired clay or concrete) and/or composite topping: (i) Inverted T-sections – Armocrete shown; (ii) Joists or X-joists – Lyncrete shown; (iii) X-joists with thin precast planks and composite in situ topping – Pierhead shown. Makers: (i) Allisons, Anglian, Armocrete, Blockspan, Bradley, Dabro, Flooring Contracts, Francis, Greenwood, HBS Self-centring, Invictus, Musker, Quikspan, Rose Green, Tilecast, Trent. (ii) Fram, Lyncrete, Pierhead (later Booth), Rapid, Trent. (iii) Pierhead.

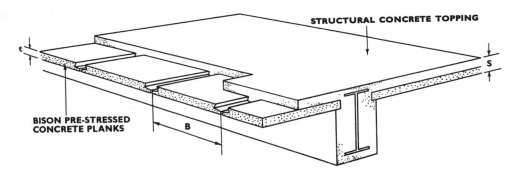

Figure III.40. Precast solid planks with/without composite topping – Bison (composite) shown. Makers: Bison, Francis, GKN, Omnia, Parcrete, Quikspan, Stotam

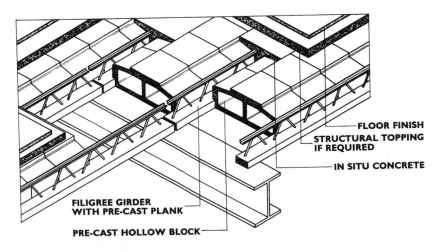

Figure III.41. Precast slabs or ribs with lattice reinforcement and in situ concrete to form composite structure, often with infill blocks between ribs – Filigree beam and block floor shown. Makers: Atlas-Omnia (with trough formers), Combric, Filigree, GKN, Katzenberger, Omnia, Sindall, Spancrete, Spanlite, Stahlton (prestressed clay block planks – see Chapter 13), Triad. Variant with wide precast planks and lightweight void formers with in situ topping: Combric Wideslab, Filigree Wideslab, Katzenberger Wideslab

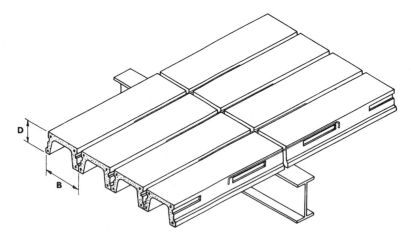

Figure III.42. Precast inverted trough sections with/without structural topping – Siegwart shown. Makers: Blatchford, Bradford, Ebor, Evercrete, Greenwood, Helicon, Invictus, Matthews & Mumby, Milbank (see Chapter 13), Siegwart, Stuarts, Truscon (sometimes used with timber bearers and floor boarding over concrete)

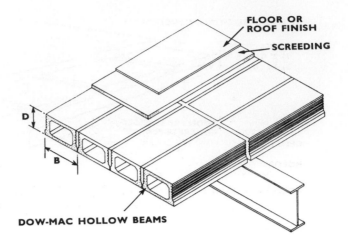

Figure III.43. Precast single-void hollow cored sections (basically rectangular profile), with/without structural topping – Dow-Mac shown. Makers: Arrow, Bison, Bradford, Corite, Croft, Dow-Mac, Ebor, Evanstone, Framlite, Girling, Greenwood, Helicon, Invictus, Matthews & Mumby, Raphcon, Siegwart, Stuarts, Viking

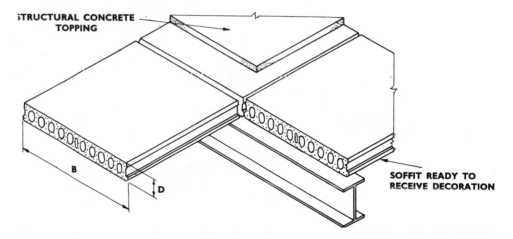

Figure III.44. Precast twin- or multi-voided hollow-cored sections (basically rectangular profile), with/without structural topping – Truspan prestressed floor shown. Makers: Bison, Dow-Mac, Evercrete, Francis, Marley, Shockcrete, Spiroll, Stotam Roth, Tembo, Trent, Truspan, Viking

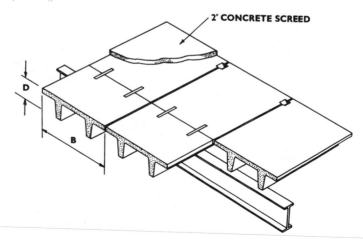

Figure III.45. Precast double tee units with structural topping – Blatchford shown. Makers: Blatchford, Dow-Mac

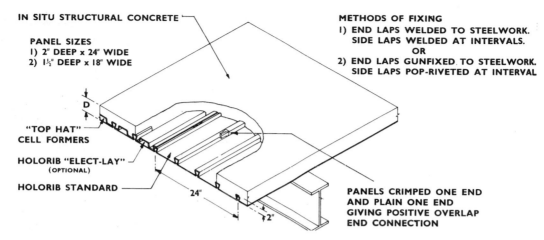

Figure III.46. In situ slabs composite with profiled metal decking permanent formwork – Holorib shown (note decking profiles vary: typically either 'dovetail' re-entrant or trapezoidal). Makers: Expamet, Holorib (later Richard Lees), Robertson Thain (Q- and QC-Floor), etc. (many more profiles and makers from 1980s onwards)

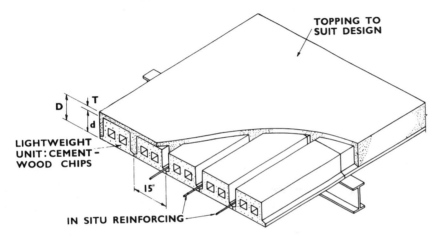

Figure III.47. In situ slabs on permanent lightweight formers (e.g. woodwool) – Chipcrete shown. Makers: Chipcrete, Gypklith, Marlith, Neolith, Spanform, Thermaflor, Wylam

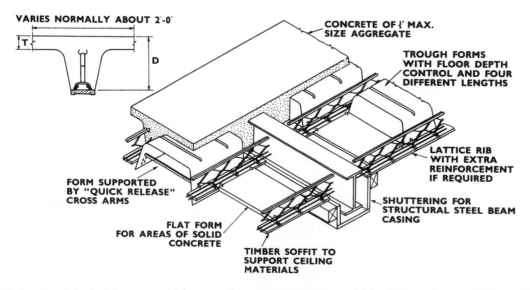

Figure III.48. In situ ribbed slabs on metal formers (permanent or removable) – Kaiser shown. Makers: BRC, Kaiser. Variant with ribbed or coffered slabs on removable non-metallic formers: GKN, Mills

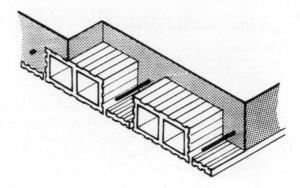

Figure III.49. In situ ribbed slabs with permanent infill blocks, the 'hollow pot' floor with hollow clay tiles, often used with thin 'slip tiles' to rib soffits to form a continuous plaster key; generic form, but particularly associated with Phorpres (London Brick Co.) blocks. Hollow clay tile makers: London Brick Co. Variant with hollow concrete blocks Smith's (with telescopic formwork for in situ ribs)

Appendix IV: assessing ancient mortars

There is often a need to determine the 'mix proportions' of old or even ancient mortars, plasters, renders, and stucco – without resorting to the costly techniques sometimes employed by archaeologists.[1]

Engineers need to know mortar 'strengths' when involved with structural alterations to old buildings; as do conservators repairing or replacing eroded parts of historic structures, and those researching the use of early materials, need to know the essential makeup of mortars. Because of the wide variety of possible binders and aggregates in these materials, together with problems relating to the limitations of chemical analysis, their assessment requires careful thought. The chemist, however, must be brought in early enough to glean valuable background information in terms of sampling and seeing the material in its context.

For visual examination, discrete pieces of material must be supplied; powdered material being virtually useless. Render materials, for example, need each layer to be separately considered. External layers in particular, due to weathering, can suffer loss of binder constituents by leaching and erosion. Colour photography is useful as a visual record, although the colours may not always be reliable.

Where mortars have been used as a bedding material, the aggregate will usually have been chosen for practical reasons. However, for visual effect as in pointing or matching mortars to types of stone or brickwork, the builder may have gone to some lengths to obtain a suitable aggregate and/or pigment.

Early colouring additives include crushed coal and ash, chalk, brick dust and naturally coloured cements (such as the pinkish-brown 'Roman' cements from the septaria of Harwich[2]). Information regarding type, shape and colour of the aggregate particles can be gained from acid wash separation (for non-calcareous types), sieve grading and visual examination with a hand lens or low-powered microscope. Detailed investigation may be made by petrographic means after preparing a thin section of the original material. The final colour of the mortar will have derived from a combination of aggregate/additives and binder.

Today, synthetically prepared pigments are available in a wide range of colours obtained from, for example, iron oxides, manganese and chromium oxides and cobalt and ultramarine blues.

Excluding Portland cement, present from the mid-19th century onwards, the most common binders[2] used were limes of non-hydraulic, semi and eminently hydraulic types. Gypsum was used mostly, but not exclusively, for indoor work. Natural cements react essentially as eminently hydraulic limes but can contain matter which, dependent on source, give rise to natural pigmentation. Binder characteristics are as follows:

- *Portland cement*: invented by Joseph Aspdin resembles Portland stone in appearance and was patented in 1824, this hydraulic cement's grey colour is roughly proportional in intensity to its iron oxide content, helping with identification in set materials.
- *Non-hydraulic or high calcium limes*: contain more impurities than modern building limes which are used mainly as workability aids, and had low proportions of soluble silica and alumina, hardening purely by drying-out and normal atmospheric carbonation.
- *Hydraulic limes*: these have chemical compositions not unlike Portland cements, the components of silica and alumina imparting the ability to set under water. Moderately hydraulic limes exhibit mid-range proportions of silica and alumina

which vary widely depending on the limestones used.

- *Magnesian limes*: essentially limes deriving from dolomitic limestones and with a capability of hydraulic properties.
- *Natural (Roman) cements*: composition of these materials is a function of the clay and mineral contents of the limestones used. Setting and hardening mechanisms resemble hydraulic limes. Because of their colours, they were used extensively for stucco and decorative features in neat form or in rich mixes with fine aggregates.
- *Gypsum*: commonly used as Plaster of Paris, the partly dehydrated form of gypsum employed as binders in mortars and plasters. Retarders were needed to give flexibility in working practice (keratin from animal hooves and horns being an early example); in early building work plaster frequently contained impurities and was used neat or with lime in the mix. Because plaster is sparingly soluble in water is more often, but not exclusively, used for indoor work. An alternative form is anhydrous gypsum (exemplified by Keene's Cement) which may need an accelerating agent.

The most common additives in non-hydraulic lime-bound materials are pozzolans, fine siliceous materials capable of reacting with lime in a mix to form calcium silicate cementing compounds. Pozzolanic materials derive from compact volcanic rocks, volcanic glasses and tuffs but where these were not available brick dust was often used. Pulverized fuel ash (pfa) from power stations is the best known modern form of pozzolan.

Where these materials are present in mortars they make identification of hydraulic of non-hydraulic limes impossible by chemical analysis, as a non-hydraulic lime/pozzolana mortar assumes a similar chemical composition, in soluble component terms, as a hydraulic lime mortar.

In addition to these additives, natural organic substances were frequently added to mortar to impart particular physical properties. These included animal fats, or blood, egg white and beeswax, while hair and straw were commonly used as reinforcement.

Such substances would generally be present in significant amounts, enabling simple inexpensive qualitative laboratory tests to identify their presence. Quantities can be determined using solvent extraction methods or comparative total organic matter once the additive has been characterised. Sometimes infrared spectroscopy is needed for positive identification. Visually, mortars do not always show the presence of additives.

Chemical analysis can be carried out using methodology based on that given in BS 4551[3], considered to be equally applicable to old lime-based mortars as to modern Portland cement ones.

Interpreting such chemical analyses without visual examination would only in a few cases lead to the correct mix being chosen.

Examples of different types of mortar analyses are as follows:

- External pointing mortar was creamy white with a discrete black surface coating. Analysis, by virtue of the high sulphate content, indicated a gypsum binder with an excess of calcium assumed to be a lime addition rather than calcium from the aggregate. The black coating consisted of a fine, essentially carbon black material with a very high lead content, which gave the conclusion that the mortar joints had been painted with black lead to protect the gypsum from weathering.
- Taken from the same building, bedding mortar showed a very different analysis with much lower content of sulphate in spite of its similar appearance. A low soluble silica ruled out the presence of hydraulic lime. Mix proportions were reported as non-hydraulic lime and sand, which assumed that the sulphate was

present as a natural impurity in the lime of aggregates rather than as a deliberate addition.

• Creamy white in appearance, old bedding mortar from a canal tunnel contained sizeable particles of unmixed lime, almost white. Aggregate contained a proportion of crushed brick debris. The high magnesium content indicated the use of a magnesium lime. Qualitative analysis of separated lumps of lime confirmed the high magnesium content but low silica. The mortar was judged to have been made with a non-hydraulic magnesian lime, a high soluble silica content being attributed to clay brick particles.

• A stucco sample was notable for its very hard dense nature and for the absence of visible aggregate particles. Analysis suggested hydraulic lime, the silica and alumina content being high. Additionally, a high soluble iron content indicated a pigment or iron-rich substance, almost certainly related to the sample's brown coloration. Composition was assumed to be of a neat natural cement with an inherent brown mineral coloration.

• Analysis of a quite firm, greenish-grey mortar gave no reaction with acids and alkalis. On heating, all of the mortar burned away. More detailed organic analysis revealed the presence of denatured protein suggesting eggs. In such an instance, bearing in mind the age of the mortar (400 years) chemical analysis could not give the exact original composition. Perusal of the parish records from June 1594 eventually solved the mystery: 'Mortar to be made up of 40 gallons of worte and 3s of eggs' (worte is unfermented beer).

Mix proportions are traditionally expressed on the basis of volume ratios, to calculate which of the bulk densities of material must be known. These values are known for modern materials, but not for the variable aggregates and aggregate mixtures used in old mortars and plasters. Thus calculations of volume proportions have to be based largely on guesswork and it is desirable to express mix proportions both by weight and volume.

Where no background information is forthcoming and several mixes are possible, alternative mixes are presented together with a 'balance of probabilities assessment' as to the most likely mix.

I have tried to demonstrate that for a relatively modest amount of work and cost, significant information can be gained about the makeup of old mortars and plasters.

As a cautionary note, it should be stressed that possession of this information is only a small step in terms of carrying remedial works such as repointing and re-rendering on old structures. The preparation of effective replacement mortars it is said, is 20% knowledge of the original mix and 80% the technique in deciding on effectively preparing the material. Deciding whether a more modern material might be the best option demands great skill and should be entrusted to those qualified to make such decisions.

References

1. An investigation of the chemical constituents of some Renaissance plasters, Newton, R.G. and Sharp, J.H, *Studies in Conservation*, 32 (1987), pp. 163–175.
2. Ashurst, J. and N. *Practical Building Conservation*.
3. BS 4551: Part 2: 1998 Analysis of mortars, screeds and plasters.

Index

Headings in italics refer to publications. Page numbers in italics refer to illustrations, a 't' suffix indicates a table of information.